Quantum Dynamics: From Theory to Applications

Quantum Dynamics: From Theory to Applications

Editor: Adaline Cerny

STATES
ACADEMIC PRESS
www.statesacademicpress.com

Published by States Academic Press
109 South 5th Street,
Brooklyn, NY 11249, USA
www.statesacademicpress.com

Quantum Dynamics: From Theory to Applications
Edited by Adaline Cerny

International Standard Book Number: 978-1-63989-456-7 (Hardback)

Cataloging-in-Publication Data

Quantum dynamics : from theory to applications / edited by Adaline Cerny.
p. cm.
Includes bibliographical references and index.
ISBN 978-1-63989-456-7
1. Quantum theory. 2. Mechanics. 3. Thermodynamics. I. Cerny, Adaline.
QC174.12 .Q36 2022
530.12--dc23

Contents

Preface

The world is advancing at a fast pace like never before. Therefore, the need is to keep up with the latest developments. This book was an idea that came to fruition when the specialists in the area realized the need to coordinate together and document essential themes in the subject. That's when I was requested to be the editor. Editing this book has been an honour as it brings together diverse authors researching on different streams of the field. The book collates essential materials contributed by veterans in the area which can be utilized by students and researchers alike.

Quantum dynamics is concerned with the exchange of energy and momentum, and the motions of systems whose behavior is governed by the laws of quantum mechanics. It is the quantum version of classical dynamics. Quantum dynamics is integral to the rapidly growing fields of quantum computing and atomic optics. In the field of mathematics, quantum dynamics is the study of the mathematics behind quantum mechanics. As a study of dynamics, this field explicitly investigates how quantum mechanical observables change over time. It also involves the study of one-parameter automorphisms of algebra of all bounded operators on the Hilbert space of observables. This book is compiled in such a manner, that it will provide in-depth knowledge about the theory and practice of quantum dynamics. From theories to research to practical applications, case studies related to all contemporary topics of relevance to this field have been included herein. As this field is emerging at a rapid pace, the contents of this book will help the readers understand the modern concepts and applications of the subject.

Each chapter is a sole-standing publication that reflects each author's interpretation. Thus, the book displays a multi-facetted picture of our current understanding of application, resources and aspects of the field. I would like to thank the contributors of this book and my family for their endless support.

Editor

1

Minimum Time in Quantum State Transitions: Dynamical Foundations and Applications

Alexandre Coutinho Lisboa and
José Roberto Castilho Piqueira

Abstract

This chapter is about the minimum time evolution between two quantum states considering the dynamics obeying either time-invariant Hamiltonians or time-varying ones. Merit figures are defined to help quantum control designers to define optimization parameters. The expressions are derived from the time-energy uncertainty relations and a practical case is studied as an example.

Keywords: merit figures, minimum time, quantum control, quantum evolution, uncertainty relations

1. Introduction

One of the most important problems in quantum control [1, 2], as well in quantum information processing and quantum computing [3, 4], is the transition from an initial state to a target state in minimum time. In fact, the existence of the *Decoherence* phenomenon in quantum systems, characterized by extremely short coherence times, presents serious difficulties on implementing quantum information devices [3, 5].

Indeed, one of the difficulties of the operational nature in quantum systems lies in the fact that they are very sensitive to the presence of the external environment, which often destroys its main quantum characteristics, which are essential, for example, for the implementation of systems processing quantum information, as well as for the viability of quantum computing. This is the problem of quantum decoherence. Thus, it is widely desirable for expedients and methods of optimal control of time minimization, applied to quantum systems, be the most efficient possible accordingly, whether of analytical nature, or of algorithmic and computational nature. However, there are physical limitations inherent by quantum dynamics, which relate to the minimum time, physically possible, so that a transition from quantum states occurs.

Therefore, it is natural to ask, what is the shortest physically possible time for a quantum state to evolve to another? The answer is provided by *time-energy uncertainty relations*. Opposing the famous quantum uncertainty relations regarding non-commuting operators, for example, position and momentum, time-energy uncertainty relations have a different mathematical and physical nature; they are deeply rooted in quantum dynamics.

The Soviet physicists Leonid I Mandelstam and Igor E. Tamm carried out the first successful theoretical approach addressing this issue, in the 1940s [6]. Under the hypothesis of time-independent Hamiltonian, Mandelstam and Tamm deduced a quantum dynamical inequality for time-energy, which sets up the temporal lower bound for a quantum state transition.

It is our goal, in the first section of this Chapter, to follow the theoretical steps of [6], translating the deductions to the modern quantum mechanical formalism, and to perform detailed analyses of the dynamical issues. Thereafter, we wish to apply the Mandelstam-Tamm time-energy inequality to a quantum system of interest, for example, the Fahri-Gutmann model of digital quantum computation [7], in order to obtain an analytical expression for the minimal time required for a state transition in such a quantum system. These analyses allow us to ultimately introduce a quantitative measure for the performance of time-optimal quantum controls [8].

In a subsequent section, we shall drop the time-independent Hamiltonian hypothesis of the original time-energy uncertainty relation and generalize it in the case of a time-dependent Hamiltonian $\hat{H} = \hat{H}(t)$. This is a type of Hamiltonian operators one finds in quantum control systems. In fact, appropriately shaped time-dependent electromagnetic fields or laser beams act as control efforts $u(t)$ [1]. The theoretical starting point for such generalization is to take into account the time evolution operator $\hat{U}(t, t_0)$ for the time-dependent Hamiltonian operator $\hat{H} = \hat{H}(t)$. Other approach to minimal time in quantum dynamical evolution is when one considers time-dependent Hamiltonians to employ the time-energy uncertainty relation obtained by Pfeifer [9].

2. Minimum time for quantum state transitions

Before tackling the control problem in quantum systems with respect to minimum time, that is, to make a given state transfer from a prescribed initial state to a desired target state as quickly as possible, one must take into account a fundamental issue of physical nature, which can be stated as follows. Given an initial state $|\psi_I\rangle$, which evolves over time according to the Schrödinger picture of quantum dynamics, that is, by means of the action of the time-evolution operator, dependent on the Hamiltonian of the system, how fast is the transition to a final state

- $|\psi_F\rangle$? In other words, what is the shortest possible time so that a transition of quantum states can occur?

This problem of a theoretical nature has been widely studied and it is closely linked to dynamical characterizations derived from time-energy uncertainty relations. Such relationships differ fundamentally from the Heisenberg's uncertainty principle (e.g., the simultaneous measurement uncertainty of the position and the linear momentum of a quantum particle), which comes from the *incompatibility* between the quantum observables, a physical fact that arises, within the mathematical and theoretical framework of quantum mechanics, from the *non-commutability* of the quantum observables involved in the measurement process. The time-energy uncertainty, on the other hand, finds its roots in quantum dynamics, as we are about to see.

From a historical point of view, since the so-called "old quantum theory," pioneered by Max Planck, Albert Einstein, Niels Bohr, among others, comprising the first two decades of the last century and firmly established, was the Planck-Einstein equation, which relates the energy and frequency of a photon through the Planck's fundamental constant h:

$$E = h\nu \tag{2.1}$$

As we know, such relationship is fundamental not only for the pioneering and groundbreaking Planck's hypothesis of quantization of radiation emission by an ideal black body, written in 1900 but also for the explanation given by Einstein for the photoelectric effect in 1905. Moreover, the atomic model introduced by Bohr in 1913, which explained the stability of the hydrogen atom by means of quantized energy levels corresponding to the stable possible orbits for the electron; postulated in addition that, when jumping to an energy level (or orbit) more or less energetic, the electron absorbed or emitted a quantum of energy, respectively, following the relationship (**2.1**), corresponding to a photon with frequency ν.

So, given a trivial variation in frequency at a given time interval, such as

$$\Delta\nu.\Delta T \sim 1 \tag{2.2}$$

providing the same account of an "uncertainty" $\Delta\nu$ for the frequency measurement of a monochromatic radiation at the ΔT time interval, by making use of the expression (**2.1**), we are led to the following:

$$\Delta E.\Delta T \sim h \tag{2.3}$$

This "uncertainty relation" deduction is eminently heuristic, although expression (**2.3**) still has experimental support within the physical conditions one has evoked to get it. Nevertheless, if one tries to generalize it as something of the form:

$$\Delta H.\Delta T \sim h \tag{2.4}$$

because the Hamiltonian corresponds to the energy of a mechanical system (classical or quantum), one will run into serious difficulties. If H represents a quantum observable in the current quantum mechanical sense, in the same way Paul Dirac had formally stated in the very first edition of his famous treatise [10], dating back to 1930, we can no longer identify the energy with the frequency of the monochromatic radiation times the Planck's constant. Moreover, expression (**2.4**) becomes inherently invalid and devoid of meaning for any quantum system when properly mathematically treated.

The Soviet physicists Leonid I. Mandelstam and Igor E. Tamm in the 1940s carried out the first successful theoretical approach addressing this quantum dynamical issue, the statement of a meaningful time-energy uncertainty relation [6].

So, let us try to present their theoretical starting point drawing on modern quantum mechanical formalism and its current notation, and finally arrive at the desired time-energy uncertainty relation through rigorous deductions combined with detailed analyses. The goal is also to modify it, in order to obtain variants of it and alternative expressions suitable for some purposes, which will become clear in the sections ahead.

Let $\overset{\frown}{R}$ and $\overset{\frown}{S}$ be two generic Hermitian operators (quantum observables). The following relationships hold [6]:

$$\Delta\hat{S}.\Delta\hat{R} \geq \frac{1}{2}\left|\left\langle \hat{R}\hat{S} - \hat{S}\hat{R} \right\rangle\right| \tag{2.5}$$

$$\frac{\mathrm{d}\left\langle \hat{R} \right\rangle}{\mathrm{d}t} = \frac{1}{i\hbar}\left\langle \left[\hat{R}, \hat{H}\right] \right\rangle \tag{2.6}$$

where $\Delta\overset{\frown}{S}$ and $\Delta\hat{R}$ are the square root of the mean square deviations (or variances) of operators $\overset{\frown}{S}$ and $\overset{\frown}{S}$, respectively, also known as "dispersion operator," or simply "standard deviation" in statistics terminology. Expression (**2.6**) is nothing but the dynamical evolution equation in the Heisenberg picture for the expectation value of observable $\overset{\frown}{R}$. If we impose $\hat{S} \equiv \hat{H}$ on (**2.5**) and apply the result of (**2.6**), the following inequality is obtained:

$$\Delta\hat{H}.\Delta\hat{R} \geq \frac{\hbar}{2}\left|\frac{\mathrm{d}\left\langle \hat{R} \right\rangle}{\mathrm{d}t}\right| \tag{2.7}$$

This expression provides, therefore, the connection between the standard deviation $\Delta\hat{H}$ of the total energy and the uncertainty in determining the energy of an isolated system, the standard deviation $\Delta\hat{R}$ (uncertainty) of some other dynamical quantity, relating them to the rate of change of the expectation value of the same physical amount.

Relationship (**2.7**) may be expressed otherwise. Since the modulus of an integral is less than or equal to the integral of the integrand modulus [11], upon performing an integration of (**2.7**) from t to $t + \Delta t$, and taking into account that $\Delta\hat{H}$ is constant, one derives inequality:

$$\Delta\hat{H}.\Delta t \geq \frac{\hbar}{2}\frac{\left|\left\langle\hat{R}_{t+\Delta t}\right\rangle - \left\langle\hat{R}_t\right\rangle\right|}{\left\langle\Delta\hat{R}\right\rangle} \tag{2.8}$$

in which the denominator of the right side of the inequality denotes the average value of $\Delta\hat{R}$ during the amount of time Δt.

It is appropriate to introduce, at this point, a special notation, ΔT, for the shortest time, during which the average value of a certain physical quantity is changed by an amount equal to the standard deviation thereof. Thus, ΔT can be called standard deviation (uncertainty) of time; making use of this notation, (**2.8**) can be rewritten as follows:

$$\Delta\hat{H}.\Delta T \geq \frac{\hbar}{2} \tag{2.9}$$

Now, let us consider a projection operator of form $\hat{\Lambda} \equiv |\psi_0\rangle\langle\psi_0|$. One can immediately notice that only one of its eigenvalues is unitary (when the projection operator is applied to the eigenstate $|\psi_0\rangle$) and all the others are zero. Thus, we have:

$$\hat{\Lambda}^2 = \hat{\Lambda}\hat{\Lambda} = \hat{\Lambda} \tag{2.10}$$

Indeed, $\hat{\Lambda}$ is an idempotent operator, as one can easily verify.

Furthermore, the expectation value $\langle\hat{\Lambda}\rangle$ may be interpreted as the probability of finding the quantum system in a specific quantum state $|\psi\rangle$, considering the initial state $|\psi_0\rangle$, since from the average value (or expectation value) definition, we have $\langle\hat{\Lambda}\rangle_\psi \equiv \langle\psi|\hat{\Lambda}|\psi\rangle = \langle\psi|\psi_0\rangle\langle\psi_0|\psi\rangle = |\langle\psi|\psi_0\rangle|^2 = P_\psi$. Of course, $\langle\hat{\Lambda}\rangle$ has its values within the interval $0 \leq \langle\hat{\Lambda}\rangle \leq 1$. According to (**2.10**), it follows that

$$\Delta\hat{\Lambda} = \sqrt{\left\langle\hat{\Lambda}^2\right\rangle - \left\langle\hat{\Lambda}\right\rangle^2} = \sqrt{\left\langle\hat{\Lambda}\right\rangle - \left\langle\hat{\Lambda}\right\rangle^2} \tag{2.11}$$

Therefore, making use of the expressions (**2.7**) and (**2.11**); then, applying operator $\hat{\Lambda}$ in the place of $\hat{R}$, we obtain

$$\Delta\hat{H}.\sqrt{\langle\hat{\Lambda}\rangle - \langle\hat{\Lambda}\rangle^2} \geq \frac{\hbar}{2}\left|\frac{d\langle\hat{\Lambda}\rangle}{dt}\right| \tag{2.12}$$

This inequality contains only one-time variable quantity, $\hat{\Lambda}=\hat{\Lambda}(t)$, and its time derivative, so that it can be integrated with respect to time. If, for example, $\hat{\Lambda}(t)=\hat{\Lambda}(0)=1$, that is, at the instant of time $t=0$, we were certain that the system was in the initial state $|\psi_0\rangle$, then it follows from (**2.12**) that, for $t\geq 0$,

$$\frac{\pi}{2} - \arcsin\sqrt{\langle\hat{\Lambda}(t)\rangle} \leq \frac{\Delta\hat{H}.t}{\hbar} \tag{2.13}$$

From (**2.13**), by using basic trigonometric properties and simple algebraic manipulations, it leads to the following expression:

$$\langle\hat{\Lambda}(t)\rangle \geq \cos^2\left(\frac{\Delta\hat{H}.t}{\hbar}\right) \tag{2.14}$$

for $t\geq 0$.

Here, an important fact should be noted. Although the projection operator $\hat{\Lambda}$ as $\hat{\Lambda}\equiv|\psi_0\rangle\langle\psi_0|$ has been defined with the outer product given by the *ket* and the *bra* corresponding to a state ψ_0, we could have also defined it, in a more general manner, such as $\hat{\Lambda}\equiv|\psi_t\rangle\langle\psi_t|$; that is, for a generic quantum state ψ_t evolved in time, for any instant of time t, so that the idempotency property of $\hat{\Lambda}$, as well as all the expressions, arguments, and previous deductions are analogous, remaining valid, thus, enjoying full generality.

From (**2.14**), two expressions relevant to our purposes shall be deduced. Since the average value of the projection operator $\hat{\Lambda}$ corresponds to the probability of observing the system at a given quantum state $|\psi\rangle$, more generally, at a certain state $|\psi_t\rangle$ evolving over time, having been (or been prepared) at an initial state $|\psi_0\rangle$ of the system, then inequality (**2.14**) shall be rewritten as

$$P_t \geq \cos^2\left(\frac{\Delta\hat{H}.t}{\hbar}\right) \tag{2.15}$$

where $P_t = |\langle\psi_t|\psi_0\rangle|^2$ is the probability of finding the system in $|\psi_t\rangle$ from $|\psi_0\rangle$; indeed, $\langle\hat{\Lambda}(t)\rangle = \langle\psi_0|\hat{\Lambda}(t)|\psi_0\rangle = \langle\psi_0|\psi_t\rangle\langle\psi_t|\psi_0\rangle = |\langle\psi_t|\psi_0\rangle|^2 = P_t$.

Finally, from (**2.15**), it can be deduced that the lowest physically possible time, or more generally, the inferior time limit,[1] physically needed to perform a transition between quantum states, under the assumption of a time-independent Hamiltonian, is given by the following inequality, which takes into account the uncertainty in determining the energy (dispersion or standard deviation of $\hat{H}$):

$$t \geq \frac{\hbar}{\Delta\hat{H}} \arccos\sqrt{P_t} \tag{2.16}$$

in which t is the necessary time for a quantum transition of states that its associated probability is $P_t = |\langle\psi_t|\psi_0\rangle|^2$.

From (**2.16**), it is immediately noticed that the time t is always a real number, despite the internal product ("bracket") of states being, in general, a complex number. Likewise, as any quantum observable, $\Delta\hat{H}$ always results in a real number. Moreover, since $P_t = |\langle\psi_t|\psi_0\rangle|^2$ is defined in [0,1], when the state normalization conditions are taken for granted, ensuring that the obtained probabilities vary always in the real interval [0,1]; it follows that the range of *arccos* function for this specific domain and, therefore, the values of t, will always be positive.

Some remarks about technological issues and its terminology are in order here. In quantum control literature, inequality (**2.16**) is also known as the "Bhattacharyya limit," after a paper by the Indian physicist Kamal Bhattacharyya, in which the author revisits the Mandelstam-Tamm's time-energy uncertainty relations, in order to apply them in decay or non-decay problems in quantum systems [12].

Inequality (**2.16**) gives a strong motivation to introduce a quantitative measure for the evaluation of the quantum control systems performance, with special interest on the time-optimal quantum control, for example, [1] and [13]. Lets then define the following *figure of merit*,

$$\eta_t \equiv \frac{t_{\min}}{t_{\mathrm{CQS}}} \tag{2.17}$$

where $t_{\min}$ is the shortest physically possible time to obtain a desired transition of quantum states, and t_{CQS} is the time by which such transition can be effectively accomplished in the *controlled quantum system*, hence, the notation chosen is [8].

In general, analytical solutions to problems of optimal quantum control are rare, and in most cases, control algorithms and numerical simulations are employed to obtain the desired results

[1] Mathematically speaking, one can say it is the *inf* or greatest lower bound (GLB) of the subset S of the physically possible times for quantum states transitions.

[1]. Indeed, in an article by Caneva and other authors [13], the Krotov algorithm is employed, for example, [14] and [15], and applied on the Landau-Zener system, as well as on a theoretical scheme of quantum information transfer in a spin chain ("One dimensional Heisenberg spin chain of length N"). The authors obtained state transition times close to the "Bhattacharyya limit," which is equivalent to inequality (**2.16**) for both systems studied [13].

Regarding the figure of merit η_t, it is easy to notice that η_t is a real number and a dimensionless quantity, defined on the interval [0,1]: $\eta_t \in \mathrm{R}: 0 \leq \eta_t \leq 1$, such that the extremal values correspond, respectively, to the cases in which the transition of quantum states has not been accomplished; in other words, the employed control algorithm did not converge to the desired target state (or mathematically speaking, it took an "infinite time"); on the other hand, there has been an ideal temporal state transfer: the control algorithm has achieved maximum efficiency in terms of the demanded time to perform the desired quantum state transition.

Combining the expressions (**2.16**) and (**2.17**), η_t can be more explicitly expressed as

$$\eta_t = \frac{\hbar.\arccos\sqrt{P_t}}{\Delta\hat{H}.t_{\mathrm{CQS}}} \tag{2.18}$$

or, alternatively as

$$\eta_T = \frac{\hbar \arccos\left|\langle\psi_{\mathrm{F}}|\psi_{\mathrm{I}}\rangle\right|}{\Delta\hat{H}.t_{\mathrm{CQS}}} \tag{2.19}$$

where $|\psi_{\mathrm{I}}\rangle$ and $|\psi_{\mathrm{F}}\rangle$ correspond, respectively, to some initial state and a certain desired final state (or target state, in control terminology).

From the dimensional analysis point of view, it is straightforward to verify that η_t, as expressed by (**2.18**) and (**2.19**), is consistent with the adimensionality requirement, since the constant $\hbar$ has dimension of action (energy × time), and the dispersion operator (or standard deviation) of the system's Hamiltonian $\Delta\hat{H}$ has natural dimension of energy.

Frequently, in practical applications, it is not always possible to obtain the "exact transition" of quantum states by the use of control algorithms and numerical simulations per se, from an initial state to the desired goal state, such as $|\psi_{\mathrm{G}}\rangle$. Instead, one looks for optimal control actions that maximizes the measure of *quantum fidelity*:

$$F = \left|\langle\psi_{\mathrm{G}}|\psi_T\rangle\right|^2 \tag{2.20}$$

or, in dual form, which minimize the amount of *quantum infidelity*:

$$I = 1 - \left|\langle \psi_G | \psi_T \rangle\right|^2 \tag{2.21}$$

that is, control actions such that, once elapsed control time T, the probability for the evolved state in time $|\psi(T)\rangle$ to be found at the target state $|\psi_G\rangle$ gets arbitrarily close to unity. Therefore, one can analyze the performance of the iterative process as a function of T and show that the algorithmic method in question is capable of producing infidelities arbitrarily close to zero [13].

3. Application to a particular quantum state transition

Now, consider the application of the time-energy uncertainty relations (**2.15**) and (**2.16**) to a specific and very important quantum state transition, namely the transition between two orthogonal states.[2] By rewriting inequality (**2.15**), in order to make explicit the probability associated to the transition from an initial state to a time-evolving state, on the left side of the expression, we have

$$\left|\langle \psi_t | \psi_0 \rangle\right|^2 \geq \cos^2\left(\frac{\Delta \hat{H}.t}{\hbar}\right) \tag{3.1}$$

where $\Delta\hat{H} = \sqrt{\langle \hat{H}^2\rangle_\psi - \langle \hat{H}\rangle_\psi{}^2}$ is the dispersion of $\hat{H}$ or the uncertainty for determining the energy.

Let us consider a quantum dynamical evolution, starting from a generic state $|\psi\rangle$, which the system dynamics leads it to the orthogonal state $|\psi^\perp\rangle$, by means of the actuation of the time-evolution operator $\hat{U}(t, t_0)$. Schematically, we have

$$|\psi\rangle \rightarrow \hat{U}(t,t_0) \rightarrow |\psi^\perp\rangle$$

with $\hat{U}(t, t_0) = e^{\frac{-i\hat{H}(t-t_0)}{\hbar}}$ —the time-evolution operator for a time-independent Hamiltonian $\hat{H}$, upon which $t_0 = 0$ can be imposed without loss of generality.

Adapting the expression (**3.1**) for such a case, results in

$$\left|\langle \psi | \hat{U}(t,t_0) | \psi \rangle\right|^2 \geq \cos^2\left(\frac{\Delta \hat{H}.t}{\hbar}\right) \tag{3.2}$$

[2] Or orthonormal states, once the normalization condition is taken for granted.

which its associated probability amplitude, $\langle\psi \mid \hat{U}(t, t_0) \mid \psi\rangle$, can be regarded as a survival amplitude of the state $\mid \psi\rangle$, evolving in time according to $\hat{U}(t, t_0)$ and $\Delta\hat{H}$, the energy uncertainty is given by

$$\Delta_\psi \hat{H} = \sqrt{\langle\psi|\hat{H}^2|\psi\rangle - \langle\psi|\hat{H}|\psi\rangle^2} \tag{3.3}$$

If a state transition from $\mid \psi\rangle$ to $\mid \psi^{\perp}\rangle$ occurs, the probability amplitude of (**3.2**) becomes zero, being able to define formally the first instant of time *t*, for which such transition takes place as follows:

$$\tau_{\psi\to\psi\perp} \equiv \inf\{t \geq 0 : \langle\psi|\hat{U}(t,t_0)|\psi\rangle = 0\} \tag{3.4}$$

Therefore, it is straightforward to conclude from (**3.2**) that $\tau_{\psi\to\psi\perp}$ is inferiorly bounded by the inequality:

$$\tau_{\psi\to\psi\perp} \geq \frac{\pi\hbar}{2\Delta\hat{H}} \tag{3.5}$$

Furthermore, the quantitative measure of temporal transfer efficiency can be defined for this specific case, figure of merit $\eta_{\psi\to\psi\perp}$ for such quantum state transfer, as

$$\eta_{\psi\to\psi\perp} \equiv \frac{\tau_{\psi\to\psi\perp}}{\tau_{\mathrm{CQS}}} \tag{3.6}$$

or, more explicitly, taking into account (**3.5**),

$$\eta_{\psi\to\psi\perp} = \frac{\pi\hbar}{2\Delta\hat{H}\tau_{\mathrm{CQS}}} \tag{3.7}$$

where $\tau_{\psi\to\psi\perp}$ is the shortest physically possible time, such that the first transition to the orthogonal state occurs, and τ_{CQS} can be stated as the time effectively spent by the controlled system or control algorithm, in order to perform such state transfer.

Finally, it is worthwhile mentioning the transitions or transfers between orthogonal (orthonormal) quantum states are of paramount importance to any schemes or devices, whether theoretical, experimental, or of technological nature, aimed at implementing quantum information processing, or quantum computing. The interested reader is referred to standard and authoritative sources like [3,4].

4. An analytical case study: The Fahri-Gutmann system

In this section, an application example is presented for illustrating the ideas and theoretical concepts developed so far. The Fahri-Gutmann system is a digital quantum computing model, which in its turn can be interpreted as a variation of the quantum search algorithm, similar to the well-known Grover's algorithm [16]. Here, we follow the steps of [17].

Let $|a\rangle$ and $|b\rangle$ be the initial and the target state, respectively, and the system's Hamiltonian defined as:

$$\hat{H} = E_a|a\rangle\langle a| + E_b|b\rangle\langle b| \tag{4.1}$$

where E_a and E_b are positive constants.

As already stressed, there can be computational difficulties to achieve the exact desired transfer of states, that is, from the initial state to the target one over the time evolution. Nevertheless, one can think of formulating the quantum control problem in a less restrictive manner, namely in terms of a state transition, as fast as possible, such that one can ensure maximum fidelity. Translating it into quantum mechanical language, we want to maximize the quantity:

$$F \equiv P_t = \left|\langle b|\hat{U}(t,t_0)|a\rangle\right|^2 \tag{4.2}$$

Firstly, let us impose $\langle a|b\rangle = s$. Without loss of generality, is assumed that s is a real number, since a phase factor can be associated to $|a\rangle$ or $|b\rangle$. Given the Fahri-Gutmann Hamiltonian expression (**4.1**), it is evident that it acts non-trivially only on the subspace spanned by the eigenstates $|a\rangle$ and $|b\rangle$, so it suffices to consider only this two-dimensional subspace. The computational basis of this quantum model is as follows:

$$\{|b\rangle, |b'\rangle\}, \text{such that} |b'\rangle = \frac{1}{\sqrt{1-s^2}}\left(|a\rangle - s|b\rangle\right) \tag{4.3}$$

which, in fact, consists of an orthonormal basis for the subspace spanned by $|a\rangle$ and $|b\rangle$.

Considering $E = E_a + E_b$, $x = E_a - E_b$, and defining the quantity:

$$\mu \equiv \sqrt{s^2 + \left(\frac{x}{E}\right)^2 \left(1 - s^2\right)} \tag{4.4}$$

We shall assume that $E_a>0$ and $E_b>0$. It is evident that $s \leq \mu \leq 1$ and $s=\mu$ if, and only if, $E_a=E_b$. In the computational basis $\{|b\rangle, |b'\rangle\}$, states $|a\rangle$ and $|b\rangle$ are represented, respectively, by the vectors:

$$|a\rangle := \begin{pmatrix} s \\ \sqrt{1-s^2} \end{pmatrix} \tag{4.5a}$$

$$|b\rangle := \begin{pmatrix} 1 \\ 0 \end{pmatrix} \tag{4.5b}$$

By defining the auxiliary constant λ as follows:

$$\lambda = s^2 - \frac{x}{E}(1-s^2) \tag{4.6}$$

the Hamiltonian $\hat{H} = E_a |a\rangle\langle a| + E_b |b\rangle\langle b|$, considering the basis given by **(4.6)**, has its matricial representation, with E being an arbitrary constant:

$$\hat{H} := \begin{pmatrix} 1+s^2-\frac{x}{E}(1-s^2) & \left(1+\frac{x}{E}\right)s\sqrt{1-s^2} \\ \left(1+\frac{x}{E}\right)s\sqrt{1-s^2} & 1-s^2+\frac{x}{E}(1-s^2) \end{pmatrix} = \frac{E}{2}\begin{pmatrix} 1+\lambda & \sqrt{\mu^2-\lambda^2} \\ \sqrt{\mu^2-\lambda^2} & 1-\lambda \end{pmatrix} \tag{4.7}$$

so that this matrix can be diagonalized as follows:

$$\hat{H} = U \begin{pmatrix} \frac{E}{2}(1+\mu) & 0 \\ 0 & \frac{E}{2}(1-\mu) \end{pmatrix} U^{-1} \tag{4.8}$$

with the U matrix given by

$$U = \frac{1}{\sqrt{2}} \begin{pmatrix} \sqrt{1+\frac{\lambda}{\mu}} & \sqrt{1-\frac{\lambda}{\mu}} \\ \sqrt{1-\frac{\lambda}{\mu}} & -\sqrt{1+\frac{\lambda}{\mu}} \end{pmatrix} \tag{4.9}$$

corresponding to the diagonalization unitary operator built up with the eigenvectors of $\hat{H}$ as given by its matrix representation (**4.7**).

After performing some cumbersome, but straightforward calculations, to diagonalize $\hat{H}$, given by (**4.7**), making use of (**4.8**) and (**4.9**), one finally obtains from (4.5a) and (4.5b) the time transition probability P_t, which can be regarded as a fidelity measure between the "search state" $e^{\frac{-i\hat{H}(t-t_0)}{\hbar}} | a\rangle$ and the target state $| b\rangle$:

$$P_t = \left| \langle b | \exp(\frac{-i\hat{H}t}{\hbar}) | a\rangle \right|^2 = s^2 \left(\left(\frac{1}{\mu^2} - 1 \right) \sin^2\left(\frac{\mu E t}{2\hbar} \right) + 1 \right) \tag{4.10}$$

where we have imposed $t_0 = 0$ on the time-evolution operator $\hat{U}(t, t_0) = e^{\frac{-i\hat{H}(t-t_0)}{\hbar}}$.

Noting that $s \leq \mu \leq 1$, it is easy to see that the maximum value of the probability P_t is

$$P_{\max} = \max_{t \geq 0} P_t = \left(\frac{s}{\mu} \right)^2 \tag{4.11}$$

indeed, it suffices to impose that the value of the function $\sin^2\left(\frac{\mu E t}{2\hbar}\right)$ must be 1 in the expression (**4.10**) to obtain (**4.11**).

In the same way, once achieving $\sin^2\left(\frac{\mu E t}{2\hbar}\right) = 1$ in (**4.10**), it can be conclude that the first instant of time t, for which maximum probability (or maximum fidelity), is given by

$$t_{\text{first}} = \inf\{t \geq 0 : P_t = P_{\max}\} = \frac{\pi\hbar}{E\mu} \tag{4.12}$$

Thus, for the Fahri-Gutmann's quantum computing model, the particular figure of merit that quantifies the time transition efficiency can be stated as

$$\eta_{\text{FG}} = \frac{\pi\hbar}{E\mu\tau_{\text{CQS}}} \tag{4.13}$$

where τ_{CQS} is the time effectively spent to drive the search state $\hat{U}(t, t_0) | a\rangle$ to the target state $| b\rangle$.

5. Quantum systems with time-dependent Hamiltonians: Two theoretical approaches to minimum time in quantum state transitions

Let us consider, a time-evolution operator $\hat{U}(t, t_0)$ characterized by a time-dependent Hamiltonian $\hat{H} = \hat{H}(t)$, such that $[\hat{H}(t_1), \hat{H}(t_2)] = 0$, $\forall\, t_1, t_2 \in I \subset \mathrm{R}^+$, $t_1 \neq t_2$; that is, the Hamiltonian operator of the quantum system of interest varies with time; however, it always commutes for any distinct instants of time t_1 and t_2, $t_1 \neq t_2$.

In quantum control, this is by far the most commonly found and studied case regarding the attempts to find control laws $u(t)$ to achieve a certain control goal of interest, which involves, for example, the minimization or maximization of dynamical variables to perform desired state transitions. Here, we are interested in minimizing transfer times between quantum states. Physically speaking, for example, a spin ½ system (e.g., an electron) subjected to a magnetic field which magnitude varies in time, but not in direction.

The time-evolution operator $\hat{U}(t, t_0)$ associated with such a time-dependent Hamiltonian is given by [18]

$$\hat{U}(t,t_0) = \exp\left[-\left(\frac{i}{\hbar}\right)\int_{t_0}^{t} dt'\hat{H}(t')\right] \tag{5.1}$$

Therefore, given an initial state $|\psi_{\mathrm{I}}\rangle$ and a final state $|\psi_{\mathrm{F}}\rangle$ (or "target state"), the shortest physically possible time (or minimum time) can be formally defined such that this transition of states occurs as follows:

$$t_{\min} \equiv \inf\{t \geq 0 : \left|\langle\psi_{\mathrm{F}}|\hat{U}(t,t_0)|\psi_{\mathrm{I}}\rangle\right|^2 = 1\} \tag{5.2}$$

where $\hat{U}(t, t_0)$ is given by the expression (**5.1**).

In an applied point of view, given the vector representations of $|\psi_{\mathrm{I}}\rangle$ and $|\psi_{\mathrm{F}}\rangle$ expressed in terms of a suitable basis (built up of linearly independent and orthonormal vectors), as well a matrix representation of $\hat{H} = \hat{H}(t)$, by calculating the integral of (**5.1**), expression (**5.2**) will provide the necessary formula to obtain the minimum time value.[3]

Now, the focus is on another possible formulation addressing this quantum dynamical issue. The goal is to directly employ a time-dependent energy uncertainty (or standard deviation) $\Delta(\hat{H}(t))$ to formally obtain an expression for the minimum time associated with transitions of states in quantum systems with dynamics governed by a time-dependent Hamiltonian $\hat{H} = \hat{H}(t)$. The Schrödinger equation taking into account such a Hamiltonian is written as

[3] In [8], we present an analytical case study of a time-dependent Hamiltonian, namely the Landau-Zener system.

$$i\hbar\frac{\partial|\psi(t)\rangle}{\partial t}=\hat{H}(t)|\psi(t)\rangle \tag{5.3}$$

with a time-evolving state $|\psi(t)\rangle$, from an initial state $|\psi_0\rangle$.

From this particular Schrödinger equation, Pfeifer proposed the following expression for a time-dependent energy uncertainty [9]:

$$\Delta(\hat{H}(t),|\psi(t)\rangle,|\dot{\psi}(t)\rangle)=\hbar\left\|\left(\hat{1}-|\psi(t)\rangle\langle\psi(t)|\right)\frac{\partial|\psi(t)\rangle}{\partial t}\right\| \tag{5.4}$$

in which we can immediately notice the dependency on the time-evolved state $|\psi(t)\rangle$, as well on its time derivative. From the original expression (**5.4**), some manipulations are carried out to deduce an alternative expression for the time-dependent energy uncertainty $\Delta(\hat{H}(t), |\psi(t)\rangle, |\dot{\psi}(t)\rangle)$, such that it may be more useful for our purposes:

$$\Delta(\hat{H}(t),|\psi(t)\rangle)=\hbar\left\|\left(\hat{1}-|\psi(t)\rangle\langle\psi(t)|\right)\left(\frac{1}{i\hbar}\hat{H}(t)|\psi(t)\rangle\right)\right\|=$$

$$\Delta(\hat{H}(t),|\psi(t)\rangle)=\hbar\left\|\left(\frac{1}{i\hbar}\hat{H}(t)|\psi(t)\rangle-\frac{1}{i\hbar}\langle\psi(t)|\hat{H}(t)|\psi(t)\rangle|\psi(t)\rangle\right)\right\|=$$

$$\Delta(\hat{H}(t),|\psi(t)\rangle)=\hbar\left|\frac{1}{i\hbar}\right|\left\|\left(\hat{H}(t)|\psi(t)\rangle-\langle\psi(t)|\hat{H}(t)|\psi(t)\rangle|\psi(t)\rangle\right)\right\|=$$

$$\Delta(\hat{H}(t),|\psi(t)\rangle)=\left\|\hat{H}(t)|\psi(t)\rangle-\left\langle\hat{H}(t)\right\rangle_{\psi(t)}|\psi(t)\rangle\right\| \tag{5.5}$$

wherein, at the first step, we replaced the time derivative of $|\psi(t)\rangle$ making use of (**5.3**), so that we could eliminate the dependency on $|\psi(t)\rangle$, the outer product $|\psi(t)\rangle\langle\psi(t)|$, and the constant $\hbar$, as well (second and third steps, respectively). As a result, we were able to express $\Delta(\hat{H}(t), |\psi(t)\rangle)$ as a functional of vector $\hat{H}(t)|\psi(t)\rangle$ and of the time-evolving state $|\psi(t)\rangle$, multiplied by the scalar quantity correspondent to the expectation value of the Hamiltonian $\hat{H}(t)$.

Now, let us consider a time-energy uncertainty relation of the form $\Delta E.\Delta t\approx\hbar$, similar to those one can find in the time-independent Hamiltonian case, in terms of stationary states of energy [18]. Afterwards, through methods and concepts from the time-dependent perturbative theory, Sakurai shows that such relation remains valid for the time-dependent case [18]. So, it becomes natural to evoke one of the Mandelstam-Tamm's time-energy uncertainty relations, namely

$$\Delta\hat{H}.\Delta T \geq \frac{\hbar}{2} \tag{5.6}$$

Furthermore, to achieve the perspective of our chain of theoretical reasoning so far, we restate it in the following way:

$$\Delta(\hat{H}(t),|\psi(t)\rangle).t \geq \frac{\hbar}{2} \tag{5.7}$$

in which the energy uncertainty is now considered as depending on a time-dependent Hamiltonian $\hat{H}(t)$, and, without loss of generality, $t_0 = 0$ in ΔT is imposed.

Finally, the minimum time for a generic quantum state transition, between an initial state $|\psi_0\rangle$, which begins to evolve over time, and a final state $|\psi_F\rangle = |\psi(T)\rangle$, can be formally defined in such a way that it can also be defined in function of an instant of time T. Its final state is achieved or characterized by

$$t_{\min} \equiv \inf\{t \geq 0 : P(t) = |\langle\psi_F|\psi(t,t_0)\rangle|^2 = 1\} = T \tag{5.8}$$

or, in terms of the maximum possible quantum fidelity, also by

$$t_{\min} \equiv \inf\{t \geq 0 : F \equiv |\langle\psi_F|\psi(t,t_0)\rangle|^2 \to 1\} = T \tag{5.9}$$

Finally, taking into account expressions (**5.7**) and (**5.5**), we have

$$t_{\min} \geq \frac{\hbar/2}{\Delta(\hat{H}(t),|\psi(t)\rangle)} \tag{5.10}$$

$$t_{\min} \geq \frac{\hbar/2}{\left\|\hat{H}(T)|\psi(T)\rangle - \langle\hat{H}(T)\rangle_{\psi(T)}|\psi(T)\rangle\right\|} \tag{5.11}$$

obtaining an alternative expression for the greatest temporal lower bound for a generic transition of states, that occurs in a quantum system with dynamics governed by a time-dependent Hamiltonian $\hat{H}(t)$, thus, without relying on the approximative hypothesis of time independency in the characteristic Hamiltonian of the system.

Author details

Alexandre Coutinho Lisboa and José Roberto Castilho Piqueira*

*Address all correspondence to: piqueira@lac.usp.br

Engineering School, São Paulo University, São Paulo, Brazil

References

[1] D. D'Alessandro. *Introduction to Quantum Control and Dynamics*. CRC Press - Taylor & Francis Group, Boca Raton, FLA, 2007.

[2] H. M. Wiseman, G. J. Milburn. *Quantum Measurement and Control*. 1st Edition. Cambridge, UK: Cambridge University Press, 2009.

[3] M. A. Nielsen, I. L. Chuang. *Quantum Computation and Quantum Information*. Cambridge, UK: Cambridge University Press, 2000.

[4] D. Bouwmeester, A. K. Ekert, A. Zeilinger (Ed.). *The Physics of Quantum Information: Quantum Cryptography, Quantum Teleportation, Quantum Computation*. 1st Edition. New York: Springer-Verlag, 2000.

[5] M. A. Schlosshauer. *Decoherence and the Quantum-to-Classical Transition*. New York: Springer Verlag (The Frontiers Collection), 2008.

[6] L. I. Mandelstam, I. E. Tamm. "The Relation of Uncertainty Energy-Time in Non-Relativistic Quantum Mechanics". *Notification of the Academy of Sciences of the USSR (Physics Series)*. v. 9, n. 1/2, 122-128, 1945.

[7] E. Fahri, S. Gutmann. "Analog analogue of a digital quantum computation". *Physical Review A*. v. 57, n. 4, 2403–2406, 1998.

[8] A. C. Lisboa, J. R. C. Piqueira. "Quantum Speed Problem: Theoretical Hints for Control". *International Journal of Quantum Information*. IJQI-D-15-00257R1. *Accepted for Publication*, 2015.

[9] P. Pfeifer. "How Fast Can a Quantum State Change with Time?". *Physical Review Letters*. v. 70, n. 22, 3365–3368, 1993.

[10] P. A. M. Dirac. *The Principles of Quantum Mechanics*. 4th Edition. Oxford, UK: Oxford University Press, 1958.

[11] W. Rudin. *Principles of Mathematical Analysis*. 3rd Edition, New York: Mc Graw-Hill, 1976.

[12] K. Bhattacharyya. "Quantum decay and the Mandelstam-Tamm time-energy inequality". *Journal of Physics A: Mathematical and General*. v. 16, 2993–2996, 1983.

[13] T. Caneva *et al.* "Optimal Control at the Quantum Speed Limit". *Physical Review Letters*. v. 103, n. 124, 240501,1–4, 2009.

[14] V. F. Krotov. *Global Methods in Optimal Control Theory*. New York: Marcel Dekker, Inc., 1995.

[15] S. G. Schirmer, P. De Fouquieres. Efficient Algorithms for Optimal Control of Quantum Dynamics: The Krotov Method Unencumbered. *New Journal of Physics*. v. 13, 073029, 1–35, 2011.

[16] L. K. Grover. Quantum Mechanics helps in searching for a needle in a haystack. *Physical Review Letters*. v. 79, n. 2, 325–328, 1997.

[17] S. L. Luo. How fast can a quantum state evolve into a target state? *Physica D: Nonlinear Phenomena.*. v. 189, n. 1, 1–7, 2004.

[18] J. J. Sakurai. *Modern Quantum Mechanics*. Revised Edition. Addison-Wesley Publishing, Reading, Mass. Company, Inc., 1994.

2

Linking Non-equilibrium Transport with the Many Particle Fermi Sea in the Quantum Hall Regime

Josef Oswald

Abstract

The communication of the electron system with the outside world at low excitation transport experiments happens by exchanging electrons at the Fermi level. We argue that the locations where this is possible in the quantum Hall effect regime are the so-called edge channels. We explain that these channels can be understood as a more general representation of the many-particle quantum Hall (QH) system close to equilibrium that allows describing transport due to non-equilibrium on a very fundamental level. Based on fundamental principles of quantum physics, a transfer matrix formulation for the local non-equilibrium electrochemical potential in a network of interconnected directed quantum channels can be used to solve the lateral distribution of the non-equilibrium excitation potentials. Instead of using the Landauer formula or other tools like the Kubo formula for addressing conductance's just a transfer matrix formulation for the transfer of electrochemical potentials is used to find the self-consistent lateral distribution of the non-equilibrium electrochemical potentials. Currents and potentials that are measured at contacts are only calculated as a post-processing step right at the contacts, and they allow calculating all the resistances and conductance's like known from QH experiments. The interplay between transport and the many-particle system is of general interest when dealing with accessing information about quantum systems. Our approach allows modeling electron systems in the QH regime for realistic sample geometries, including inhomogeneities, like present in real samples or that are forced by gate electrodes as well as the random disorder potential. We combine our network model for transport with the Hartree–Fock method that allows the inclusion of realistic screening effects as well as the magnetic field-dependent enhanced g-factor.

Keywords: network model, quantum Hall effect, magneto-transport, conductance quantization, non-equilibrium transport, exchange interaction, Hartree-Fock approximation, many particle quantum system

1. Introduction

Even more than 35 years after discovering the integer quantum Hall effect [1], it is still a lively discussed topic at many conferences. In the quantum Hall (QH) regime, the many-particle quantum nature of the electron system can be studied from mesoscopic to nearly macroscopic length scales. The challenge in modeling electron transport in this regime is that the many-particle quantum states of the electron system exist as stationary states at thermal equilibrium, while transport is driven by introducing deviations from the equilibrium. Even for the QH experiments that happen close to thermal equilibrium, a direct modeling of current flow does not fit the world of stationary states. The general purpose of quantum Hall effect experiments is to unveil the physics of the stationary many-particle quantum states behind. This implies that such experiments should happen close to the thermal equilibrium and therefore they are mostly performed also at lowest achievable temperatures. However, any experiment in quantum physics is a disruption of the quantum states under investigation and that is true also for the quantum Hall experiments. On this background, the major question arises to what extend experiments can provide information about the underlying physics of the stationary many-particle states. At the same time, one has to consider that the stationary state under investigation itself cannot be directly used for modeling current flow. Stationary states do not exhibit any time-dependent observables, except while destroying them. This fundamental dilemma can be seen as the playground for the numerous theoretical approaches to the quantum Hall effects in order to compare theory and experiment. Various theoretical approaches focus on finding a basis for relating steady state properties of the quantum system to experimental properties without finally being able to model directly the current flow like it happens in the various experiments. One most prominent theoretical approach is the scaling theory [2], that is based on the localization picture of the quantum Hall effect and that relates the localization length of the stationary quantum states to experimental properties like the width of the quantum Hall plateau transition regimes that can be affected by experimental conditions like temperature [3] and measurement frequency [4]. Despite the great success of this theory, that finally proofs the quantum nature of the physics behind; the current flow itself is not addressed by this method. A further step in this direction is the usage of local conductance's that are extracted from the network of quantum channels evolving for the steady state. However, local conductances are finally introduced even for such most successful and sophisticated methods like, for example, the Kubo formula [5], while any local quantities associated with local current flow is finally a violation of the many-particle physics, as will be discussed in the next section.

2. Single-particle picture versus many-particle states

One of the most important aspects for the interpretation and modeling of experimental investigations is the fact that electrical measurements in the quantum Hall regime deal in fact with single particles in terms of counting carriers when measuring currents. On this background, it is not that surprising that also the interpretation and the way of thinking about the quantum Hall system is dominated by the single-electron picture. Anyway the human ability for thinking, intuition and finally picturing processes that take place in the electron system is limited to the imagination of single electrons, or at least single quasi-particles. From the general point of view, any electrical measurement as performed in the quantum Hall system has to be understood as an extraction and/or injection of single electrons from and to a many-particle quantum system at the contacts, while any imagination of the movement of single electrons between those contacts in the electron system must consequently fail and is finally forbidden. However, there still exist very useful pictures and models that are based on single electrons moving along the quantum Hall systems like, for example, the edge channel picture [6] that has proven to be very successful especially for the interpretation of QH experiments. Last but not least the scientific community needs such restricted pictures in order to communicate and deal with the subject. Therefore, it is a special challenge to find out to what extend such single-electron pictures may be used without seriously violating the many-particle physics behind, but also finding out, why such "wrong" pictures still give right answers and where the limits for the applications of these "wrong pictures" are. In the following, we finally try to extend the limits of these "wrong" single-particle pictures by a modified interpretation of the single-electron picture.

As already mentioned, one of the most prominent and useful single-electron pictures in the QH physics is the so-called edge channel picture that gives useful results even when considering the single electrons as interaction free. Edge channels act as one-dimensional directed quantum channels. However, before having a closer look to this model, we should have to look at some basic aspects about one-dimensional channels in general.

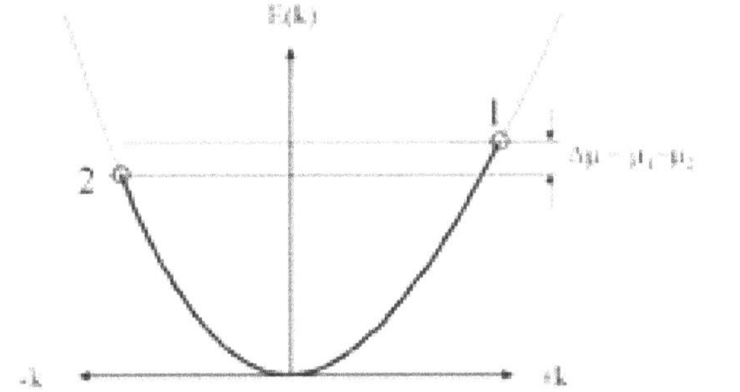

Figure 1. Band structure of a one-dimensional channel with parabolic k-branches for forward and backward direction; labels 1 and 2 indicate the quasi-Fermi levels for the case of an injected non-equilibrium by opposite contacts. The bold parabolic branches indicate the filled states in k-space, which are different for the two k-branches at non-equilibrium.

Figure 1 schematically shows the band structure of a one dimensional channel. It consists of two branches in k-space, one for the forward and one for the backward direction of the channel. This picture is often used to introduce the one-dimensional quantized conductance, and it allows getting the correct value of $G = 2^{e^2}/h$ for spin degenerate electron channels on the basis of semiclassical arguments. The main assumption for this is that only those electrons contribute to the current that are occupying states in the energy interval between μ_1 and μ_2. However, despite the correct result, this picture gets in serious conflict with the many-particle nature of the electron system if one takes that assumption for true. Already the interpretation of this simple picture seems to be crucial, and it is worth to find out why to some extend the main results are still valid and if we need to change the way of thinking how the single electrons in the channel carry the current.

It is well known that according to the above semiclassical model, a net current flows, if the two branches in k-space are driven out of thermal equilibrium by applying different voltages to the contacts that inject carriers from opposite sides via the opposite k-branches. However, a common misinterpretation in this context is now to say that all electrons on opposite lying states in k-space do not produce a net current flow because of k-cancellation; only those electrons occupying the states between μ_1 and μ_2 produce a net current that can be measured in experiments. If thinking this way, we attribute single individual electrons to single quantum states; or at least, we attribute different groups of electrons to sit on different regions in k-space. That is clearly in contradiction with the many-particle picture, because that interpretation suggests that only a small fraction (group) of (necessarily distinguishable) electrons carry the current. According to the many-particle picture, all electrons and single-electron states should lose their individuality and it should not be allowed to attribute any subset of electrons to any subset of states. Nevertheless, the consequent further application of this interpretation leads still to the correct value of the one-dimensional conductance. This fact makes it difficult to realize if the limits of interpretation are already violated or not. We identify this as the main source of many controversial discussions in the community. But how can we think about electrons acting in any system without the ability of attributing subsets of electrons to particular regions, states, or considering different electrons for different actions?

In order to reconsider this problem for trying to find an appropriate re-interpretation, we would like to draw the attention back to the very basics of the many-particle picture. It is well known that the many-particle states are composed by single-particle states by superposition while mixing up all different configurations for the occupation of different states by different electrons. This means that all single electrons lose completely any relation or association with any particular states and they get complete delocalized in space as well as in energy. This happens while preserving the negative Eigenvalue of the operator for exchanging any two single electrons of the system, like demanded for Fermions. Just as a reminder, the composition for a two-electron state in this context reads as follows:

$$|\psi\rangle = \frac{1}{\sqrt{2}}\{|\psi_1,\psi_2\rangle - |\psi_2,\psi_1\rangle\} \tag{1}$$

Here, $|\psi_1, \psi_2\rangle$ means that electron 1 occupies state 1 and electron 2 occupies state 2, while $|\psi_2, \psi_1\rangle$ means electron 1 occupies state 2 and electron 2 occupies state 1. By superposition of the different configurations, the electrons lose their identity and their association with a particular single-electron state and the negative sign in the superposition ensures the required change of sign when interchanging any two electrons as required for Fermions. For more than two electrons and more than two states, this composition is done in real space representation by applying the so-called Slater determinant for mixing up all possibilities of attributing all electrons to all available single-electron wave functions. This technique is applied also in context with the Hartree–Fock [7] approach as used in our modeling below. At this point, only the fundamental consequences should be brought back to attention. The crucial point is that the human way of thinking is always bound to the imagination of single electrons and the fundamental question arises if there exists a more appropriate way of intuitive thinking about single electrons that is more suitable for the many-particle physics. This question is important, because most of the controversial discussions in the scientific community arise from the way, how different people obtain their mostly intuitive understanding in terms of single electrons, which anyway has to fail at some point for the many-particle system. Therefore, we have to re-interpret the single-electron aspect on the background of the many-particle system. For this purpose, we have to answer two fundamental questions that have to be correctly answered also by any interpretation in terms of single electrons [8]:

I. *What is the ground state energy of the many-particle system?*

II. *Which of the supposed single electrons are sitting right at that ground state energy?*

According to the basic principles of quantum physics, the ground state energy has to be defined in terms of an "observable" that needs an appropriate operator for measuring that physical quantity. This "measurement" has to be based either on a real experiment or at least on a "Gedankenexperiment." On this basis, the two questions can be re-formulated as follows:

ad I) At which energy can we extract any single electron from the system without destroying the ground state of the system left behind, after just taking an electron, or just adding an electron?

It quickly becomes clear that a single electron has to be extracted from or added to the Fermi level if subsequent relaxation processes of the electron system should be avoided. If we, for example, pull out an electron at lower energy than the Fermi energy, an excited hole would have been left behind that subsequently relaxes. If adding an electron at higher energy, it is also clear that we would create an excited state that still has to relax into the ground state subsequently. The only way to put in or take out an electron without the need of subsequent spontaneous relaxation is if that happens right at the Fermi energy. The Fermi energy therefore appears as the lowest possible energy at which the electron system can communicate with the outside world, as, for example, done by a transport experiment at low temperatures and low driving currents. Consequently, the Fermi energy should be defined as the (observable) ground state of the many-particle systems.

ad II) What is the probability for "catching" any particular single electron of the system at the Fermi energy?

We consider the electron system in the ground state as a black box, and we randomly take out and put back electrons at the Fermi level according to I. Knowing how the many-particle state is constructed [see e.g. Eq. (1)], it becomes immediately clear that we can find each individual electron of the system with same probability sitting at any electronic state, which applies of course also for the Fermi level! So if one particular electron out of N electrons can be found at the Fermi level with probability $^1/_N$, this can only mean that all single electrons sit right at the

Fermi level. In this case, the "observation" has to overrule some possibly contradicting intuition.

In contrast to the most common way to look at the many-particle electron system in a single-electron representation, we consequently have to consider all single electrons sitting right at the Fermi energy.

Except this maybe somehow counter intuitive fact which we have to accept some how, we now can go on to think in terms of single electrons. In this context, it is interesting to reconsider the conductance of the one-dimensional channel according to **Figure 1**: All electrons are considered to sit at the top occupied energy level, that is, twofold degenerate for the +/− k-branches in the case of equilibrium. Additionally, the "single electrons" are also delocalized between these degenerate levels because of the many-particle character of the whole system. The single-particle wave functions are consequently just plane waves at k_F for *all* electrons, which still compensate each other at thermal equilibrium, meaning that the quasi-Fermi levels in the two branches of the k-space are the same (see **Figure 2** left) and the counter-propagating identical electron waves create a stationary periodic charge distribution along the channel.

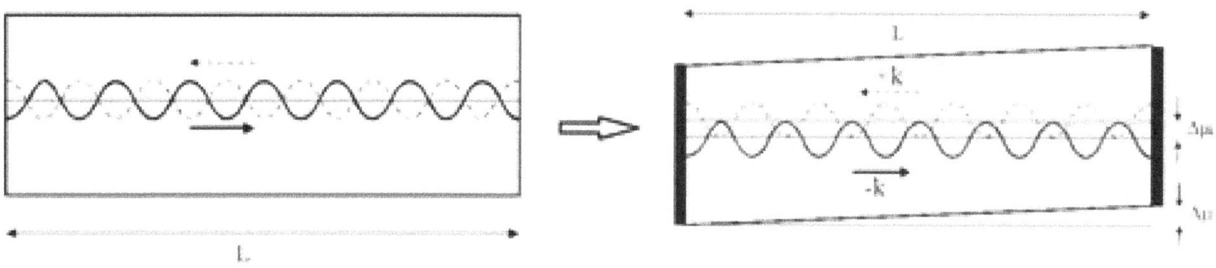

Figure 2. Left: Counter-propagating electron waves from states in the opposite k-branches at Fermi level for equilibrium. Right: Counter-propagating electron waves from states in the opposite k-branches at quasi-Fermi levels that are slightly shifted against each other by a small potential difference introduced by opposite contacts.

If we now introduce a vanishing small non-equilibrium Δμ between the two k-branches by applying a vanishing small potential difference at the associated contacts, the two counter-propagating single-electron wave functions face a slight energy difference that affects the result of their superposition (see **Figure 2** right). This energy difference affects the usually not observable universal time dependence of stationary quantum states that reads as follows:

$$\psi(x,t) = \psi(x) \cdot e^{-i\frac{2\pi \cdot E}{h}t} \tag{2}$$

However, this time dependence becomes relevant in case of superposition. The superposition of the counter-propagating waves with slightly different energy E_I and E_{II} leads now to a beating effect in time that causes a lateral movement of the whole charge distribution that is associated with the superposition that originally resembles just a standing wave (that now starts to move). The beating frequency ν therefore derives as $\nu = \Delta\mu / h$ which finally leads to a non-stationary time-dependent state that now creates charge transport. This fact can be seen as a most elementary step of composing a non-stationary state by the superposition of two stationary states at slightly different energy. It is easy to convince oneself that during a single beating cycle, the original charge distribution is restored, while the charge distribution as a whole moved on by exactly one period of the initial charge distribution. This represents a certain transferred charge from one contact to the other. We know that a single electron on the Nth quantum state in space representation is build up by N charge maxima that are distributed along the channel. Since they shift laterally by one period during a single beating cycle, the transferred charge relates to a fraction of $\Delta q = q_e/N$ of the charge of one single electron. However, due to the many-particle aspect and the Pauli principle, we know that we have also a total of N electrons in the whole system that subsequently fills up all lower states. Since we have to consider all N electrons equally at the Fermi energy, they consequently contribute all equally in parallel. All together we get therefore a charge transfer of N times the fraction Δq of charge. This gives in total charge per beating cycle of $\Delta Q = N \cdot \Delta q$. This finally appears to be the charge of a single electron q_e. As a consequence, we obtain the fundamental result that the complete electron system transfers exactly the charge of one single electron during each beating cycle from one contact to the other. Assuming spin degeneracy, we just have to multiply by a factor of 2 and we get $I = 2 \cdot \nu \cdot q_e$ as the total current which easily allows deriving an associated conductance:

$$G = 2 \cdot \frac{e^2}{h} \tag{3}$$

In this way, we end up with the well-known quantized value for the conductance of a spin degenerate one-dimensional channel. However, it results from a slightly different way to look at the many-particle systems in term of single electrons that strictly avoids attributing individual electrons to individual single-electron states, but at the same time including all electrons of the system.

If we reconsider now the implications for transport experiments, it becomes clear, that for low-excitation experiments close to the thermal equilibrium, the electron system is only accessible in regions where the spectrum of the single electrons has states close to the Fermi level. These Fermi edge electrons take the role as a representation of the whole electron system that, however, can be accessed close to the ground state only at the Fermi level. Therefore, these electrons seem to appear just at locations, where there are states at the Fermi level in the single-electron picture that make up the quantum channels or so-called edge channels in the QH regime. However, it is a common misinterpretation to conclude that current carrying electrons exist only in those channels. Moreover, according to the composition of the many-particle states, it is even forbidden to attribute single electrons to special regions, energies, and states. The channels are only regions where one can communicate with the electron system close to thermal equilibrium, while the electrons themselves remaining de-localized over the whole system [8]. We may picture the transport channels as the dip of the iceberg that sticks out of the Fermi sea, but still belonging to the one and only many-particle system. Doing low-excitation transport is like walking on the tip of the iceberg, and it crucially depends on the topology of the ground on the dip.

The consequences of these considerations in this paragraph will become clearer if applying them to the edge channel picture in the next paragraph.

3. The edge channel picture

Since the edge channel picture [6] is a widely known model, only a short qualitative overview of the main aspects should be given in this chapter. However, we need this chapter in order to make clear the consequences of the above discussion and the implications for our modeling below. The main ingredient of how edge channels are created is that for the Schrödinger's equation a strong homogeneous magnetic field appears as an effective parabolic confinement potential that acts in addition to a possibly present electrical potential. A strong magnetic field in this context is present, if the magnetic length is small as compared to the typical length scales of the electrical potential or in other words, if the electrical potential varies slowly on the length scale of the cyclotron radius. However, the centre coordinate for the magnetic contribution to the confinement potential is not fixed in space and has therefore a free centre coordinate for the Schrödinger's equation (see e.g. Ref. [9]). Furthermore, for realistic magnetic fields of several Tesla and sample parameters as used in real experiments, the quantization due to the magnetic confinement dominates strongly over the contribution of the electrical field. In this regime, the quantization due to the magnetic field leads to the well-known Landau levels, just like without electrical field. It appears as a good approximation for the energy spectrum that the local electrical potential simply shifts the magnetically split LLs parallel in energy. This is shown in **Figure 3**.

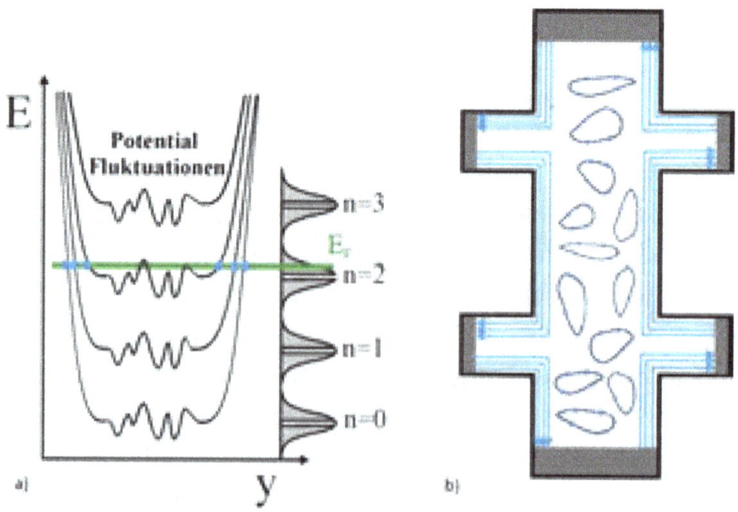

Figure 3. (a) Schematic representation of the laterally varying Landau levels across the bulk in the transverse y-direction that mimic the lateral electrical potential fluctuations; at interceptions with the Fermi energy, there appear directed channels that resemble loops around potential hills or valleys and edge channels along the sample edge near the sample boundaries (blue color); (b) schematic representation of the Hall geometry with the bulk area in top view with the edge channels along the edge and loops (magnetic bound states) in the bulk region.

In this regime, the states at the Fermi energy (E_F) extend only along equipotentials, where the LLs intercept E_F, while in direction transverse to the equipotentials, they are localized on the length scale of the cyclotron radius. These equipotentials resemble directed channels along the edges of the electronic system and create directed loops in the bulk region that might also couple opposite edges by tunneling across magnetic bound states [10]. Conduction through the bulk becomes possible, if the bulk states, that mostly create localized loops, extend over the sample size. That happens, if the Fermi level is close to the centre of the potential fluctuations, which also means the centre of the disorder broadened LL. Whenever a bulk current is possible, we get into the regime between QHE plateaus and whenever the bulk is insulating, we are in the plateau regime. This aspect is used by the Chalker–Coddington network model [11], which has been introduced in order to calculate the localization length that is decisive for the bulk states whether they contribute to the spectrum of the localized states or they belong to the spectrum of de-localized states within the DOS peaks of the LLs.

Figure 4 schematically summarizes the results of the so-called localization picture of the QHE [12] that finally leads to the scaling theory of the QHE [2]. The essential point of the scaling theory is that the localization length of an electronic state close to the centre of the Landau level scales according to a particular exponent of the inverse energy relative to the LL centre.

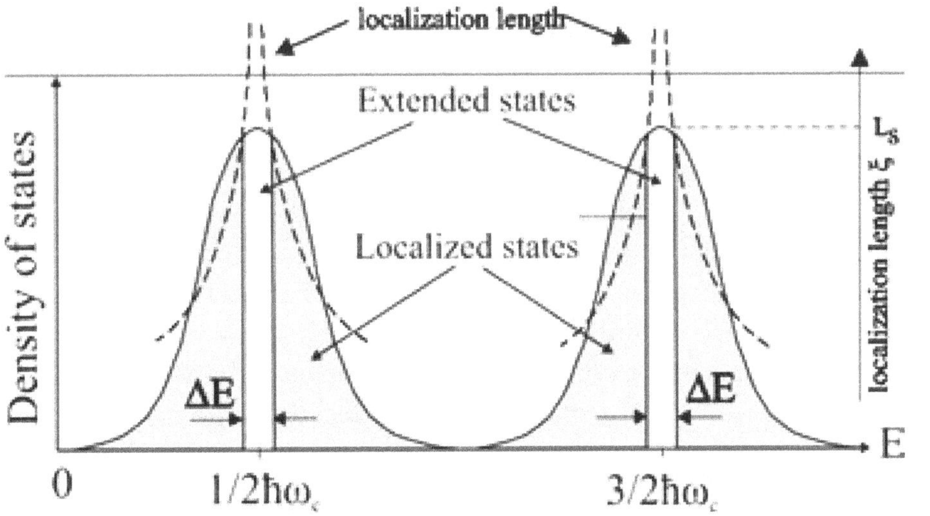

Figure 4. Schematic representation of the localization picture of the integer quantum Hall effect. It illustrates the density of states (DOS) for the two lowest Landau levels (LL). The energy interval ΔE indicates that part of the DOS around the centre of the LL, in which there exist delocalized states with respect of the sample size. The shaded tails of the DOS are explained to contain localized states. The dashed lines indicate schematically the dependence of the localization length ξ on the energy of the states relative to the LL centre.

This suggests that at zero temperature the localization length would diverge to infinity for one particular energy right at the centre of the LL. For finite sample size, any localization length larger than the sample size produces consequently delocalized states that open up a window of de-localized states around the LL centre. But also a finite temperature opens up such a window because a finite temperature brings in a phase uncertainty that destroys quantum-driven localization mechanisms like, for example, Anderson localization for localization lengths above certain values. Intensive experimental investigations of this scaling behavior as well as theoretical investigations confirmed that the scaling exponent results from quantum percolation, which can be seen as the main message for the ongoing chapter.

In conclusion of the scaling theory, the experimentally obtained scaling exponent suggests that bulk transport is indeed driven by quantum percolation. We also want to emphasize at this point that the Chalker–Coddington network model is a powerful tool for successfully dealing with the localization problem, but definitely it deals with the stationary states of the system at the Fermi level at equilibrium. The mapping onto the scaling behavior of experimentally investigated quantities is the most important contribution of the CC model for understanding the QHE, but it does not directly explain the excess current flow as it happens in real experiments.

4. Introducing non-equilibrium

From the previous section, we conclude that the channels of non-vanishing DOS at the Fermi level which develop at the intersections of the LLs with the Fermi energy are indeed quantum channels. Now, we need a model that captures the behavior of these channels if introducing non-equilibrium like in real experiments.

In **Figure 3**, there has been already shown a QHE sample with edge channels, as introduced by Büttiker [6]. The blue lines are the edge channels that appear as intersections of the LLs that follow the uprising potential at the edges of the sample. Topologically opposite pairs of channels belong to a single channel that follows the edge all around the sample. However, current contacts at the ends break up the loop, leaving two opposite directed channels, but they have still to be understood as parts of one common channel (still created by one and the same many-particle electron system). These opposite directed channels can be seen as the two opposite branches in k-space of an ordinary one-dimensional channel as depicted in Section 2. From this point of view, such a pair of opposite located edge channels have still to be understood as one single one-dimensional channel that in total is able to contribute the quantized conductance. The most important criterion is that one must not consider the edge currents as real currents located just at the edges while current flowing just in narrow channels. Using the arguments given in Section 2, the edge channels are not more than the locations on the sample area, at which one can communicate with the electron system close to equilibrium by transport experiments (the channels are the possible walk ways on the dip of the "many-particle iceberg"). But it is not allowed by quantum physics to even think that single electrons are moving just there semiclassically like in a narrow classical conductor. In other words, the so-called edge current is not a current flowing at the edges; it is carried by the edge channel pair as a whole that is just a representation of the whole electron system. According to the many-particle character of the whole electron system, all electrons are delocalized in between, which means that the path for single electron is completely unknown between the sample boundaries, as it should be in a quantized system. On this background, the historic battle between the so-called edge channel picture and bulk-current picture of the QHE has no real meaning and is just a result of overstretching the single-electron picture as discussed in Section 2.

If we look at the bulk region, the interceptions with the Fermi energy produce mostly closed loops around hills or valleys of the potential fluctuations which would encircle those in opposite directions. Physically, those loops are the same like the edge channels which run all around the sample, except that edge channels will be interrupted by contacts, while in the bulk region these channels stay to be closed loops. The only way to connect to those loops is either by tunneling if such loops come close to each other at potential saddles, or if such loops finally start to merge, while the Fermi level touches the saddle energy.

The plateau regime of the quantum Hall effect is perfectly described by the edge channel picture that in this context means indeed the channels that run along the edge, while the bulk channels do not contribute. However, the physically most interesting and challenging problem is the transition regime between QH plateaus, at which also a bulk current is flowing. The representation of this bulk current is therefore is the major goal of the ongoing discussion.

A fundamental building block of our approach is the tunneling regime between loops at the potential saddles. Such saddles exist in real electronic structures in a random potential, but as will be shown subsequently they can be used also as the nodes or building blocks of a network approach in order to discriticize a realistic random potential distribution.

For this purpose, we represent such a saddle by some kind of circuit element that consists of two counter-propagating channels that get close to each other and couple, for example, by tunneling.

Figure 5 shows the situation of two counter-propagating channels, while they get close at the potential saddle and can couple by tunneling. The channels are part of closed loops that can couple also at other saddles to other loops as will be shown below. The most important aspect now is how a tunneling current between the channels has to be handled. In this context, our discussion in Sections 2 and 3 play a major role. There we argued that the counter-propagating channels can be seen as those regions, at which transport close to thermal equilibrium is possible, but all channels belong to the same many-particle quantum state as a whole. If we forget about the association with particular channel pairs unlike done in **Figure 5**, we get two possibilities for building channel pairs, as would be the case if considering the encircled node as a black box. The one is like shown in **Figure 5**, but there is also the possibility to build a pairs with channels 1 → 4 and 3 → 2. Considering the saddle as black box both combinations are possible on equal basis and therefore have to be just different representations of the same problem. If we now consider an introduced non-equilibrium at input channels 1 and 3 that means that the input potentials V_1 and V_3 are set to be different, the associated currents have to serve consistently both possibilities to build channel pairs. Using channel pair 1 → 2 and 3 → 4, they represent a horizontal current transport in **Figure 5** from the right to the left by $I_{xx}=(\mu_1-\mu_4)\cdot {}^{e}/_{h}$ that is also $I_{xx}=(\mu_2-\mu_3)\cdot {}^{e}/_{h}$. Using the notation for the quantum Hall effect I_{xx} would serve as sample current, $(\mu_1-\mu_4)$ and $(\mu_2-\mu_3)$ would be measured like Hall voltages, while $(\mu_1-\mu_2)$ and $(\mu_4-\mu_3)$ would be measured like longitudinal voltage drops. However, considering 1 → 4 and 3 → 2 as channel pairs, they represent a current in vertical direction

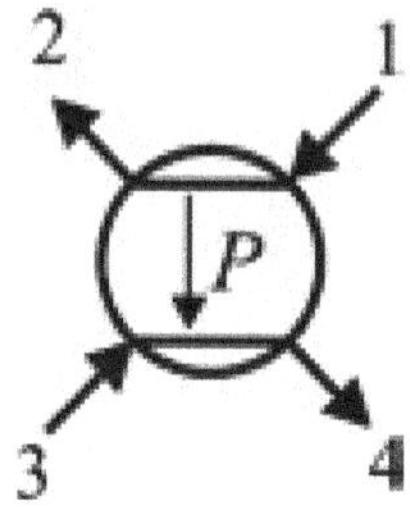

Figure 5. Representation of the situation at a saddle point that can be used as a node of a network. The channels 1 → 2 and 3 → 4 pass each other and the tunneling rate between them is represented by the factor $P = R / T$ where R means the tunneling probability and $T = 1 - R$.

$I_{yy} = (\mu_1 - \mu_2) \cdot {}^{e}/_{h}$ that is also $I_{yy} = (\mu_3 - \mu_4) \cdot {}^{e}/_{h}$. For the quantum Hall notation, in this case, I_{yy} would serve as sample current, $(\mu_1 - \mu_2)$ and $(\mu_4 - \mu_3)$ would be measured like Hall voltages, while $(\mu_1 - \mu_4)$ and $(\mu_2 - \mu_3)$ would be measured like longitudinal voltage drops. In this context, we can interpret the saddle situation in **Figure 5** as a single-channel quantum Hall sample, for which we define the longitudinal voltage drop in horizontal direction as $V_{xx} = V_1 - V_2$ that is also given by $V_{xx} = V_4 - V_3$ and the Hall voltage $V_{xy} = V_1 - V_4$ that is also given by $V_{xy} = V_2 - V_3$. Using further the definition of the longitudinal resistance $R_{xx} = {}^{V_{xx}}/_{I_{xx}}$ and the Hall resistances $R_{xy} = {}^{V_{xy}}/_{I_{xx}}$, we can write the following equations:

$$R_{xx} = \frac{R}{T} \cdot \frac{h}{e^2} \tag{4}$$

$$R_{xy} = \frac{h}{e^2} \tag{5}$$

The above equations use also the fact that the relation between the currents ${}^{I_{yy}}/_{I_{xx}}$ is identical to the associated relation between the probabilities for single carriers that means ${}^{I_{yy}}/_{I_{xx}} = {}^{R}/_{T}$ and we define $P = {}^{R}/_{T}$. There is another interesting fact to be checked: A current I_{xx} and a longitudinal voltage V_{xx} imply a dissipated power of $P_{xx} = I_{xx} \cdot V_{xx}$ in the node. However, since a particular choice of the channel pairs is just one of two possible equivalent choices for the same electron system, the other choice of tunneling current I_{yy} and transverse (Hall) voltage V_{xy} with dissipated power $P_{yy} = I_{yy} \cdot V_{xy}$ should represent the same power dissipation $P_{yy} = P_{xx}$, which is indeed the case and can be easily checked. Turning the node by 90°, which means flipping the x and y directions leads just to a replacement of P by ${}^{1}/_{P}$. The most important aspect is that we can use this node as a kind of building block to build a more complex circuit as a network. Also in real, it happens that several loops get arranged in the bulk region of a real sample that may get interconnected by tunneling. Therefore, as the next step we use a periodic arrangement in two directions to fill the space in two dimensions. This is schematically shown in **Figure 6** below.

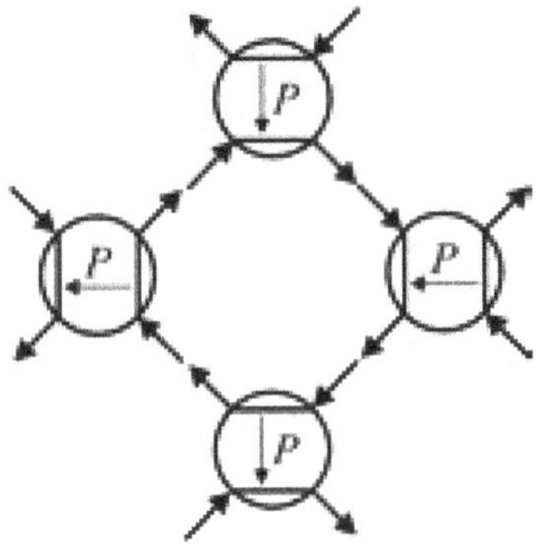

Figure 6. Arrangement of the nodes in Figure 5 in order to resemble a single closed loop in the bulk that allows coupling to adjacent loops in order to fill up the whole area.

From **Figure 6**, one can easily imagine that it is possible to fill up the space with any number of adjacent loops that may fill the bulk region. However, besides the Eqs. (4) and (5), we need a more compact relation that allows obtaining the output potentials of the nodes in terms of the input potentials. For this purpose, we rewrite Eqs. (4) and (5) in terms of potential differences, which means $(\mu_1-\mu_2)=(\mu_4-\mu_3)=\left(R/T\right)\cdot(\mu_1-\mu_4)=\left(R/T\right)\cdot(\mu_2-\mu_3)$. After some steps of derivations, we can show that for each single node, the potentials of the input channels are translated to the potentials of the output channels by some kind of transfer matrix:

$$\begin{pmatrix}\mu_2\\ \mu_4\end{pmatrix}=\begin{pmatrix}T & R\\ R & T\end{pmatrix}\cdot\begin{pmatrix}\mu_1\\ \mu_3\end{pmatrix} \tag{6}$$

This provides a basis for a numerical approach to find the lateral distribution of the potentials over the whole network. It should be emphasized at this point that these are potentials that have to be understood as the deviations from the equilibrium potentials that would be all the same in thermal equilibrium, which of course is the trivial solution for the network. For a non-trivial solution, we need at least two channels of the whole network to be fixed at different potentials, so that the remaining network has to adjust self consistently to these non-equilibrium potentials.

Considering the single nodes of the network, the tunneling current between adjacent loops depends on the relative position of the Fermi energy and the saddle energy, while the smoothness of the saddle potential serves an important parameter. However, this also means that a formulation in terms of the filling factor should be possible too. In Ref. [13], this problem has been considered in detail and formulations have been presented for the dependence on the Fermi energy as well as the dependence on the filling factor. The main results will be discussed in the following and for details please refer to the cited literature [13].

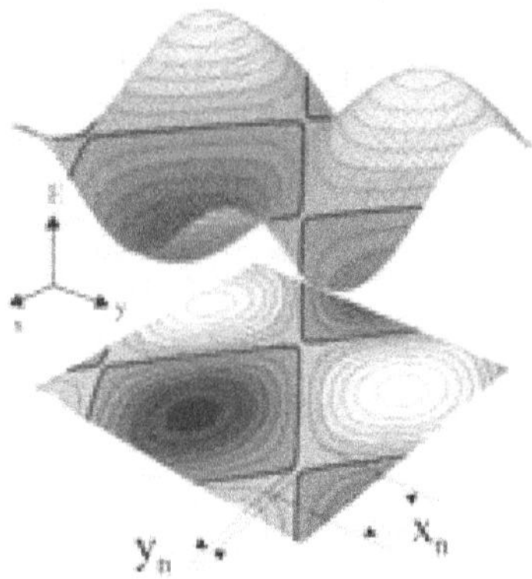

Figure 7. Saddle potential that is produced by a two-dimensional Cosine function; x_n and y_n are the tunneling distance between loops, if they come close to each other in either x or y direction.

As a first step, the situations have been analyzed for saddle potentials that are created by a two-dimensional Cosine function. However, the choice of this special function is not a restriction to a special case, because tunneling becomes important only if the Fermi level gets close to the saddle energy at which the opposite channel can get close to each other. In this regime, the saddle potential can be represented already in very good approximation by the second-order saddle curvature, which is very well parametrized by a Cosine function in both directions. **Figure 7** shows a part of the potential modulation near such a saddle. The bold line for this case represents the channels at a Fermi energy that is slightly below the saddle energy, at which the loops encircle the potential valleys and get close to each other in y-direction by y_n. If the Fermi level moves slightly above that saddle energy, it is easily seen that the loops will encircle potential hills in opposite direction and they get close to each other in x direction by x_n, which resembles a node that is turned by 90°. On this basis, the following equation has been found in Ref. [13]:

$$P = exp\left[\pm\frac{L^2E_F}{e\cdot V_0}\cdot\frac{e\cdot B}{h}\right] \tag{7}$$

In this equation, E_F is the Fermi energy relative to the saddle energy, L is the period, and V_0 is the amplitude of the representing Cosine function that has been taken into account up to the second-order Taylor expansion. If the Fermi energy crosses the saddle energy, the argument of the exponent changes sign which automatically means that $P \rightarrow {}^{1}\!/_{P}$ and at the same time, the node turns by 90°, as already mentioned above. Considering one square of length L, that is, one full period L of the Cosine function in both directions, the above equation can be mapped onto the filling factor scale [13]:

$$P = exp\left[-\Delta\nu\cdot L^2\cdot\frac{e\cdot B}{h}\right] \tag{8}$$

$\Delta\nu$ is the difference of the filling factor relative half filling. The above equation suggests that at half filling at $\Delta\nu=0$ exactly half of the area defined by L^2 is occupied with electrons.

As can be seen in **Figure 8b**, at exactly half filling when the Fermi level meets the saddle energy, exactly half of the area appears as shaded, which means filled with electrons. If the Fermi level is slightly above the saddle like shown in **Figure 8a**, a small shaded gap opens right at the saddle that indicates that the shaded area starts to dominate, and at the same time, the shaded areas get interconnected to each other across the saddle, while the unshaded areas shrink to isolated droplets. The opposite happens if the Fermi level is slightly below the saddle energy like shown in **Figure 8c**, because in this case, a gap of empty states now opens up at the saddle that now interconnects the unshaded areas and leaving the shaded areas as isolated droplets. This behavior is in agreement with the so-called bulk-current picture of the QHE regime that is explained as a mixture of a so call QH liquid phase (shaded area) and an insulating phase (unshaded areas) [14]. Due to the bulk-current picture, a plateau transition happens, if the one phase takes over from the other and this happens around the half-filled LL. In context with **Figure 8** and Eq. (8), we get the quite simple and universal looking equation $P=\exp[-\Delta N]$, where ΔN is the number of electrons that are either in excess or missing from an exactly half-filled Landau level on a square defined by the period L.

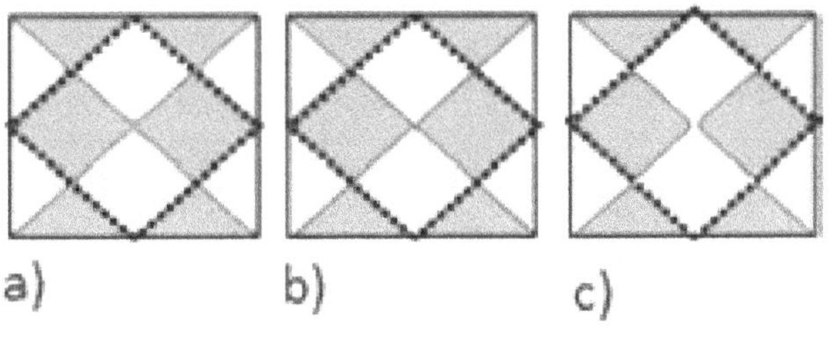

Figure 8. Top view of a complete period L containing a central saddle at (a) Fermi energy slightly above saddle energy, (b) Fermi energy exactly at saddle energy, (c) Fermi energy slightly below saddle energy. The areas filled with electrons are shaded; the unshaded areas are empty.

A major necessary further step is now the introduction of a disorder potential. This is achieved by the superposition of a random potential to the so far regular saddle potentials. That leads to the effect that no longer all nodes behave the same. Looking at the filling factor representation, a random potential also leads to a lateral fluctuation of the local filling factors at the positions of the nodes.

Including an additional long range disorder potential, the saddle energies get nonuniformly distributed and a partly transmission and reflection at the nodes appears only at those locations where the saddle energy stays close to the Fermi energy. This usually happens only for a small fraction of the nodes in the network, like indicated schematically in **Figure 9**. All nodes (saddles) with energy far from the Fermi energy have either full transmission (*T = 1, R = 0*) or full reflection (*T = 0, R = 1*), depending on whether the Fermi energy is above or below the saddle energy. In this way, the network guides the channels around potential fluctuations and only those nodes at which the saddle energy gets close to the Fermi energy become physically active as tunneling junctions (red arrows in **Figure 9**). In this context, our network, which we call non-equilibrium network model (NNM) can be understood as a method to approximate arbitrarily shaped magnetic bound states and it also finds the "hot spots" (active saddles of the disordered potential).

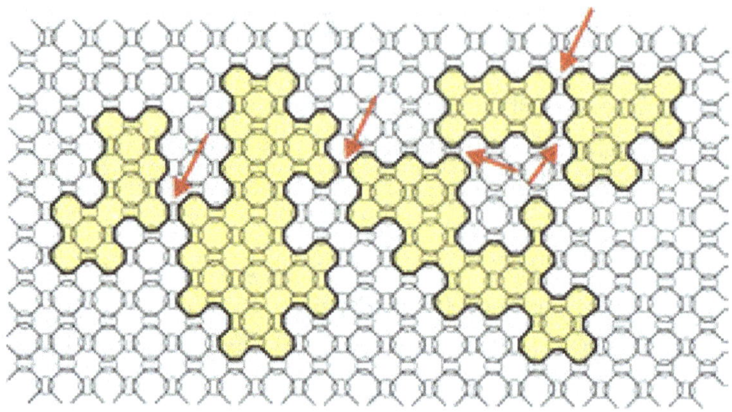

Figure 9. Schematic representation of the effect of a random potential that is superimposed on the regular periodic grid of saddles. The yellow areas indicate regions of saddle potentials below the Fermi energy that lead to higher carrier density and hence local filling factors larger than ν=0.5, while the white areas indicate larger saddle potentials above the Fermi level leading to lower carrier density and hence local filling factors lower than ν=0.5. At the boundaries between these regions we cross the situation of half-filled LLs, where the Fermi energy meets saddles and as discussed already before, while crossing the saddle energy the involved nodes turn by 90°. This can be seen by either following a particular row or column of nodes if we enter or leave the yellow colored area.

The NNM as introduced so far is a standalone model for describing the non-equilibrium transport that means the lateral distribution of the disruption from thermal equilibrium. The physics, how the screened potentials or local filling factors are obtained as input information for the nodes, is not directly addressed by the NNM. There exist different levels for approaching the underlying electron system from semiclassical screening effects that neglect exchange interaction to a full treatment as many-particle system on, for example, the Hartree–Fock level. While for the semiclassical model, the screened electrostatic potential can be used as an input for the network, this might not be sufficient for the Hartree–Fock treatment, because the exchange contribution might dominate over the pure classical electrostatic part of the carrier distribution, and therefore, there is good reason to use the network representation in terms of the local filling factors.

Another very important aspect of our non-equilibrium network model is that it addresses finally only the lateral distribution of the non-equilibrium excitation potentials without explicitly relating to quantities like local currents or local conductivities, which would be in violation of the many-particle physics, because of implying an existence of individual electrons in the system. The currents of the non-equilibrium are not addressed while solving the problem. Only as a post-processing step, the currents are calculated just at the contacts where the system communicates with the outside world. Therefore, our network model is compatible also with the demands of many-particle physics, and it is independent of the type of carriers that show up as quasi-particles just at the contact in a post-processing step. In this post-processing step, the type of carriers become relevant only by the way how the current balance at the contact has to be done. For electrons, this is the usual quantized conductance times the potential difference. At this point it should be reminded that also for the usual electrically confined one-dimensional channels the quantized conductance $G = 2 \cdot e^2/h$ appears only in context with the presence of metallic contacts. However, there is no reason why the same network model should not be used for fractional channels exactly in the same way, where the only modification would be that the current balance at the contact has to be done for the fractional charges of composite fermions. This is the subject of future work.

5. Implications

Many implications of the various approaches and arguments as given above can be found retrospective already in the literature, and therefore, the next paragraph is devoted to numerous already existing implications. For better readability in the above modeling sections, only some of those citations concerning primary resources are contained. It is also worth to mention that the presentation as it appears above is not the chronological process of a recent development. The chronological process contains many partly independent steps collected over almost 20 years of research that finally have been but together by the time. Therefore, there already exists a huge punch of related literature, in which supporting implications for many of the different features and aspects of our model can be found. In this section, we will address part of this additional background literature for the reader to gather some further insights of different aspects, however, without being able to make a full review that would definitely go beyond the boundaries of this chapter.

Let us start directly with Eq. (8) that is one of the major ingredients of our model. Almost 20 years ago, a similar equation has been found to be necessary in order to meet specific symmetry conditions of quantum Hall measurements [15]. That equation has been applied to a quantum Hall sample as a whole and the typical behavior now turns out to apply also for a single node of our network model. Even without creating a full network, there have been attempts to put together an equivalent circuit for a quantum Hall sample. Each channel pair of the whole sample has been handled like a single node of our network, but only for the channel pair representing the top LL backscattering ($P > 0$) has been assumed. The total sample has been modeled by parallel connection of the suitable number of edge channel pairs [16]. Although there have been strong indications that tunneling may play an important role, the particular shape as an exponential function of P versus filling factor ν at that time was found initially just by symmetry requirements. A similar function had been found also by Shahar et al. [17], who needed it for interpretation of experimental data in context with deviations from scaling theory that the authors have been observing. Their results have been inspiring in the sense that the scaling theory is an important aspect of the QHE, but maybe not directly the driving force for the existence of the QHE. A first version of the network model has been formulated on the basis of local filling factors [18], which in turn is based on a special earlier version for addressing the Hall insulator [19]. The tunneling process was explicitly addressed later when looking for a better way to introduce the disorder potential that was expected to be dominated by the screened electrostatic potential [13]. Using just a semiclassical approach to screening for modeling and shaping the samples structures has already proven to be sufficient for successfully modeling many features of quantum Hall samples in terms of all geometrical aspects like length and distance between contact arms [20], effects of gate electrodes [21, 22], effects due to the different equilibration lengths [20], effects due to the number of contact arms and non-local effects [23], inhomogeneous magnetic fields [24] and also most complex sample geometries like an anti-Hall bar structure that is embedded in a Hall bar [25–27]. The latter demonstrates the appearance of simultaneous independent Hall effects for the inner and outer boundaries of the same electron system that is supplied in parallel by different independent constant current sources. That example was at the same time a demonstration of the flexibility of our network model in terms of addressing most complex experimental conditions.

But even much earlier, we have found indications of such a unique behavior when coupling quantum channels that carry different electrochemical potentials, which now turn out to look similar like the formulations for our network model. That behavior has been found by investigating magneto-transport in quasi-3D systems that do not show quantum Hall effect. In this context, we have performed magneto-transport experiments and modeling for the so-called wide parabolic quantum wells [28, 29]. These structures are no longer two-dimensional because the sub-band splitting is smaller as compared to the LL splitting, which means that those structures are somehow in the regime between 2D and 3D. Although there have not been observed any quantum Hall plateaus, relations between longitudinal and transverse (Hall) resistance have been found that in a retrospective view also support the relations between the chemical potentials as discussed above. The wide parabolic quantum wells at large filling factors had been created by the Narrow Gap Semiconductor PbTe and instead of quantum Hall plateaus magneto-resistance fluctuations have been found at temperatures below 100 mK and at small current of the order of a few nA. The temperature and current dependence of these fluctuations that also appeared in a so-called non-local contact configuration even at macroscopic sample size [30] indicated the involvement of quantum channels [31]. In this context, the relations between the potentials of channels that meet at non-equilibrium situation seem to be valid also for that regime far from the standard quantum Hall regime in the clean two-dimensional case. That can be seen as a hint that the basic principles addressed in the previous sections seem to be more general and hold also beyond the standard quantum Hall regime. There are also other examples from the literature, in which a behavior similar to that in a single node was found by Haug et al. [32]. However, that was derived just for the special case of a gate initiated scattering experiment in the QHE regime and has not been considered further from a more general point of view. Another interesting connection can be found with the work of Chklovskii et al. [33], where the authors deal with the conductance of the central compressible bulk channel. Such a compressible bulk channel appears in the transition regime between conductance plateaus, while successively depopulating the LLs by either increasing the magnetic field or successively narrowing a channel by spit gates until the density in the center falls below the density of a fully populated LL. Besides the screening properties, which the authors handled as a pure classical electrostatic problem, they use some kind of semi empirical formula in order to translate the electron density in the central compressible bulk stripe into a conductance of that stripe:

$$G = \frac{e^2}{h} \cdot \nu(0) \tag{9}$$

Here, $\nu(0)$ means the filling factor in the center of the compressible bulk channel at $\nu(0) < 1$. This Eq. (9) has been extracted from a paper of Beenakker et al. [34] who deals with edge channel transport in the fractional quantum Hall regime. Equation(9) can be seen as an appropriate approach to give a correct interpolation between conductance steps. At this point, we want to extend the discussion of our model in order to demonstrate that our approach captures the main features of the approach of these authors as well. We make a simplification and use only a single node of our network in order to capture the situation of two opposite channels that merge in the centre of a sample. This situation has been already addressed in Ref. 13 (there in context to **Figure 14**). It can be also understood if considering just a single node like in **Figure 5**. Without backscattering (P = 0), the quantized conductance of the node for transmission from right to left is $G = e^2/h$. This value would be measured only if terminals 2 and 3 of the node are connected together at an ideal metallic contact on the left and terminals 1 and 4 are connected to another ideal metallic contact of the right. If applying a potential difference between those contacts, a current according to that quantized conductance e^2/h would flow. If we now merge the channels in the centre of the nodes, reflection and transmission get equal probability, which means $R = T = 0.5$ and $R/T = 1$. According to Eq. (4), this means that the junction itself represents the universal conduction e^2/h that adds another universal conductance in series to the backscattering free case, which means that now we have connected in series two times the universal conduction which gives in total:

$$G = \frac{e^2}{h} \cdot 0.5$$

This fits exactly Eq. (9) for $\nu(0) = 0.5$, which means a half-filled LL. Obviously, this is also in agreement with Eq. (8), for which the exponent vanishes right at a half-filled LL. The advantage in our case is that we can extend our model to a network in order to capture further details like length, width, and also inhomogeneities of the metallic bulk stripe, while the authors of Ref. [33] have to restrict them self to special cases and assumptions. Another point is that the approach of Chklovskii et al. provides no handle for extending their approach beyond semiclassical treatment of the stripes, and therefore, they neglect many body effects that have been found to play an important role in the quantum Hall regime. Since we consider current transport from a most general point of view and taking into account facts related to many-particle physics, the application of our transport model has no such restriction and can be applied also to a modeling based on many body effects, as will be demonstrated in the next section.

The fact, that there can be found numerous implications for the basic behavior that is represented by the nodes of the network, suggests that there are more general aspects behind that might be interesting also for communities outside the quantum Hall community. That was finally the motivation to propose this chapter as a very fundamental step toward introducing some kind of dynamic in terms of stationary net current flow close to thermal equilibrium while starting from the stationary many-particle ground states. In the final chapter, we present some most recent results for the quantum Hall regime on the basis of a many-particle system by applying the Hartree–Fock approximation for finding the stationary ground state and introducing the presented treatment of non-equilibrium for obtaining the longitudinal and Hall resistances as a function of magnetic field.

6. Recent results

In the previous section, we have presented a selection of examples for implications of our model that can be already found in the literature that give support for the results of our modeling. This is done mainly just by referring to the literature without reproducing the results in detail. Therefore, the reader is advised to refer to the cited literature for further details. In this paragraph, we want to present most recent results by taking into account the many-particle physics up to the Hartree–Fock level including disorder. The particular numerical Hartree–Fock approximation as used here has been outlined in several publications, and therefore, the reader is also advised to look up further details in the given Refs. [35–37]. Here, we will just present the results for two examples.

6.1. Imaging of condensed quantum states

The first example deals with a many-particle system that contains only few electrons that get localized in some model potential and which do not directly contribute to transport. This system is considered as embedded in a conducting environment and the impact of the charge distribution of such a localized state on the tunneling barrier of a nearby quantum point contact (QPC) is taken into account. Such a QPC can be understood as a saddle potential that acts as a charge detector for nearby charge distributions. In this case, the nodes of the network for transport at the QPC are controlled by the local potentials based on Eq. (7). For demonstration, an artificial confining potential ring (see **Figure 10**) has been created and filled with eight electrons. These eight electrons are treated as a many-particle system using the Hartree–Fock method and as shown in **Figure 11** the charge density resembles a periodic structure along the ring that can be understood also as a one-dimensional Wigner crystal. Here, just the main results are presented and for further details please refer to Ref. [38]

The eight confined electrons have been treated as a many-particle system separately from the conducting environment that has been represented by a semiclassical approach based on Eq. (7) of the NNM. The presence of the charge distribution in the ring was taken into account by using the recalculated Hartree potential of the charge distribution in the ring at the potential saddles of the QPCs. The transmission of the QPCs, that have been modeled using the network model as described above, has been calculated, while a repulsive potential distortion was superimposed at different locations in the ring area. The potential distortion represents the moving tip of a so-called scanning gate microscope [39]. The effect of the tip on the condensed quantum state affects the charge distribution, which affects the nearby saddle potentials that finally modulates any tunneling current in the environment nearby the ring. If a current in the environment of the ring passes those saddles, this current give a response if the SGM-tip hits the charge maxima of the condensed state. If the total sample current is recorded as a function of tip position, the response pattern creates something like an image of the charge distribution of the quantum state.

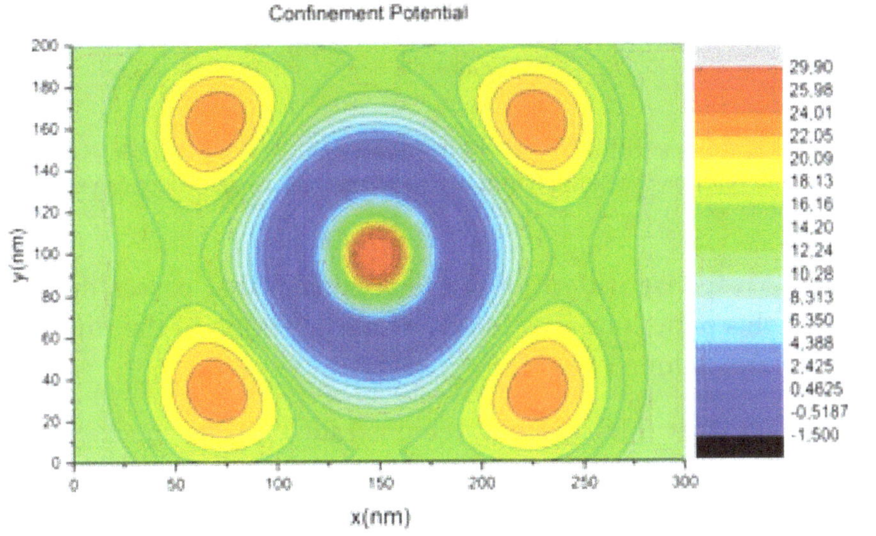

Figure 10. Contour plot of the bare ring-shaped confinement potential with a diameter of about 100 nm and a depth of about 15 meV as compared to the saddle energies of the QPCs.

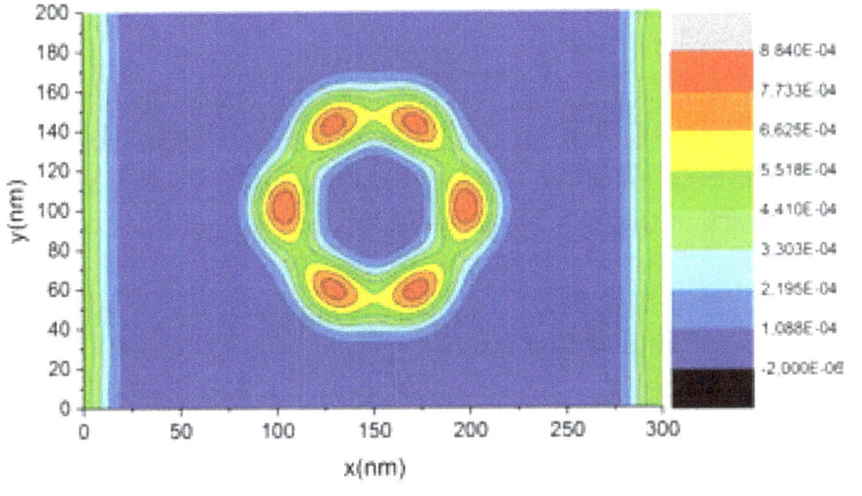

Figure 11. Contour plot of the charge distribution obtained from the Hartree–Fock calculation in multiples of 10^{14}cm^{-2} for a magnetic field of B = 2.5 Tesla. In this regime, the electron system appears fully spin polarized.

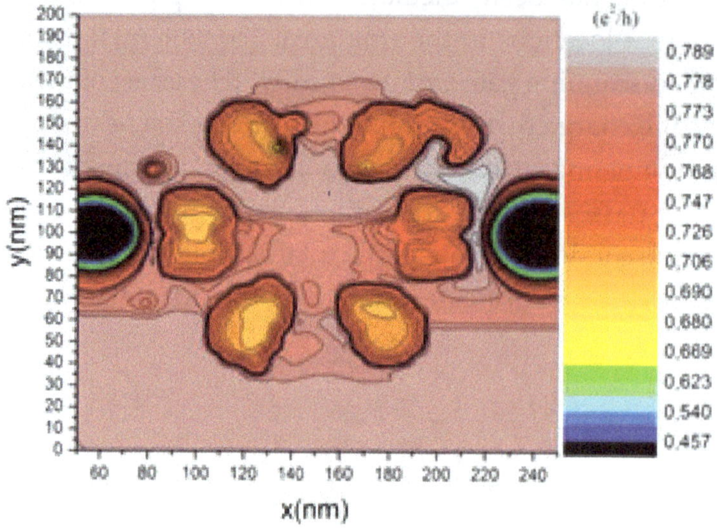

Figure 12. SGM response pattern as a contour plot of the two-point conductance (in multiples of e^2/h) of a nearby QPC in the tunneling regime as a function of SGM tip position. Tip parameters: circular Gaussian shape of repulsive height 1 mV and half-width diameter of 20 nm.

Figure 12 shows the response pattern of the total current that passes the nearby potential saddles and indeed this pattern images the charge distribution in the ring, even though the electrons in the ring are excluded from transport.

It is important to note that this is just a demonstration of the mechanism that transfers information from a decoupled localized many-particle quantum state to a current response of a nearby electron system. A more realistic structure requires a much larger system size of up to several microns length that so far is out of reach for our simulation capabilities. Therefore, in our example, the space beyond the QPC's that carries the environmental current can be understood to be folded back to the same area of the localized states and thus is used twice as some kind of second layer in the model, but assuming that this current bypasses the localized states. The necessary extensions of the model are the subject of ongoing work which needs also to acquire more computing power.

Anyway, this example demonstrates the basic mechanism that allows the SGM-method to create images of charge distribution from condensed quantum states and serves at the same time as an example for the application of the network model for transport where the driving force is coulomb interaction that can be addressed by using Eq. (7) for controlling the nodes of the network.

6.2. Quantum transport at large filling factors

The previous example for a system of just few electrons is far from a fully filled Landau level, and the interaction with the environment was still just coulomb interaction that can be handled with the network model on the basis of Eq. (7) as demonstrated. Now, we turn to the extreme opposite regime of many electrons that fill up several LLs. That means that now we have of the order of 500 electrons in the system instead of eight electrons that is addressed by the

Hartree–Fock model. We use a random potential at an average carrier density of 2×10^{11} cm^{-2}. Because of the many-particle physics at higher-filled LLs, we have to expect that the usage of the Hartree potential for the network will be insufficient because we expect a significant impact from the exchange interaction. In this case, the better representation of the electronic system for the NNM is therefore the local filling factor that is controlled by both the Hartree and the exchange interaction. Once more we want to bring to attention that also Beenakker [34] had found the local filling factor to be relevant for the transport behavior also in the fractional quantum Hall effect, where many-particle physics plays the dominant role. We have compared both methods and plotted the results on top of each other. Exchange-driven effects are lost if using just Eq (7) that also means that enhanced spin splitting is lost because that results from the exchange-driven enhancement of the g-factor. In order to get the right position of the peaks in QHE, one has to introduce the exchange-enhanced g-factor into the NNM, while no such adjustments are needed if using the filling factor version based on Eq. (8). By choice of using Eq. (7) or (8) for the simulation and comparison with experiment, we have a tool to distinguish between effects that are either strongly affected by exchange interaction or not. From our simulations, we get the strong indication that exchange effects are dominating almost all transport regimes that involve more than a single-filled LL even at moderate magnetic fields between B = 1–2 Tesla. That has a significant impact on screening effects that turn out to be strongly reduced as compared to semiclassical modeling as based on, for example, a Thomas-Fermi approach. Investigations concerning the screening properties in context with many-particle physics are in progress.

In the following, we present an example for the simulation of QHE measurements of a disordered sample. The sample is of mesoscopic size of 600 nm length and 400 nm width. We used two degrees of disorder and compare also the different calculation methods, the full many-particle approach in combination with the filling factor version of the NNM as well as the Hartree version.

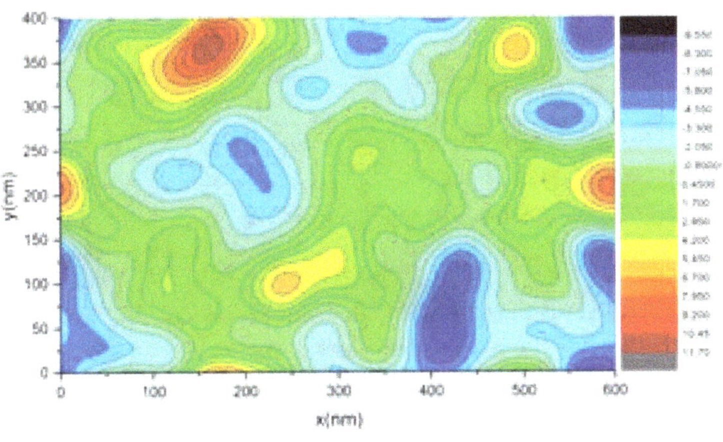

Figure 13. Disorder potential for the case of strong disorder as used for the HF simulation of the bulk region. The colors represent the potential in mV as given in the color map.

Figure 13 shows the bulk potential distribution for the case of strong disorder as used for the Hartree–Fock calculation of the bulk region. Alternatively also a low disorder potential was used that makes use of the same random distribution of Gaussian peaks in order to produce the random potential, but the Gaussians are just half of amplitude and half of width that ends up in a weaker disorder potential. In all cases, an average fixed carrier density of 2×10^{11} cm^{-2} has been used, that results in 480 electrons in the sample area, that get distributed.

The lateral carrier distribution obtained from the Hartree–Fock calculation is shown in **Figures 14** and **15** which show the carrier distributions for spin 1 and spin 2 separately. This is important since it turns out that the many-particle system avoids as good as possible to have partly filled spin 1 and spin 2 LLs at the same time. The calculation was done at magnetic field B = 2.3 Tesla that means a total filling factor close to $\nu = 3.5$. This means automatically that the higher second spin 1 LL remains partly filled, while the lower second spin 2 LL is fully occupied. As can be seen in **Figure 14**, this average filling factor of $\nu = 1.5$ of the spin 1 LLs is achieved by creating areas that are close to filling factor $\nu = 2$ in red and filling factor $\nu = 1$ in blue. That regions are separated by green stripes around filling factor $\nu = 1.5$. An indicated enhanced occurrence of filling factors close to of $\nu = 1.5$ is the subject of ongoing investigations. According to Eq. (8) these green-colored stripes will create stripes of good transmission of the excitation potential in the network model and represent at the same time also the so-called compressible stripes, while the other regions remain insulating. Looking at **Figure 15** for the spin 2 electrons, one can see that the dominating part of the sample area remains close to filling factor $\nu = 2$ which therefore do not contribute to transport that also will have only a weak contribution to screening.

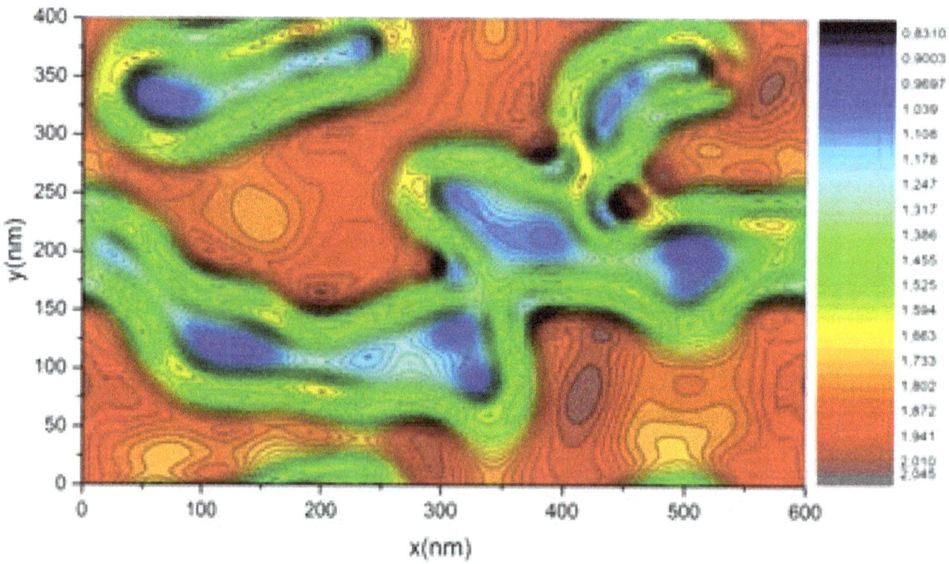

Figure 14. Lateral carrier distribution of the spin 1 electrons, mapped on the filling factor scale. The color mapping ranges over three colors from blue to green to yellow that spans a filling range from about $\nu = 1$ to $\nu = 2$, which means that green appears around filling factor $\nu = 1.5$.

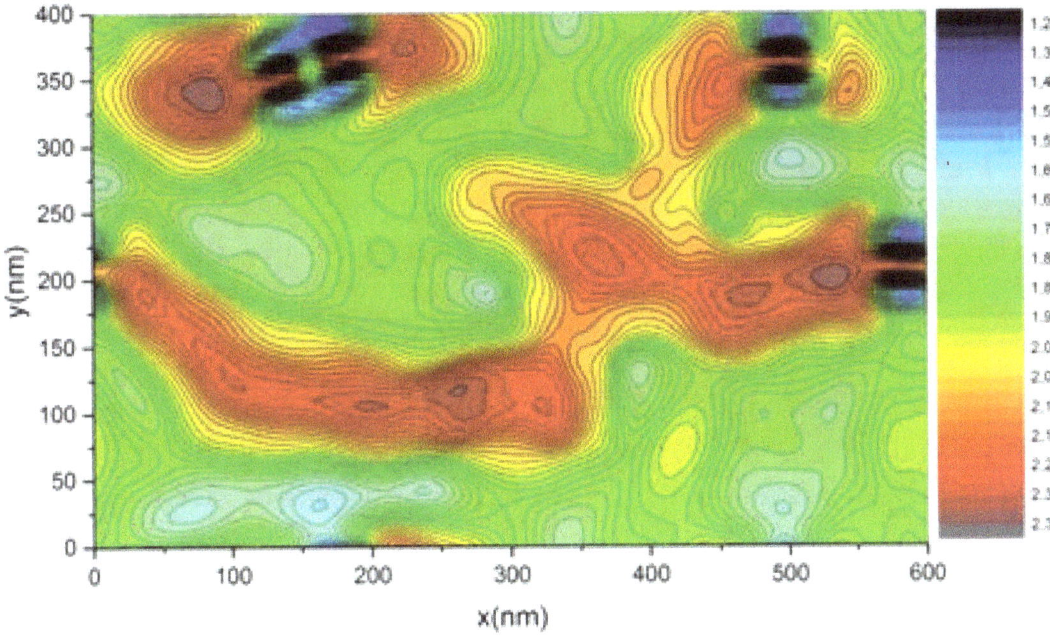

Figure 15. Lateral carrier distribution of the spin 2 electrons, mapped on the filling factor scale. The color mapping ranges over three colors from blue to green to yellow that spans a filling range from about $\nu = 1.5$ to $\nu = 2.3$, which means that green appears close filling factor $\nu = 2$.

The whole calculation has been repeated for different magnetic fields, and the network model has been used to solve the lateral distribution of the excitation potential. For this purpose, the non-equilibrium potential has been set to different fixed values at the supposed current contacts on the left and right boundaries.

In **Figure 16**, there is shown the solution for the lateral distribution of the excitation potential for the same situation like in **Figure 14**. By comparing with **Figure 14**, the transmitting stripes that evolve in the network follow the green stripes. The color represents the potentials that those stripes pick up at the boundaries they are coming from and they may mix up a locations where channels of different excitation potentials meet. By comparing with **Figure 13**, one can see that those stripes follow just roughly the contours of the disorder potential, but deviate from the details. This is a clear indication that besides the Hartree contribution also the exchange interaction is important. After finding the lateral distribution of the excitation potential as shown in **Figure 16**, only directly at the current contacts along sides 1 and 2, the currents are calculated by using Eq. (2) for each channel pair that is arriving at each point of the metallic contacts. Voltage probes have been created at locations 3–6 at the boundaries by just placing a metallic contact, but keeping the potential numerically floating, which means that no current is extracted or injected at those locations. A Hall voltage can be obtained, for example, by taking the difference of the excitation potentials between contacts 3 and 5 or 4 and 6. If dividing by the obtained sample current, we get the Hall resistance that is plotted as a function of the magnetic field. The same can be done for the longitudinal resistance by taking first the potential differences between contacts 3 and 4 or 5 and 6 and dividing by the total current.

Figure 16. Screenshot of the NNM showing the lateral distribution of the injection potential in the network model. The high potential appears in red and is injected from the left, and the low potential appears in green and is injected from the right. The chirality of the edge channels appears as clockwise. Expect the top spin1 LL, all lower LLs create narrow channels at the network boundaries that experience no backscattering.

In this way, the simulation follows the same procedure like any real experiment that measures voltages and currents and only then the values are translated into resistances, without calculating a total resistance or conductance in terms of butting together elements of any kind of equivalent circuit. In this way, we avoid to use any quantities that might be related to local behavior in terms of conductance or local Ohm's law for the electron system and thus consequently avoid any steps that could be used to attribute single electrons to particular locations or areas of the sample. Constant colour of the stripes in **Figure 16** indicates that there is no current flow within those stripes because of the missing gradient of the chemical potential. The current therefore has to be attributed to the mostly dark regions between stripes of different potential. The current path within these regions remains unknown, like demanded by a quantum system. If stripes of different potential meet the appearing potential gradient (crossing over the yellow colour) between them indicates dissipation in this region. Sometimes numerical noise can also lead to yellow colour in usually de-coupled (black) regions, which, however, does not harm the total current balance.

Figure 17 shows the magnetic field-dependent two-point conductance for strong and low disorder potential and also for the two different methods of incorporating the Hartree–Fock solution into the NNM. Using just the Hartree part of screening of the random potential on the basis of Eq. (7), we have the possibility of manually taking into account exchange-enhanced spin splitting or not. When neglecting the enhanced spin splitting only the even conductance plateaus appear (dashed blue). If introducing the enhanced g-factor manually all plateaus appear correctly also in the Hartree version (red). However, the transition regime looks different if looking at details. Especially, the $\nu = 5$ plateau of the Hartree version (red color) is almost wiped out because of a strong overlap of the spin split $\nu = 5$ LL (see also below). The strongest deviations in the transition regime are between the case of strong (pink color) and weak disorder (green color), which is not really surprising. For strong disorder, there appear additional overshoots and dips, which result from the fact, that the strong disorder for this small sample dimensions acts already as an inhomogeneity.

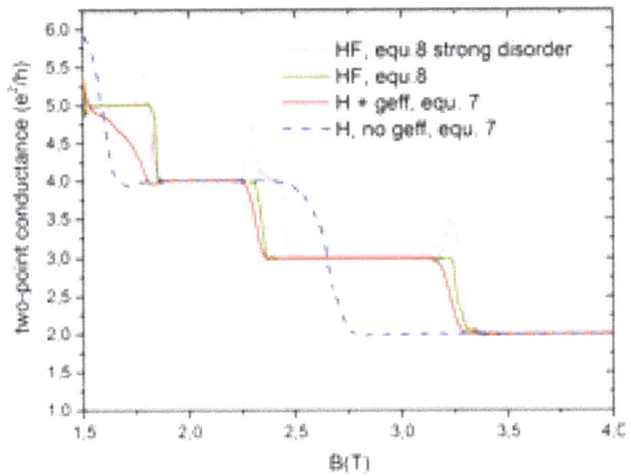

Figure 17. Total sample current mapped out as two-point conductance as a function of the magnetic field. The simulation was done for the structure shown in Figure 13 for the shown strong disorder potential and the not shown weaker disorder potential. HF means that the full Hartree–Fock contribution has been used in the NNM by using the local filling factor on the basis of Eq. (8); H means that the full Hartree–Fock carrier distribution has been obtained as well, but from that only the Hartree potential has been recalculated in order to get its contribution to screening of the random potential and Eq. (7) has been used in the NNM. If using just the Hartree part of screening on the basis of Eq. (7), the LLs have to be calculated separately on the basis of a semiclassical model directly in the NNM, which also requires to use the g-factor the for spin splitting as a semi empirical parameter; the label geff means that a semi empirical enhanced g-factor of geff = 14 has been used, while no geff means that enhanced spin splitting has been neglected.

If extracting potential differences between Hall voltage probes and calculating the Hall resistances for the same cases as in **Figure 17** as a function of magnetic field, we obtain **Figure 18**.

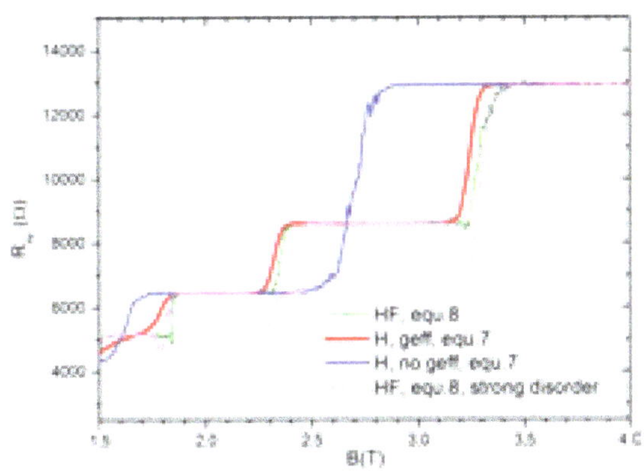

Figure 18. Hall resistance for the same parameters as described in Figure 17.

The Hall plateau values for all cases are of course the same, also the plateau transitions appear almost at the same magnetic fields, just the transition itself is different for the different calculations. An exception is again the non-spin split case (blue) where only the even Hall plateaus appear. The slight fluctuations in the transition regime comes from the fact that also the Hartree part of the Hartree–Fock solution fluctuates with changing magnetic field, and this is communicated to the NNM as well. When using the enhanced g-factor in the higher magnetic field regime (red), the agreement between the two versions is quite good. Only at low magnetic fields, there is again a strong deviation between the Hartree and the HF version because the semiclassical representation of the LLs in the Hartree version of the NNM overestimates the LL overlap between different spins. This results in a wiped out ν = 5 plateau of red curve. But we know that the Hartree–Fock solution tries to avoid different partly filled spin LLs at the same time, and therefore, the ν = 5 plateau of the HF versions is still well developed (green). For the strong disordered and almost inhomogeneous sample, we get strong fluctuations and over shoots in the plateau transitions (pink)

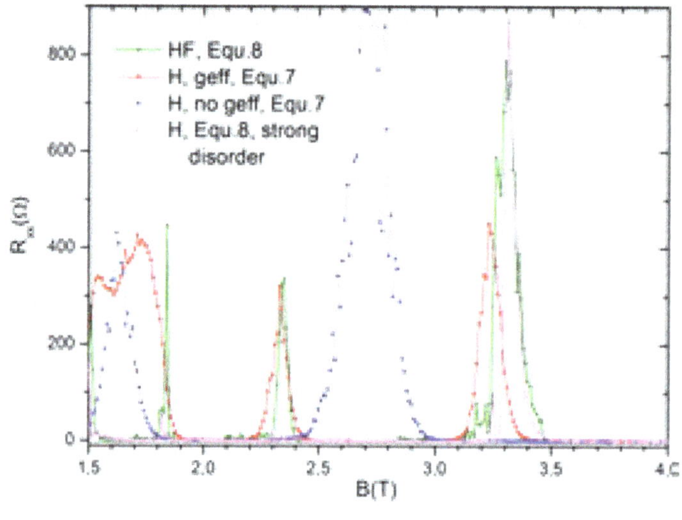

Figure 19. Longitudinal resistance versus magnetic field for the same cases as explained in context with Figure 17.

Figure 19 finally shows the longitudinal resistance for the different cases. The overall shape is covered with strong magneto-resistance fluctuations for all cases, as one expects from QH samples of mesoscopic size. The peak positions of the Hartree version with correctly set enhanced g-factor (red) matches quite well the HF-version (green), again except at fields below B = 2 Tesla. Here, the Hartree version (red) produces a strongly overlapping double peak that results from the overlapping spin 1 and spin 2 levels of the third LL. Exchange interaction counteracts this overlap, and therefore, there appear two separated sharp peaks for the HF version (green), one at B = 1.5 Tesla that corresponds to $\nu = 5.5$ and the other at 1.85 Tesla that corresponds to $\nu = 4.5$. For strong disorder, the R_{xx} peaks (pink) are almost missing in the field range below 3 Tesla. That can be understood by the fact, that the strong disorder produces already an inhomogeneity that can lead to the fact, that not all edge channels reach also the voltage probes at the boundaries, as e.g. can be seen in **Figure 16**. This results in the fact that any dissipation that is generated by the two inner stripes is not monitored by the contacts at the boundaries, and therefore, the associated R_{xx} peaks are missing. The blue dashed curve again represents the Hartree version without enhanced spin splitting, which means that the R_{xx} peaks appear at odd filling factors ν = 5 and ν = 3, where both overlapping spin 1 and spin

2 levels of LL 3 and LL2 are taken into account as half filled, which is not the case in the HF solution because of the enhanced spin splitting.

7. Summary

Based on a fundamental discussion of the many-particle aspects of the electron system, we have presented a modified way to look at the many-particle electron system in terms of a single-particle picture that considers channels for current flow from a more general point of view. While quantum channels like the edge channels in the QHE regime can be identified as locations on the sample area where current exchange for transport experiments is possible, one must not attribute any current flow exclusively to those channels like for classical wires; instead, the whole electron system is understood to carry the sample current, while it is not possible to track individual electrons through the electronic system. Instead of directly addressing current flow, we describe the lateral transmission of the experimentally introduced non-equilibrium electrochemical potentials between such channels on the bases a network approach. This network is used to find a solution for the lateral distribution of the injected non-equilibrium without needing to consider explicitly current flow or local conductances. Conductances come into play only in context with experiments when extracting electrons just directly at contacts by relating the applied non-equilibrium potentials and the resulting flow of electrons through these contacts. Only the lateral distribution of the non-equilibrium potentials is found by our network model on the basis of an iterative procedure, and currents are only obtained as a post-processing step just directly at the current contacts where the electrons are extracted like also in real experiments! This makes our transport model also applicable for systems that forbid the way of thinking in terms of single electrons moving through the electron system, like this is the case in many-particle systems. This has been demonstrated by applying our transport model to the many-particle electron system for which realistic results have been achieved by simulating quantum Hall measurements for disordered samples of mesoscopic size. This makes our network model also applicable for the fractional quantum hall regime, which is the topic of future work.

Acknowledgements

JO acknowledges support from Austrian Science Foundation FWF Project P19353-N16. JO thanks R. Römer for his important cooperation in context with implementing the Hartree-Fock method for the non-equilibrium network model, as well as for substantial discussions and providing access to computing facilities via partial support from EPSRC Grant EP/J003476/1 and provision of computing resources through the MidPlus Regional HPC Centre Warwick UK (EP/K000128/1).

Author details

Josef Oswald

Address all correspondence to: josef.oswald@unileoben.ac.at

Institute of Physics, Montanuniversität Leoben, Leoben, Austria

References

[1] Vonklitzing, K., Dorda, G., Pepper, M. New method for high-accuracy determination of the fine-structure constant based on quantized Hall resistance. Physical Review Letters. 1980; 45(6): 494–497.

[2] Huckestein, B. Scaling theory of the integer quantum Hall-effect. Reviews of Modern Physics. 1995;67(2): 357–396. doi:10.1103/RevModPhys.67.357

[3] Hohls, F., Zeitler, U., Haug, R. J. Hopping conductivity in the quantum Hall effect: Revival of universal scaling. Physical Review Letters. 2002; 88(3): 036802. doi:10.1103/PhysRevB.88.036802

[4] Engel, L. W., Shahar, D., Kurdak, C., Tsui, D. C. Microwave frequency-dependence of integer quantum Hall-effect—Evidence for finite-frequency scaling. Physical Review Letters. 1993; 71(16): 2638–2641. doi:10.1103/PhysRevLett.71.2638

[5] Kubo, R. Statistical-mechanical theory of irreversible processes. 1. General theory and simple applications to magnetic and conduction problems. Journal of the Physical Society of Japan. 1957; 12(6): 570–586. doi:10.1143/jpsj.12.570

[6] Buttiker, M. Absence of backscattering in the quantum Hall-effect in multiprobe conductors. Physical Review B. 1988; 38(14): 9375–9389.

[7] Roothaan, C. C. J. New developments in molecular orbital theory. Reviews of Modern Physics. 1951; 23(2): 69–89. doi:10.1103/RevModPhys.23.69

[8] Oswald J. On the Landauer formula and the edge channel picture of the integer quantum Hall effect. In: Miura, N., Ando, T., editors. Proceedings of the 25th International Conference on the Physics of Semiconductors, Pts I and Ii; Sep 17–22, 2000; Osaka, Japan. Springer Proceedings in Physics; 2001. pp. 977–978.

[9] Sohrmann, C., Oswald, J., Roemer, R. A. Quantum percolation in the quantum Hall regime. Quantum and Semi-Classical Percolation and Breakdown in Disordered Solids. 2009; 762: 163–193. doi:10.1007/978-3-540-85428-9_6

[10] Hashimoto, K., Sohrmann, C., Wiebe, J., Inaoka, T., Meier, F., Hirayama, Y., Roemer, R. A., Wiesendanger, R., Morgenstern, M. Quantum Hall transition in real space: From

localized to extended states. Physical Review Letters. 2008; 101(25): 256802. doi:10.1103/PhysRevLett.101.256802

[11] Chalker, J. T., Coddington, P. D. Percolation, quantum tunnelling and the integer Hall-effect. Journal of Physics C: Solid State Physics. 1988; 21(14): 2665–2679. doi: 10.1088/0022-3719/21/14/008

[12] Kramer, B., Kettemann, S., Ohtsuki, T. Localization in the quantum Hall regime. Physica E-Low-Dimensional Systems & Nanostructures. 2003; 20(1–2): 172–187. doi: 10.1016/j.physe.2003.09.034

[13] Oswald, J., Oswald, M. Circuit type simulations of magneto-transport in the quantum Hall effect regime. Journal of Physics-Condensed Matter.2006; 18(7): R101–R138. 18(7): R101–R138. doi:10.1088/0953-8984/18/7/r01

[14] Ruzin, I., Feng, S. C. Universal relation between longitudinal and transverse conductivities in quantum Hall-effect. Physical Review Letters. 1995; 74(1): 154–157

[15] Oswald, J., Span, G., Kuchar, F. Universality in the crossover between edge-channel and bulk transport in the quantum Hall regime. Physical Review B. 1998; 58(23): 15401–15404

[16] Oswald J. A new model for the transport regime of the integer quantum Hall effect: The role of hulk transport in the edge channel picture. Physica E. 1998; 3(1–3): 30–37

[17] Shahar, D., Hilke, M., Li, C. C., Tsui, D. C., Sondhi, S. L., Cunningham, J. E., Razeghi, M. A new transport regime in the quantum Hall effect. Solid State Communications. 1998; 107(1): 19–23

[18] Oswald, J., Homer, A. A new network model for the integer quantum Hall effect. Physica E-Low-Dimensional Systems & Nanostructures. 2001; 11(4): 310–322

[19] Oswald, J., Homer, A., Ganitzer, P. A network model for the Hall insulator. Microelectronic Engineering. 1999; 47(1–4); 31–33

[20] Oswald J. The role of edge channel equilibration for the R_{xx}—Peak height of the integer quantum Hall effect. In: 15th International Conference on High Magnetic Fields in Semiconductor Physics, SEMIMAG, 5–9 Aug, 2002, Oxford, UK, CD-ROM

[21] Oswald, J., Uiberacker, C. Gate controlled separation of edge and bulk current transport in the quantum Hall effect regime. Journal of Low Temperature Physics. 2010; 159(1–2): 180–183

[22] Oswald, J., Uiberacker, C. Gate Controlled Narrowing of the Quantum Hall Effect Plateau Transitions. In: Goll, G., Lohneysen, H. V., Loidl, A., Pruschke, T., Richter, M., Schultz, L., Surgers, C., Wosnitza, J., editors. International Conference on Magnetism (ICM 2009), Jul 26–31, 2009, Karlsruhe, GERMANY. Journal of Physics Conference Series. 2010; 200. doi:10.1088/1742-6596/200/1/012153

[23] Oswald, J., Uiberacker, C., Stecher, C. A Numerical Study of Non-local Magneto Transport Effects in Quantum Hall Device Structures. In: Muraki, K., Takeyama, S.,

editors. Horiba International Conference: The 19th International Conference on the Application of High Magnetic Fields in Semiconductor Physics and Nanotechnology. Journal of Physics Conference Series. 2011; 334. doi:10.1088/1742-6596/334/1/012020

[24] Oswald, J., Uiberacker, C. A Numerical Study of Magneto Transport in 2D Electronic Systems in the Presence of Non-Uniform Magnetic Fields. In: Ihm, J., Cheong, H., editors. Physics of Semiconductors: 30th International Conference on the Physics of Semiconductors; Jul 25–30, 2010, Seoul, South Korea. AIP Conference Proceedings 1399; 2011.

[25] Oswald, M., Oswald, J., Mani, R. G. Voltage and current distribution in a doubly connected two-dimensional quantum Hall system. Physical Review B. 2005; 72(3): 035334

[26] Oswald, J., Oswald, M. Magnetotransport in a doubly connected two-dimensional quantum Hall system in the low magnetic field regime. Physical Review B. 2006; 74(15): 153315

[27] Uiberacker, C., Stecher, C., Oswald, J. Microscopic details of the integer quantum Hall effect in an anti-Hall bar. Physical Review B. 2012; 86(4): 045304

[28] Oswald, J., Span, G. Novel non-local behaviour of quasi-3D wide quantum wells. Semiconductor Science and Technology. 1997; 12(3): 0268–1242

[29] Oswald, J., Span, G., Homer, A., Heigl, G., Ganitzer, P., Maude, D. K., Portal, J. C. Anomalous magnetotransport in wide quantum wells. Solid State Communications. 1997; 102(5): 391–395. doi:10.1016/s0038-1098(97)00017-3

[30] Oswald, J., Heigl, G., Span, G., Homer, A., Ganitzer, P., Maude, D. K., Portal, J. C. Conductance fluctuations in PbTe wide parabolic quantum wells. Physica B. 1996; 227(1–4): 360–362. doi:10.1016/0921-4526(96)00442-5

[31] Homer, A., Ganitzer, P., Span, G., Oswald, J. A novel transport model for the metal-to-insulator transition in quasi-three-dimensional wide quantum wells. Superlattices and Microstructures. 1999; 25(1–2): 191–196. doi:10.1006/spmi.1998.0636

[32] Haug, R. J., Kucera, J., Streda, P., Vonklitzing, K. Scattering experiments in two-dimensional systems in the presence of quantizing magnetic-fields. Physical Review B. 1989; 39(15): 10892–10900. doi:10.1103/PhysRevB.39.10892

[33] Chklovskii, D. B., Matveev, K. A., Shklovskii, B. I. Ballistic conductance of interacting electrons in the quantum Hall regime. Physical Review B. 1993; 47(19): 12605–12617. doi:10.1103/PhysRevB.47.12605

[34] Beenakker, C. W. J. Edge channels for the fractional quantum Hall-effect. Physical Review Letters. 1990; 64(2): 216–219. doi:10.1103/PhysRevLett.64.216

[35] Roemer, R. A., Sohrmann, C. Hartree-Fock interactions in the integer quantum Hall effect. Physica Status Solidi B: Basic Solid State Physics. 2008; 245(2): 336–343. doi: 10.1002/pssb.200743321

[36] Sohrmann, C., Roemer, R. A. Kubo conductivity in the IQHE regime within Hartree-Fock. Physica Status Solidi C: Current Topics in Solid State Physics. 2008; 5(3): 842–847. doi:10.1002/pssc.200777586

[37] Sohrmann, C., Roemer, R. A. Compressibility stripes for mesoscopic quantum Hall samples. New Journal of Physics. 2007; 9. doi:10.1088/1367-2630/9/4/097

[38] Oswald J., Römer R. A. Imaging of condensed quantum states in the quantum Hall effect regime. Physics Procedia. 2015; 75: 314–325. doi:10.1016/j.phpro.2015.12.038

[39] Sellier, H., Hackens, B., Pala, M. G., Martins, F., Baltazar, S., Wallart, X., Desplanque, L., Bayot, V., Huant, S. On the imaging of electron transport in semiconductor quantum structures by scanning-gate microscopy: Successes and limitations. Semiconductor Science and Technology. 2011; 26(6): 064008. doi:10.1088/0268-1242/26/6/064008

3

Unitary Approaches to Dissipative Quantum Dynamics

Matteo Bonfanti and Rocco Martinazzo

Abstract

We describe in detail a general system–bath strategy for investigating the quantum behavior of small systems interacting with complex environments. In this approach, a simplified heat bath is used as a surrogate for realistic environments, and explicit, unitary quantum simulations of the "universe" (the system *plus* the bath) are performed by means of high-dimensional wave-packet techniques. In this chapter, we describe the underlying Hamiltonians and the related reduced dynamical descriptions, show how to recast real-world problems into this form, introduce some of the methods currently used to deal with high-dimensional quantum dynamics, and present the results of this strategy when applied to numerous problems of physicochemical interest.

Keywords: System–bath dynamics, Multi-configuration time-dependent methods, Generalized Langevin equation, Brownian motion, Effective modes

1. Introduction

Recent years have witnessed an ever growing interest in dynamical processes that occur in complex environments, for example, ground- and excited-state molecular reaction dynamics in condensed phase, charge and excitation energy transfer in organic functional materials and biomaterials, and elementary processes at the gas–solid interface [1–6]. Their importance is hardly overemphasized, because of the key role they play in fields as diverse as catalysis, optoelectronics, nanotechnology, biochemistry, and astrophysics, just to mention a few.

The common structure of these problems—a relatively simple system that can be measured and manipulated and that is coupled to an environment only partially under control—has long been subjected to thorough theoretical investigation. Since Einstein's [7] seminal work on

Brownian motion, many important analytical results have been obtained regarding the statistical description of the effect that a large medium—being it a surface or a solvent—has on the dynamics of the small system of interest [8]. The environment, usually designated as the "bath," is seen to exert two different kinds of force, a friction, and a stochastic force. They result, respectively, in dissipation and fluctuations in the system dynamics and represent two opposite but intimately related effects that ultimately lead to the establishment of equilibrium. This has been made apparent since Langevin formulated the first sound description of an open-system dynamics in 1908, with his equation of motion (the Langevin Equation, LE) and its Generalized version (GLE) [8–10].

These ideas have been extensively deepened in classical mechanics in the following years and thoroughly validated in both model and realistic systems by several numerical experiments, that is, explicit molecular dynamics (MD) simulations of a "universe" (the system *plus* the bath) comprising a huge number of degrees of freedom [11].

In quantum dynamics, the situation is considerably more complicated. A *brute-force* approach is out of reach because of a well-known exponential scaling problem; hence, much of the research in this field focused on reduced dynamical descriptions and aimed at obtaining reasonable master equations for the system density operator [12]. In these open-system quantum dynamical approaches, the degrees of freedom of the environment are traced out and the system undergoes a dissipative, non-unitary dynamics. These approaches are exact in limiting cases only, since most often strong assumptions are needed to obtain manageable equations in a closed form. Among these, the Markov approximation is often invoked (and an effective coarse graining of the dynamics performed) on the basis that the environment correlations last much less than the characteristic time of the system dynamics, a condition that not always holds in practice. Lifting these constraints is possible, for instance, with the help of auxiliary density matrices, but a price of an enormous increase of complexity.

An intermediate possibility between the impractical *brute-force* approach and the limited reduced dynamics is the so-called system–bath approach, whereby one introduces a surrogate for the environment and explicitly describes its degrees of freedom in the dynamics. In this effective description, the bath is a collection of simple systems (*e.g.*, harmonic oscillators [13–15]) and evolves with a relatively simple dynamics. Hence, it is possible to exploit the progress that has been made in the last 20 years or so in propagating high-dimensional wave packets in time [16]. These wave-packet approaches make the unitary evolution of the universe (the system *plus* the bath) computationally feasible for a large number of coordinates, in many cases large enough to mimic true dissipative environments. The expectation values of interest can then be extracted from the full dynamical evolution using the relevant system operators, and thermal effects can be handled by sampling the mixed initial state of the universe.

In this chapter, we present the work done in the last few years in mapping a physical problem of interest into a system–bath model—the so-called independent oscillator (IO) model—and in solving such model with multi-configuration wave-packet approaches. The dissipative processes investigated range from small amplitude, damped vibrations in model anharmonic systems to "real-world" problems such as hydrogen atom sticking on graphene.

The chapter is organized as follows: Firstly, we describe the IO model in the framework of classical and quantum statistical mechanics, with a focus on its relationship to the generalized Langevin equation and on the role played by the so-called spectral density (SD) of the environmental coupling $J_0(\omega)$ [13,14,17–19]. In addition to the well-established results, we include some recent developments that improved the numerical appeal of the model [20–28]. Secondly, we address the problem of mapping a complex (realistic) dynamics into an IO model and deriving the appropriate SD [29–31]. We focus in particular on *dynamical* approaches that suit well to the current practice of accessing dynamical information using *on-the-fly* simulations, thereby bypassing the need of computing and fitting accurate model potentials. The third part of this contribution deals with the problem of the dynamical evolution of a small system coupled to its IO bath. We describe the now standard and numerically exact multi-configurational time-dependent Hartree (MCTDH) approach [16,32,33], as well as related approximate approaches which better suit to the description of a bath in the IO form. In particular, we discuss in some detail the local coherent-state approximation (LCSA), where the bath evolution is described by a number of Hartree products of pseudo-classically evolving coherent states [34–36]. All these approaches are variational and, as such, they share a number of highly desirable features which will be described in some detail. Finally, we present the results of some numerical investigations on both model and realistic systems. Issues, such as vibrational relaxation, decoherence, and scattering, have been extensively investigated in model systems (a harmonic or a Morse oscillator coupled to an oscillator bath) [27,29,34–39] and will be summarized in the following. Work on "real-world" problems is still at its infancy, but already offers some notable examples. For instance, we have recently settled a long-standing issue concerning chemisorption of hydrogen atoms on graphene and obtained the first fully quantum and numerically converged results for the probability that the atoms stick on the surface [40,41]. We describe these first exciting results and further provide an outlook of the application of our strategy to other challenging physicochemical problems.

2. Independent oscillator models

The IO Hamiltonian is a popular and extremely powerful tool to study the dynamics of an open system in a quantum setting [13–15]. Here, we discuss its connection with the generalized Langevin equation, emphasizing the role played by the SD in the mapping between the two [18,19]. Later, we introduce an effective mode transformation that casts the IO bath into a linear-chain form which suits well to truncation schemes [20–22,25,26,28].

2.1. Generalized Langevin equation and SD

The generalized Langevin equation describes the dynamics of a Brownian particle in both the classical and the quantum (Heisenberg) setting. It is a stochastic equation for a system degree of freedom s of mass m subjected to a deterministic potential V, a random Gaussian force ξ and a friction term, determined by a memory kernel $\gamma(t)$,

$$m\,\ddot{s}(t) + m\int \gamma(t-\tau)\,\dot{s}(\tau)\,d\tau + V'\big(s(t)\big) = \xi(t) \tag{1}$$

Causality of the memory kernel ($\gamma(t) = 0$ for $t < 0$) has important implications for the analytic properties of its Fourier transform $\tilde{\gamma}(\omega)=\int\gamma(t)e^{i\omega t}dt$ when continued to the upper-half complex ω-plane. The SD of the environmental coupling, $J_0(\omega)$, is defined with the help of the real part of $\tilde{\gamma}(\omega)$

$$J_0(\omega) = m\omega\,\Re\tilde{\gamma}(\omega) \tag{2}$$

and fully determines the memory kernel by virtue of the Kramers–Kronig relations, namely through

$$\gamma(t) = \frac{2}{\pi m}\Theta(t)\int_0^\infty \frac{J_0(\omega)}{\omega}\,cos(\omega t)\,d\omega \tag{3}$$

where $\Theta(t)$ is the Heaviside step function. It further determines the stochastic process $\xi(t)$, provided is Gaussian, by virtue of a fluctuation–dissipation (FD) theorem of the second kind, [1]

$$\gamma(t) = \frac{2}{\pi m}\Theta(t)\int_0^\infty \frac{J_0(\omega)}{\omega}\,cos(\omega t)\,d\omega \tag{4}$$

here written for a quantum environment (the classical result follows in the limit of high temperatures, $\beta = (k_B T)^{-1} \to 0$ and takes the form $\xi(t)\xi(0) = \Theta(t)m\gamma(t)/\beta$). Hence, all the environment-related terms included in the GLE are uniquely defined by the SD.

Once $J_0(\omega)$ is known, it can be used to construct an IO Hamiltonian[2]

$$H^{IO} = \frac{p_s^2}{2m} + V(s) + \sum_{k=1}^{\infty}\left[\frac{p_k^2}{2m_k} + \frac{m_k\omega_k^2}{2}\left(x_k - \frac{c_k}{m_k\omega_k^2}f(s)\right)^2\right] \tag{5}$$

that can be made (quasi) equivalent to the GLE above by appropriately choosing the harmonic oscillator (HO) frequencies and the coupling coefficients. In Eq. (5), the system degree of freedom s is coupled to a collection of harmonic oscillators (x_k, p_k) of mass[3] m_k and frequency ω_k. The system–bath coupling is a linear function of the bath coordinates, whereas its dependence on the system coordinate is here specified through the function $f(s)$, which typically

---

[1] Here and in the following ... denotes an average over the canonical equilibrium.

[2] This is also known as Caldeira–Leggett Hamiltonian, after the seminal work by Caldeira and Leggett on the effects of dissipation on quantum tunneling [17].

[3] In the following we will adopt, without loosing generality, the same mass for all the oscillators, *i.e.* $m_k \equiv \mu$ for all k, where μ is a numerically convenient value.

complicates the GLE (by introducing state-dependent friction) but is often necessary on physical grounds. The simple linear coupling which makes Eq. (5) equivalent to the standard GLE of Eq. (1) is appropriate to cases where only the near equilibrium configurations of the system are explored ($s \approx 0$), but is clearly limited because it describes a coupling which steadily increases when moving the system out of its equilibrium position. Thus, for instance, in previous scattering calculations using a model Morse potential, a coupling function with a finite limit for $s \to +\infty$ [27,36],

$$f(s) = \frac{1-e^{-\alpha s}}{\alpha} \tag{6}$$

(but yet such that $f(s) \approx s$ for $s \approx 0$ was used to correctly describe the asymptotically free system.

Notice that in Eq. (5), a system potential counter-term $f(s)^2$ appears which balances the distortion induced by the system–bath coupling and ensures the thermodynamic stability of the Hamiltonian [14]. Indeed, in the form given in Eq. (5), the bath adds only quadratic terms (the sum on the r.h.s.) to the system Hamiltonian, thereby guaranteeing a lower bound to the energy spectrum for any reasonable system potential.

The equivalence between the two dynamical formulations [Eqs. (1) vs. (5)] is established when the coupling coefficients sample the SD $J_0(\omega)$ of the problem, for example, for evenly spaced bath frequencies $\omega_k = k\Delta\omega$, when the coefficients are set according to

$$C_k = \sqrt{\frac{2}{\pi}\,\mu\,\omega_k\;\Delta\omega\;J_0(\omega_k)} \tag{7}$$

It rigorously holds for a finite time only, determined by the size of the bath in Eq. (5). For longer times, Eq. (1) keeps on describing a dissipative dynamics, whereas the Hamiltonian dynamics of Eq. (5) displays the consequences of discretely sampling the bath frequencies. More precisely, the equivalence is guaranteed up to the Poincaré recurrence time $t_P = 2\pi\ /\ \Delta\omega$ of the finite system, which thus needs to be set larger than the any time scale of interest. This has to be done by choosing the appropriate discretization $\Delta\omega$, compatibly with a reasonable number of oscillators N and a high frequency cutoff $\omega_c = N\Delta\omega$ of the bath. The latter determines the smaller time that can be resolvedt$_c = 2\pi\ /\ \omega_c$; higher frequencies, if present, can always be absorbed in a *mass* renormalization term provided we are not interested in times smaller than t_c.

Unlike the starting GLE, the IO Hamiltonian of Eq. (5) can be quantized by applying standard quantization rules. Furthermore, the relatively trivial dynamics followed by the harmonic oscillators of the bath makes the use of standard time-dependent wave-packet approaches to the dynamics possible. As long as a system–bath Hamiltonian can be effectively mapped into a GLE, this represents a powerful and general methodology to tackle an open-system quantum dynamical problem.

2.2. Chain representation of the bath

Recent work has shown that the IO Hamiltonian of Eq. (5) can be expressed in an alternative representation which is particularly suited to truncation schemes or hierarchical descriptions of the dynamics [20,23,24,27,42]. The idea is incredibly simple and reads as follows. Eq. (5) naturally introduces a collective or "effective" mode $X_1 = \sum_{k=1}^{N} c_k x_k / D_0$ (here $D_0^2 = \sum_{k=1}^{N} c_k^2$ is just a kind of normalization constant) that allows one to write the interaction term as

$$H_{\text{int}} = -\sum_{k=1}^{N} c_k x_k f(s) = -D_0 X_1 f(s) \tag{8}$$

The definition of this effective mode fixes the first "column" of an orthogonal transformation of the original bath coordinates into a new set of coordinates, otherwise arbitrary. The rest of the transformation matrix can be fixed by requiring that the "residual bath" is in normal form. In this way, Eq. (5) becomes an equivalent IO Hamiltonian for the *s plus* X_1 degree of freedom, coupled to a bath of $N-1$ oscillators. The coupling *only* occurs through X_1, and allows one to define a new function $J_1(\omega)$, that is, the SD felt by the mode X_1 as a consequence of its interaction with the residual bath. In the continuum limit, this procedure can be indefinitely iterated to define a sequence of effective modes X_1, X_2 … X_M … coupled in a linear chain form and a corresponding sequence of SDs $J_1(\omega)$, $J_2(\omega)$ … $J_M(\omega)$ … which describe the residual bath "felt" by each mode. The sequence of SDs is determined by a simple recurrence relation which can be started with $J_0(\omega)$ [23,24], that is, $J_{n+1}(\omega)$ only requires the previous SD $J_n(\omega)$ and two of its functionals

$$J_{n+1}(\omega) = \frac{D_n{}^2 J_n(\omega)}{|W_n^+(\omega)|^2} \tag{9}$$

Here, $W_n^{+(\omega)}$ is the limit on the real axis (from above) of the "Cauchy transform" of $J_n(\omega)$,

$$W_n(z) = \frac{1}{\pi} \int_{-\infty}^{+\infty} d\omega \frac{J_n(\omega)}{\omega - z} \tag{10}$$

and

$$D_n{}^2 = \frac{2}{\pi} \int_0^{\infty} \omega J_n(\omega)\, d\omega \tag{11}$$

determines the coupling coefficients between the nth and the $(n+1)^{\text{th}}$ effective modes. The linear chain form of the Hamiltonian further requires the frequency Ω_{n+1} of the $(n+1)^{th}$ effective mode,

$$\Omega_{n+1}{}^2 = \frac{2}{\pi D_n{}^2}\int_0^\infty \omega^3 J_n(\omega)\, d\omega \tag{12}$$

and reads, for mass-scaled coordinates, as

$$H^{chain} = \frac{p_s{}^2}{2m} + V(s) + \Delta V(s) - D_0 f(s) X_1 + \sum_{n=1}^{\infty}\left[\frac{1}{2}P_n{}^2 + \frac{1}{2}\Omega_n{}^2 X_n{}^2 - D_n X_n X_{n+1}\right] \tag{13}$$

Here, $\Delta V(s)$ is the potential counter-term

$$\Delta V(s) = \frac{1}{2}m\delta\Omega_0{}^2 f(s)^2 m\delta\Omega_0{}^2 = \frac{2}{\pi}\int_0^\infty \frac{J_0(\omega)}{\omega}\, d\omega \tag{14}$$

And $\{(X_n, P_n)\}_{n=1\ldots\infty}$ is the set of effective modes and their conjugated momenta. Equivalently, Eq. (13) can be rearranged to explicitly display its thermodynamic stability [29]

$$H^{chain} = \frac{p_s{}^2}{2m} + V(s) + \sum_{n=1}^{\infty}\left[\frac{P'_n{}^2}{2\mu} + \frac{\mu\omega_n^2}{2}(X'_n - \beta_n X'_{n-1})^2\right] \tag{15}$$

by introducing new coordinates, $X'_0 = s$ and $X'_n = X_n/\sqrt{\mu}$ for n=1,2.., the *bare* effective mode frequencies $\omega_n{}^2 = \Omega_n{}^2 - \delta\Omega_n{}^2$ (where $\delta\Omega_n{}^2$ is defined analogously to $\delta\Omega_0{}^2$ above, Eq. (14), but in terms of J_n and μ) and the coefficients $\beta_1 = D_0/\sqrt{\mu}\omega_1{}^2$ and $\beta_n = D_{n-1}/\omega_n{}^2$ for $n > 1$.

One interesting issue concerning Eq. (13) is whether a limiting residual SD exists and, in that case, which forms takes $\lim_n J_n$. It can be readily shown [23] that, if the limit exists, it is the quasi-Ohmic SD given by the Rubin dissipative model. Numerical tests have further shown that this convergence is fast enough that inclusion of a relatively small number of effective modes in an enlarged system makes the resulting dynamics effectively Markovian. Thus, the chain transformation provides a powerful tool for describing non-Markovian phenomena by means of Markovian master equations (applied to enlarged systems).[4] Further analytical work rigorously proved the convergence conditions of the sequence of spectral densities and established the connection between the chain construction and the moment problem, the theory of orthogonal polynomials, and the Padé approximants [21].

One further issue on the effective mode construction is of much practical interest and concerns its role in defining approximate representations of the bath [24]. In fact, the construction of the linear chain amounts to "unroll" the memory kernel in time—any excitation initially localized in the system necessarily moves sequentially along the chain starting from its end attached to the system. This is contrast to what happens with Eq. (5), where the coupling pattern is appropriate for a frequency resolution of the kernel. As a consequence, truncated or Markov-closed chains with n effective modes can be shown to exactly reproduce $\gamma(t)$ to the fourth order

[4] Strictly speaking such "Markovian reduction" rigorously holds in classical mechanics only; in a quantum setting the very definition of Markovian dynamics is still debated. Thus, one should better refer here to an "Ohmic embedding".

in time, up to an irrelevant constant of order $\gamma(0)/n$ [24]. Apart from its conceptual significance, this property proves to be extremely useful for numerical simulations too, since it allows one to single out those bath modes which are most important for the system dynamics. In this representation, in fact, one can easily identify "primary modes" that need to be treated at a high correlation level and "secondary modes" that can be left weakly correlated or even uncorrelated. As it will be shown in the following with some numerical examples, this simple prescription offers the opportunity of tackling long-time issues in explicit dynamical studies of system–bath problems [27].

3. Mapping of a complex system to an IO model

The possibility of using the IO Hamiltonian for simulations of "real-world" systems relies on the existence of a general strategy to derive a GLE from a given microscopic model. In the past, the fundamental problem of mapping in an exact way a reduced dynamics into a GLE was addressed by many authors,[5] but seldom checked in realistic physicochemical problems [43,44]. A substantial contribution in this direction has been recently given by Ivanov and co-workers [30,31], who proposed a classical MD-based methodology that makes use of the combined information of two correlation functions to extract the SD of the bath, and validated it in realistic molecular problems. Their technique is similar in spirit to the one that will be described in the following and that we independently devised at the same time. Though of more limited applicability, the latter has the advantage on relying *only* on simple dynamical information and does not require any *a priori* knowledge of the system potential. It thus suits well to an *ab initio* MD determination of the environmental forces. This inversion procedure (from the dynamics to the SD that generated it) is briefly described in this section, along with an account of its numerical performance [29]. In concluding this section, we further address the problem of going beyond the simple IO model in order to capture anharmonic effects of the environment [29,39].

3.1. Inversion of classical dynamics

When the Brownian motion along s is limited to near-equilibrium configurations, the bare system potential is harmonic, $V(s)=\frac{1}{2}m\omega_0^2s^2$, and the SD determines not only the correlation function of the environmental fluctuations [see Eq. (4)], but also the frequency-dependent autocorrelation function of the position,

$$\mathcal{C}(\omega) = \int_{-\infty}^{+\infty} e^{i\omega t}C(t)dt = \int_{-\infty}^{+\infty} e^{i\omega t}\langle s(t)s(0)\rangle dt \qquad (16)$$

namely through

[5] The problem is essentially classical in nature, since the statistical properties of the bath (when subsumed in the spectral density $J_0(\omega)$ are the same for both the classical and quantum GLE.

$$\frac{1}{2}\omega\tilde{C}(\omega) = \frac{k_B T}{m}\,\mathrm{Im}\left(\frac{1}{\omega_0{}^2-\omega^2-i\omega\gamma(\omega)}\right) \tag{17}$$

that follows from Eq. (1) by harmonic analysis. This equation relates the reduced dynamics of the system to its coupling with the environment and can be "inverted" to give an analytic expression for $J_0(\omega)$ in terms of $\tilde{C}(\omega)$. This can be accomplished by introducing the retarded correlation function $C^+(t)=\Theta(t)C(t)$ and by exploiting the analytic properties of its Fourier transform (see Ref. [29] for the detailed derivation). The resulting formula for $J_0(\omega)$ is as follows

$$J_0(\omega) = \frac{k_B T}{2}\frac{\omega\tilde{C}(\omega)}{|\Gamma^+(\omega)|^2} \tag{18}$$

Where $\Gamma^+(\omega)=\lim_{\epsilon\to 0^+}\Gamma(\omega+i\epsilon)$ is the limit on the real axis from the u.h.p. of the function,

$$\Gamma(z) = \frac{1}{\pi}\int_{-\infty}^{+\infty}\frac{\omega\tilde{C}(\omega)/2}{\omega-z}\,d\omega \tag{19}$$

that is, the Cauchy transform of the function $f(\omega)=\omega\tilde{C}(\omega)/2$. Here, the displacement autocorrelation function $C(t)$(or, equivalently, the velocity autocorrelation) (t)(0)$=-d^2C(t)/dt^2$ is readily available from equilibrium MD and is the only dynamical information required.

3.2. Numerical tests

In Ref. [29], we tested the inversion procedure of Eq. (18) on some model systems to elucidate how its effectiveness is affected by the presence of an harmonicities and/or by a "Debye" cutoff frequency in the environment. A number of IO Hamiltonians of the kind of Eq. (5) and its variants were used to generate the dynamics, with a reasonable choice of the parameters that mimicked typical molecular problems. Some key alternatives were considered (*e.g.*, a harmonic *vs.* an anharmonic system oscillator, Ohmic *vs.* non-Ohmic baths, small *vs.* large oscillator frequency [compared to the Debye cutoff], etc.), and the position correlation function $C(t)$ was computed with MD and then inverted according to Eq. (18). Some results of these numerical tests are displayed in **Figure 1**.

The main conclusions of such numerical analysis can be summarized as follows. When the system is a harmonic oscillator, the transformation perfectly recovers the original SD up to the bath Debye frequency ω_D (1000 cm^{-1} in our numerical tests). For higher frequencies, the spectrum is not identically zero but rather shows an increasing baseline due to the numerical implementation of the Cauchy transform[6] of Eq. (19). Further, when the system frequency ω_s lies *above* ω_D, the SD features a single Lorentzian peak, which stands out from the background and is robust against variations of the bath and/or the temperature. This is the numerical representation of a Dirac-δ contribution[7] that follows from Eq. (18) when ω_s lies outside the support of $J_0(\omega)$. Anharmonicity has two strong effects in this case: on the one hand, it causes the appearance of higher harmonics at a frequency $2\omega_s$; on the other hand, it induces a broadening of the Dirac-δ signal that is about twice as large as the spectral width of the bath (*i.e.*, the appearance of combination frequencies $\omega_s \pm \omega_B$). Importantly, the temperature has a dramatic effect on the computed SD: At "high" temperature, and in contrast to the HO case, the remnant of the δ-peak undergoes substantial broadening, in striking contrast with the expectation of a temperature-independent SD.

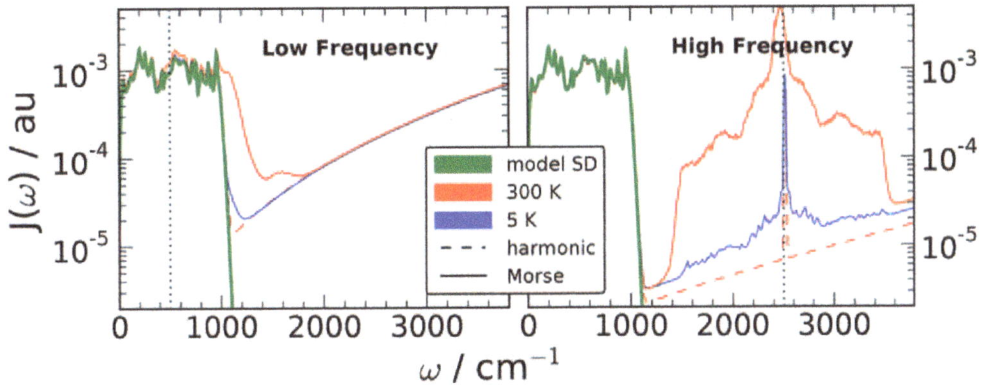

Figure 1. The spectral densities (atomic units) obtained by "inverting" the dynamical information of the position correlation function according to Eq. (18) (color coded as in that figure) are compared to the original non-Markovian SD used to define the models (green lines). Results are shown for different temperature—50 and 300 K—and for different choices of the system potential—harmonic and Morse—with an intrinsic frequency either above or below the Debye frequency ω_D of the bath. Dotted lines mark the frequencies of the system oscillators.

The presence of these features warns against blind application of the transformation of Eq. (18), particularly when the system frequency is larger than ω_D. The temperature-dependent background in the SD at high frequencies is unphysical and reflects just the anharmonicity in the system. However, provided such effect is clearly identified, no real problem arises in modeling since the anharmonicity in the system potential can always be included in the IO Hamiltonian by using the appropriate potential V(s).

3.3. Nonlinear extensions of the IO Hamiltonian

As mentioned in the previous section, system anharmonicity poses no real problem to modeling, and only generates a spurious high-frequency coupling that can be minimized by working at low enough temperature. In realistic situations, however, structured features in the spectral region $\omega > \omega_D$ are expected quite generally also from the failure of the bilinear coupling model and/or of the harmonic approximation of the bath oscillators. In both cases, the apparent coupling at frequencies above the Debye bath cutoff does have a *physical* origin,

[6] Numerical evaluation of Eq. (19) requires the introduction of a high-frequency cutoff. The problem arises when using an "unbiased" cut-off frequency well above the spectral range of interest ($\omega_c = 4000\text{cm}^{-1}$ in the simulations) and can be easily amended by setting ω_c equal to the bath Debye frequency (if known).

[7] A fictitious coupling to the bath appears here because numerically the autocorrelation function needs to be damped.

and thus the question arises whether $J_0(\omega)$ at such frequencies can be a surrogate for a more complicated coupling model. This is a rather intricate issue related to the general problem of whether a mapping of the system dynamics to a GLE exists and how it can be realized in practice (see, e.g., Ref. [30] for a recent, in-depth analysis of this issue). In general, real molecular oscillators are not harmonic and the system–bath coupling is not bilinear—especially if highly excited vibrational states are being probed—two factors which are hard to disentangle at finite temperature. In these situations, the present "dynamical" approach, when considered in the low-T limit above, can only provide the small-amplitude expansion of the coupling term and needs to be integrated with some empirical knowledge about the interaction between the system and the environment that could guide the formulation of an extended IO model.

In Eq. (5), our definition of IO Hamiltonian, we have already included a coupling which is not linear with respect to the system coordinate. It involves a shape function $f(s)$ that can be used to modulate the strength of the coupling to the bath in the state space of the system, thereby giving rise, as already mentioned, to state-dependent friction. Such extension seems to be a necessary (and simple) modification to address realistic situations; for instance, if s is the height of an adsorbate above a surface, the coupling should vanish for large s and exponentially increase for small$s$, as indeed occurs with Eq. (6).

As for the bath, the consideration of a nonlinear coupling poses more problems. In general, an exponential interaction model seems to be appropriate in typical physicochemical situations, where relaxation occurs as a consequence of close encounters between the molecular system of interest and the atoms of the environment. One simple *ansatz* of this kind is the replacement of $f(s)\sum c_k x_k$ in Eq. (5) with

$$\sqrt{\mu} D_0 e^{\alpha X'_1} \frac{e^{\alpha X'_1 - 1}}{\alpha} f(s) \tag{20}$$

Such coupling can be justified in the context of the linear-chain representation of the bath (see Section 2.2) and follows from Eq. (15) upon replacing the first harmonic term of the series with a Morse potential with the same frequency. Here, α^{-1} is an empirical parameter that describes the characteristic length of the interaction and, for consistency, should be of the order of the atom dimensions. The thus-defined exponential model makes use of the spectral properties of the *proper bath* only (*i.e.*, for frequencies $\omega < \omega_D$) to introduce *multiphonon* relaxing pathways already at the lowest order in perturbation theory, as a simple calculation of the Fermi's golden-rule rate shows. It is further simple enough to be easily handled with the same numerical methods to be described below and can thus be considered beyond the limits of applicability of perturbation theory.

As a last possibility, there may be realistic situations in which the dynamics of the bath *per se* shows strong anharmonic effects. For such purpose, in Ref. [39], a new IO Hamiltonian was proposed in which a bath of Morse oscillators was nonlinearly coupled to the system, according to

$$H^{\mathrm{Morse}} = \frac{p_s{}^2}{2} + V(s) + \sum_{k=1}^{N}\left[\frac{p_k{}^2}{2} + D_k\left[(1 - e^{-\alpha_k x_k}) - \frac{c_k}{2D_k\alpha_k}s\right]^2 - D_k\right] \tag{21}$$

where, as before, (s, p_s) and (x_k, p_k) denote the coordinates and momenta of the system and of the bath degrees of freedom, respectively. In Eq. (20), the parameters of the Morse potentials, D_k and α_k, can be specified by fixing the oscillator frequencies and ensuring a uniform number of bound states Λ over the bath oscillators. Similarly, the coupling coefficients can be chosen to sample a given SD according to Eq. (7).

The bath dynamics entailed by Eq. (21) can be highly nonlinear, especially for those low-frequency oscillators which undergo large amplitude motion.[8] In Ref. [39], this Hamiltonian was used to show that the anharmonicity of the bath induces in fact nontrivial variations in the (quantum) vibrational dynamics and in the corresponding relaxation rates.

4. Techniques for high-dimensional wave-packet dynamics

As is well known, the brute-force numerical solution of the Schrodinger equation rapidly runs into troubles when increasing the system dimensions. This is clearly seen by a simple scaling argument: With N degrees of freedom, if each of them requires on average a basis functions (or grid points), the total number of vector components is a^N and that of any operator (matrix) is $a^N \times a^N$. For $a = 6$ and double precision (complex) arithmetic, this means $\sim 10^{2N}$ *byte* for just storing the vector representing the wave function, an incredibly large number for all but the smallest N. Hence, it became clear soon that developing *approximate* methods was a necessity for facing large-dimensional quantum problems, and much effort has been spent to this end since the dawn of molecular quantum dynamics. The ideal method should be accurate enough to provide reliable information on the dynamics but also sufficiently cheap to be used for extensive sampling of "initial conditions."

Clearly, one central point to address when devising a method that could fulfill the above requirements is how to define approximate equations of motion. According to our experience, the most general strategy relies on the use of a variational principle that, apart from being physically transparent and mathematically sound, endows the resulting scheme with nontrivial properties. These features have important consequences in practice (*e.g.*, norm and energy conservation issues are settled at the outset) and will be discussed in some detail in the next subsection. In the rest of the section, a number of multi-configurational methods of increasing simplicity (and decreasing computational costs) will be introduced, with a focus on the dynamics at $T = 0K$ where a wave-function suffices. Finite temperature situations can be handled (at least in principle) by applying the same methodologies to the realizations used to sample the mixed (initial) state of the whole system.

[8] In principle, such model also describes energetic processes that irreversibly modify the environment, a phenomenon that can be mimicked by the dissociation of one or more oscillators.

4.1. Variational principle and Hamiltonian flows

The time-dependent variational principle is usually stated in terms of the Dirac–Frenkel condition

$$\langle \delta\Psi | i\hbar\partial_t - H | \Psi \rangle = 0 \tag{22}$$

where Ψ is the approximate solution sought for at a time t (the "trial" wave function which lies in some specified variational manifold) and $\delta\Psi$ is an arbitrary variation (an arbitrary vector tangent to the manifold in Ψ). The condition does lead to the Schrödinger equation when the manifold extends over the whole Hilbert state of the system[9] and admits a very simple interpretation, which is the best seen when multiplying Eq. (22) by $-i/\hbar$. In fact, $-i/\hbar H\Psi$ is easily recognized to be the *exact* time derivative for the system state Ψ, and thus Eq. (22) is seen to be a condition on the "error" vector $\left(\partial_t - \partial_t^{exact}\right)\Psi$. As such, it has a trivial geometrical solution: $\partial_t\Psi$ needs to be the *projection* of the exact time derivative onto the tangent space, a recipe that guarantees that the state vector keeps on staying on the manifold during the time evolution.

The above analysis shows that the time-dependent variational principle provides a *local-in-time* approximation to the system dynamics: It represents the best one can do *in the short run*, for the given state *at time t*. It does not offer, though, any guarantee that the solution $\Psi(t)$ at a finite time t is "close enough" to the exact solution for some specified initial state $\Psi(0)$, not even that is the best one can do with the specified manifold at that time.[10] However, the expectation is that if the trial manifold is flexible enough to include much of the state space spanned by the exact solution during its journey, the variational solution will remain a rather good approximation *at any time*, and one can thus exploit the nice features associated with the variational principle.

Energy and norm conservation follow immediately from Eq. (22) under quite mild conditions. For the energy, just notice that $\partial_t\Psi$ is always an allowed variation (and assume H be self-adjoint, of course). Norm conservation follows similarly when the manifold is scale-invariant, since in that case, Ψ itself is an allowed variation. A much deeper result emerges though when the Dirac–Frenkel condition is replaced by the (quasi) equivalent lowest action principle[11] $\delta S = 0$ and can be formulated as follows: Any variational quantum method, under quite mild regularity conditions, can be recast in the form of a symplectomorphism on a symplectic manifold [45]. Here, the relevant action functional is determined by the (real) Lagrangian

---

[9] In fact, this is the *condicio sine qua non* for the existence of a variational principle.

[10] In fact, the best approximation would just be the point on the manifold that lies closest to the exact solution at time t (whose identity may further depend on the adopted metrics).

[11] This is the *true* variational principle, *i.e.* a stationarity condition of some cleverly designed functional (something which Eq. (22) is *not*). However, the two formulations can be shown to be equivalent under quite mild conditions that are usually satisfied in practice.

$$L = \frac{i\hbar}{2}\frac{\langle\Psi\vee\partial_t\Psi\rangle - \langle\partial_t\Psi\vee\Psi\rangle}{\langle\Psi\vee\Psi\rangle} - \frac{\langle\Psi\vee H\vee\Psi\rangle}{\langle\Psi\vee\Psi\rangle} \quad (23)$$

We expand somehow on this issue in the following since it does not seem to be widely appreciated, despite its deep significance and its potential practical utility. To this end, let us first briefly introduce the concept of Hamiltonian flows and symplectic manifolds [46]. A symplectic manifold is a differentiable manifold equipped with a *closed*, *non-degenerate* 2-form ω. In a coordinate system x^i, it can be written as $\omega = \sum_{i,j>i}\omega_{ij}dx^i dx^j$, where dx^i are the fundamental 1-form—that is $dx^i(v) = v^i$ for any tangent vector v in a given point x—and the product of differentials is the so-called "wedge" product. Non-degeneracy means in practice that ω_{ij} is non-singular everywhere in the manifold,[12] and this allows one to set up a one-to-one map between tangent and *co*-tangent vectors (1-form). That is, for a given 1-form $\alpha = \sum\alpha_i dx^i$, there exists an associated vector field X_α such that $\alpha(v)=\omega(X_\alpha, v)$, and its *flow*, defined by the curves $\acute{x}^i = X_\alpha^i$. Then, given a smooth function H (which can be called a Hamiltonian) and its 1-form dH, the flow induced by its associated vector field X_H (what can be called a Hamiltonian flow) conserves the function itself, $dH(X_H) = \omega(X_H, X_H) = 0$. Closedness ($d\omega = 0$ where d is the "exterior" derivative) means that these properties can be "transported" over the manifold and guarantees that the symplectic form ω itself is invariant under any Hamiltonian flow (formally $L_{X_H}\omega=0$, L_Y being the Lie derivative along the vector field Y. This forms the basis for Liouville's theorem).

With this premise in mind, we (smoothly) introduce a set of real variational parameters x, forming a coordinate system in the sample space.[13] In terms of this parametrization, the Lagrangian reads as

$$L = \sum_{i=1}^{n} \acute{x}^i Z_i(x) - H(x) \quad (24)$$

Where $H(x) = \Psi(x) \vee H \vee \Psi(x)/\Psi(x) \vee \Psi(x)$ and $Z_i = \frac{i\hbar}{2\langle\Psi \mid \Psi\rangle}\left(\left\langle \Psi \mid \frac{\partial\Psi}{\partial x^i}\right\rangle - \left\langle\frac{\partial\Psi}{\partial x^i} \mid \Psi\right\rangle\right)$. The latter are the components of a 1-form $\alpha = -\sum Z_i dx^i$ that can be differentiated to give a closed 2-form, $\omega = d\alpha = \sum\omega_{ij}dx^i dx^j$. Provided $\omega_{ij} = \partial Z_i/\partial x^j - \partial Z_j/\partial x^i$ is non-singular, such 2-form is non-degenerate and thus represents a symplectic structure. In fact, the variational equations of motion take the form

$$\sum \acute{x}^i \omega_{ij} = \frac{\partial H}{\partial x^j} \quad (25)$$

or, equivalently, for a generic tangent vector v,

---

[12] This condition restricts the analysis to even dimensional manifolds.

[13] That is, we set $\Psi = \Psi(x)$, where **x** are, *e.g.*, the real and the imaginary parts of a set of complex parameters specifying the wave function. Though not necessarily finite in number or numerable, it is conceptually easier to think of a large but finite number of parameters.

$$\sum \acute{x}^i \omega_{ij} v^j = \omega(\acute{x}, v) = \sum \frac{\partial H}{\partial x^j} dx^j(v) = dH(v) \tag{26}$$

Hence, if ω is a symplectic form, the "variational flow" *is* the Hamiltonian flow of the Hamiltonian $H(x)$, namely $\acute{x} = X_H$. Under such circumstances, the equations of motion can also be written with the help of the Poisson brackets

$$\acute{x}^i = \{H, x^i\} \tag{27}$$

which are defined by $\{f, g\} = \omega(X_g, X_f)$ for any two smooth functions f and g, or, equivalently, by

$$\{f, g\} = \sum \frac{\partial f}{\partial x^i} \frac{\partial g}{\partial x^j} \xi^{ij} \tag{28}$$

(where ξ^{ij} is the matrix inverse of ω_{ij}) when a coordinate system is introduced.

The importance of the symplectic structure thus described is hardly overemphasized. Apart from its possible consequences on fundamental issues such as the emergence of classicality, or the coexistence of quantum and classical worlds, it offers in practice the possibility of introducing *robust* propagation schemes in solving the variational equations of motion. These symplectic propagators would not only conserve energy (and norm) but also the whole symplectic structure, a property that might be of great help when numerically investigating the emergence of irreversibility in Hamiltonian systems like the ones described by Eq. (5).

4.2. MCTDH, G-MCDTH, LCSA, and related methods

In this section, we give a brief account of (wave packet) quantum dynamical methods that have been applied in the past to system–bath problems of the kind discussed in this Chapter. The presentation is necessarily limited and, following authors' personal experience, focuses on the so-called multi-configurational methods only. In these methods, the wave function is written as a superposition of "Hartree products" of single-particle states (or single-particle functions, *spf's*), and both the coefficients of such superposition and the single-particle states (or some of them) are variationally optimized. The accuracy, the numerical complexity and the target problems of the method strongly depend on the choice of *spf's*, that is, whether they are fully flexible, constrained to a given functional form or frozen. As a result, several different methods exist which stem from the same multi-configurational *ansatz*.

Among the various possibilities, the conceptually simplest method is the *multi-configuration time-dependent Hartree* (MTDH), developed decades ago by Meyer et al. [32]. The *ansatz* is a straightforward expansion of the wave function in terms of (orthonormal) single-particle states,

$$\Psi\rangle = \sum_J A_J \left|\phi_{j_1}^{(1)}\right\rangle .. \left|\phi_{j_k}^{(k)}\right\rangle .. \left|\phi_{j_F}^{(F)}\right\rangle \tag{29}$$

where the sum runs over the possible configurations labeled by the multi-index $J = (j_1,.. j_N)$and $|\phi_j^{(k)}\rangle$ for $j = 1,.. N_k$ is a set of orthonormal states for the kth "particle." Here the "particle" can be either a single degree of freedom or a collection of modes, according to what is called a "mode combination." The corresponding variational manifold is obtained by varying the complex amplitude coefficients of the superposition and the single-particle states, which are assumed to be orthogonal to each other but otherwise fully flexible in their respective Hilbert space. The orthogonality condition proves to be a key strength of the method, since it guarantees that the configurations entering Eq. (29) are orthogonal to each other.

The equations of motion follow from the application of the Dirac–Frenkel condition to the above MCDTH *ansatz*, under the orthonormality constraints on the *spf's* which are typically introduced by means of arbitrary Hermitian matrices $g^{(k)}$ (one for each mode) that fix the evolution of the *spf's* in the "occupied" space. The important equations of motion are for the evolution of the *spf's* in the "unoccupied" space and for the evolution of the amplitude coefficients. Their derivation is straightforward, though lengthy, and the result can be summarized as follows. The amplitude coefficients satisfy a kind of matrix form of the Schrödinger equation in the basis of configurations

$$i\hbar \acute{A}_J = \sum_L \langle J|H|L\rangle A_L - i\hbar \sum_{k=1}^{F} \sum_{l=1}^{N_k} g_{j_k l}^{(k)} A_{J^k(l)} \tag{30}$$

only corrected for the "*gauge*" terms arising from the orthonormality constraints. Here, $L = (l_1,.. l_F)$ is a multi-index analogous to J above, L and J are shorthands for the configurations, j_k is the kth index of J, and $J^k(l)$ is the same multi-index J with l replacing j_k. The orbital equations, on the other hand, are mean-field like[14]

$$i\hbar(1 - P_k)\left|\acute{\phi}_j^{(k)}\right\rangle = (1 - P^k)\sum_{l,m=1}^{n_k} {\rho_k}^{-1}{}_{jl} H_{lm}^{(k)} |\phi_m^{(k)}\rangle \tag{31}$$

and involve the projection P_k onto the space spanned by the *spf's* of the kth mode, the mean-field operators $H_{lm}^{(k)} = \Psi_l^{(k)} | H | \Psi_m^{(k)}\rangle$ and the inverse of the density-matrices[15] ρ_k, $(\rho_k)_{lm} = \Psi_l^{(k)} | \Psi_m^{(k)}\rangle$, here written in terms of hole wave functions, that is, $|\Psi_m^{(k)}\rangle = a_m^{(k)} \vee \Psi\rangle$, where $a_m^{(k)}$ annihilates the state m of the kth mode. Details on the method and its numerical implementation can be found in the literature [16,32,33].

Some general considerations are in order. The MCDTH method does *not* solve the exponential scaling problem of quantum dynamics, but considerably alleviates it since replaces a potentially large number of (static) basis functions with a smaller set of "dynamically optimized"

[14] This expression is generally written as an explicit equation for the orbitals, by adding the above mentioned projection onto the occupied space of the spf velocity.

[15] Strictly speaking, ρ_k is the *transpose* of the 1-particle density for the kth mode, in the *spf's* basis.

elements. As such, it is extremely flexible and allows a systematic search of the convergence of the solution with respect to the length of the expansion of Eq. (29). In fact, provided large enough computational resources are available (how large depends of course on the problema at hand), the MCTDH solution can be made numerically *exact*.

A second issue concerns the kind of problems MCTDH may handle. The method is "general purpose" and can tackle arbitrary problems, provided the interaction terms between modes can be reasonably described as (sum of) products of terms involving one mode at a time. This is due to the appearance of the mean-field operators above, whose evaluation requires "tracing" over potentially many degrees of freedom. Apart from this, there exists no limitation in the form of the system Hamiltonian and indeed, MCTDH has been applied with success to a very large number of different problems. The application to system–bath problems to be described below represents just one possible problem where the method applies; further applications can be found in the original research papers and in the extensive review literature [16]. Here, we just mention that a user-friendly, highly efficient, general MCDTH code which takes arbitrary Hamiltonians as input is freely available upon request to the author [47].

A second class of multi-configurational methods is represented by the *Gaussian-multi-configuration time-dependent Hartree* (G-MTDH) developed a while ago by Burghardt et al. [48]. It is still a general-purpose method that can handle different kinds of quantum dynamical problems, and it is obtained from Eq. (22) by fully or partially replacing the flexible *spf's* with Gaussians. As a result, the equations of motion for the Gaussians become classical-like with considerable saving of memory and computer time (in fact, one propagates the few parameters needed to define the Gaussians), at the expense of introducing overlap matrix elements between them.

Though the method has several variants depending on the number of Gaussians introduced, it was originally formulated for system–bath-like problems, where one easily identifies primary modes (to be described at the high, fully flexible level) and secondary, less important modes that can be managed with moving Gaussians. In that case, the equations of motion for the amplitude coefficients and for the fully flexible *spf's* of the primary modes are similar to those of the MCDTH above, with minor modifications only, whereas a new set of first-order differential equations appear for the Gaussian parameters [48].

Along this line of thought, LCSA was *specifically* designed as a *local coherent-state approximation* [34] to the dynamics of system–bath Hamiltonians of the general form

$$H = H^{\text{sys}} + H^{\text{env}}(..q_k p_k..;s) \tag{32}$$

where H^{env} is an "environment" Hamiltonian (comprising the coupling with the system) which is supposed to be *local* in system coordinate(s) s and approximately *harmonic* in the bath degrees of freedom ($..q_k, p_k..$). In this case, the shaping of the wave function relies on the fact that (i) the coupling to the bath is *local* in system coordinates, and (ii) the bath is approximately *harmonic*. Upon introducing a set of system discrete variable representation (DVR) states[16] one expands the wave function as

$$\Psi\rangle = \sum_\alpha C_\alpha \vee \xi_\alpha\rangle \vee \Phi_\alpha\rangle \tag{33}$$

where ξ_α is a DVR set for the subsystem coordinates, and Φ_α are the resulting *local* bath states (one for each grid point α used to cover the relevant system configuration space). The latter are then written as products of HO coherent states (CSs), that is,

$$\Phi_\alpha\rangle = z_\alpha^1\rangle \vee z_\alpha^2\rangle..\vee z_\alpha^F\rangle := Z_\alpha\rangle \tag{34}$$

and, as a result, the bath dynamics is described by a set of coupled, *pseudo-classical* trajectories $z_\alpha^k = z_\alpha^k(t)$, one for each bath degree of freedom k *and* system grid point α. The system dynamics, on the other hand, is contained in the time evolution of the amplitude coefficients[17] C_α.

Equations of motion can be derived from the Dirac–Frenkel condition, using C_α and z_α^k as dynamical variables [34]. When using conventional phase factors for the CSs, they take the following form. The "system equation" is a kind of Schrödinger–Langevin equation

$$i\hbar \acute{C}_\alpha = \sum_\beta H_{\alpha\beta}^{\text{damp}} C_\beta + v_\alpha^{\text{eff}} C_\alpha \tag{35}$$

in which the elements of the system DVR Hamiltonian are *damped* by the overlap between bath states, $H_{\alpha\beta}^{\text{damp}} = H_{\alpha\beta}^{\text{sys}} \langle Z_\alpha \vee Z_\beta \rangle$. The local, effective potential, $v^{\text{eff}} = v^{\text{lmf}} + v^{\text{gauge}}$, contains a "local mean-field" potential

$$k\,*, z_\alpha^k, ..; s_\alpha$$
$$..z_\alpha$$
$$v_\alpha^{\text{lmf}} = \langle Z_\alpha \vee H^{\text{env}}(s_\alpha) \vee Z_\alpha \rangle = H_{\text{ord}}^{\text{env}} \tag{36}$$

(here, $H_{\text{ord}}^{\text{env}}$ is the environment Hamiltonian operator expressed in terms in $a_k^\dagger$, a_k and normally ordered, that is, with all $a_k^\dagger$'s on the left of a_k's) and a *gauge* potential

$$k\,*\,\acute{z}_\alpha^k$$
$$z_\alpha$$
$$\text{Im}$$
$$v_\alpha^{\text{gauge}} = -i\hbar \sum_{k=1}^F \langle z_\alpha^k | \acute{z}_\alpha^k \rangle = \hbar \sum_{k=1}^F \tag{37}$$

---

16 These are highly localized objects in configuration space which underlie the use of any "grid". See Refs. [49,50] for a formal introduction.

17 Rigorously speaking, the system reduced density matrix ρ also requires the overlap between bath state, $\rho_{\alpha\beta} = C_\alpha C_\beta$ $Z_\beta \vee Z_\alpha$ in the underlying DVR.

which can be explicitly written down with the bath equations below. The bath equations are *pseudo-classical*

$$k *,..; s_\alpha$$
$$..z_\alpha^k z_\alpha$$
$$i\hbar C_\alpha \acute{z}_\alpha^k = \sum_\beta H_{\alpha\beta}^{\text{damp}} (z_\beta^k - z_\alpha^k) C_\beta + C_\alpha \frac{\partial H_{\text{ord}}^{\text{env}}}{\partial a_k^\dagger} \tag{38}$$

and contain both a classical (local) force ($- i/\hbar C_\alpha$ times the second term on the r.h.s) and a quantum one ($- i/\hbar C_\alpha$ times the first term on the r.h.s) coupling the CSs of the *same* degree of freedom at *different* grid points. The latter is essential for a quantum, though approximate, description of the bath dynamics. For a detailed derivation of the equations see Ref. [34], and notice that, in general, $[a, f_{ord}(a^\dagger, a)] = \partial f_{ord}(a^\dagger, a)/\partial a^\dagger$.

This concludes the description of the original LCSA method. Several variants are possible (*e.g.*, replacing the DVR states with energy eigenstates or fully flexible system states) and can be found in the original literature [35,36]. Also, the closely related CC-TDSCF method [49] in which the CSs are replaced by fully flexible functions has been shown to provide essentially the same results as LCSA [36], thereby showing the soundness of the CS approximation for the (local) bath dynamics.

In fact, among the features of LCSA, one key strength of the method is that it reduces the bath dynamics to classical-like evolution, with a number of trajectories that scales linearly with the bath dimensions. This means that the method itself has a power-low scaling with such dimensions, the exponent of this scaling depending (eventually) on the interaction between bath modes. For bath modes coupled to the system only [as in Eq. (5)], *linear* scaling has been observed and model simulations with tens of thousands of bath degrees of freedom performed on modest computers. This good scaling is in common with mixed quantum-classical methods, which, however, fail to correctly represent the system–bath correlations.

Coupled trajectories also arise in a number of closely related approaches, namely the coupled coherent-state method of Shalashilin and Child [50,51] and the G-MCTDH method mentioned above. The latter, in fact, is strongly connected with LCSA and reduces to it as a limiting case (see Appendix B of Ref. [34]). The main difference between the two is that in LCSA *all* the configurations are orthogonal to each other, as a consequence of the presence of a different DVR state in each of them. This leads to considerable simplifications in the resulting equations, at the price of a reduced accuracy.

Finally, one interesting property about the pseudo-classical description of the bath degrees of freedom is that it suits well to induce *dissipative* dynamics into the total system. This can be accomplished by adding a suitably designed friction coefficient η to the bath equations, mimicking the presence of a *secondary* (infinite, memory-less) bath. More formally, it can be shown that applying the LCSA approximation to a system+bath+secondary bath configuration, a classical approximation to the secondary bath dynamics, and standard assumptions (Ohmic bath in the continuum limit) a friction coefficient appears in the LCSA equations for the system +bath degrees of freedom[18] (see Appendix A of Ref. [34]). This possibility has been exploited, especially in conjunction with the need of removing numerical instabilities of the method without altering the system dynamics.

5. Applications

Here, we present some numerical applications of the IO model, starting from simple simulations of a Brownian anharmonic oscillator—used as a testing ground for new dynamical methodologies [34,36–38] and/or for different representation of the Hamiltonian [27]. The last part of the contribution will be devoted to a "real-world" application, namely the hydrogen atom dynamics on the graphene surface [40,41].

5.1. Model systems

We consider here a model Hamiltonian describing an anharmonic (Morse) oscillator coupled to a heat bath. A typical problem considered in this context is the small amplitude, damped motion of the oscillator. The initial state is taken in product form, with the bath in its ground state (to mimic relaxation at $T = 0K$), and the system slightly displaced from its equilibrium position.

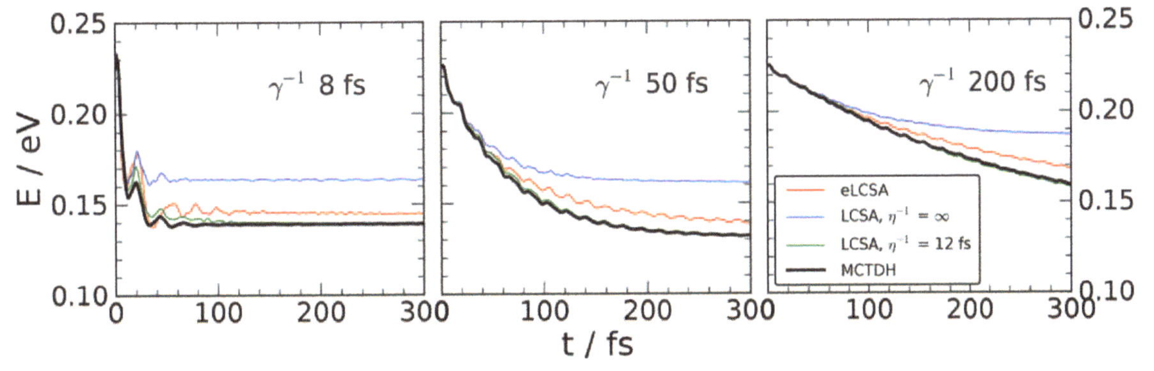

Figure 2. The small amplitude relaxation problem described in the text, for an Ohmic bath with relaxation time γ^{-1} = 850, 200 fs (panels from left to right). The energy of the system is computed with standard LCSA (blue line), LCSA coupled to a secondary bath with η^{-1} = 12 fs (green), and eLCSA (red), and compared with the MCTDH benchmark (black line).

This type of simulations was used in Ref. [34,36] to test the performances of different quantum dynamical approximations (the LCSA and its energy-local version, eLCSA). In **Figure 2**, the results of these techniques are shown along with benchmark MCTDH results, for different Ohmic spectral densities sampled with a bath of 80 oscillators. A Markovian exponential decay of the energy was found for all but the strongest coupling case, where some energy oscillations

[18] The same applies to finite temperature cases where, as expected, both a friction *and* a fluctuating term appear in the LCSA equations of each realization.

are clearly evident. The graph illustrates the main problems of standard LCSA, an inherent numerical instability related to saturation of the bath. These problems are solved in either its "damped" version [34] or with eLCSA [36]. The good agreement between LCSA and MCTDH is impressive, especially in light of the timing of the calculations (for LCSA only 2–3 min on a standard PC).

Similarly, in Ref. [27], the small amplitude relaxation of the Morse oscillator was used to illustrate the advantages of the chain transformation (Section 2.2). Here, MCTDH was used and different degrees of correlation were introduced along the chain, namely a small number of oscillators were described by a full, many particle expansion, whereas the rest of the chain was treated with one *spf* per mode. In this way, we exploited the strengths of the linear-chain representation of the bath to enlarge the *physical* time window of the simulation (*i.e.*, to increase the recurrence time) at a computational cost which scaled only linearly with the chain length [27].

Some results for small amplitude relaxation with the bath in linear-chain form are reported in **Figure 3**, for different structured SDs. The agreement with the benchmark is rather satisfactory and, as expected, the minor discrepancies were removed by increasing the correlation level. This is true both for the average system energy and for more detailed quantities like the position correlation functions.

The Morse oscillator was also used to model a dissipative scattering event, one in which the system is initially asymptotically free and moves toward the potential well where energy exchange with the bath occurs. Typically, in the interaction region, the wave-packet splits into two parts: One gets trapped in the well and fully relaxes on the long run, while the other returns to the asymptotic region. The first fraction, possibly resolved over the collision energy of the incoming wave packet, defines the "sticking" probability (having in minds problems where the bath represents a surface and a projectile sticks to it).

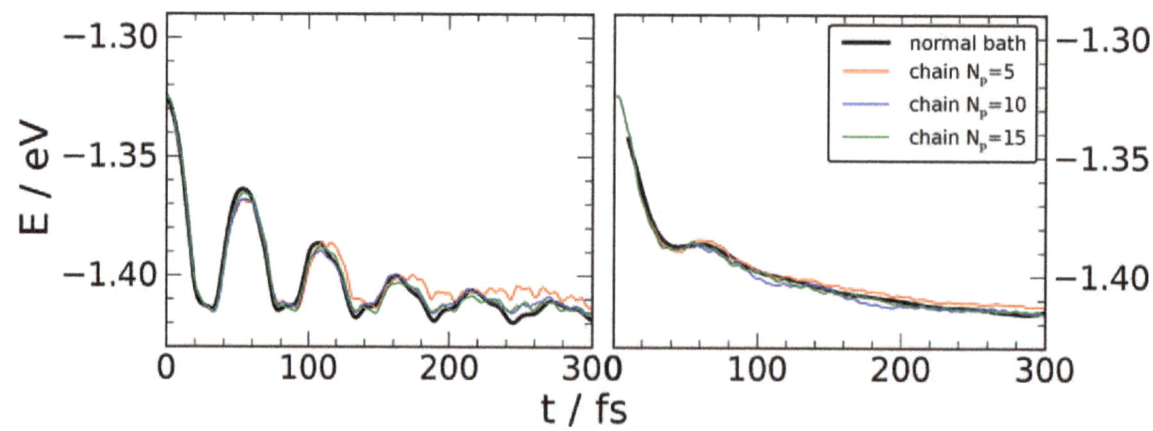

Figure 3. Small amplitude motion for the two non-Ohmic models of Ref. [27]. Computed system energy is shown for chain baths of increasing level of correlation, specified by N_p, that is, the number of fully correlated effective modes of the chain (red, blue, green lines for N_p = 5, 10 and 15, respectively). Also shown for comparison the benchmark obtained with the bath in normal form, as in Eq. (5) (black line).

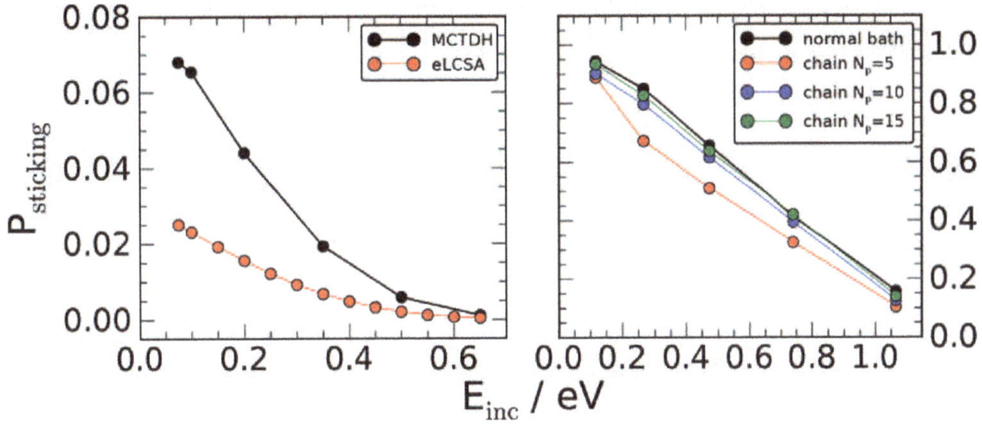

Figure 4. Sticking probability as a function of the incident energy, for the Morse scattering problem described in the main text. Left panel: Results for eLCSA (red circles) are compared to the MCTDH benchmark (black circles) for an Ohmic model with $\gamma^{-1} = 1$ ps. Right panel: The structured SD of Ref. [27] is adopted, and results are shown for baths in chain form with three different correlation schemes of increasing N_p, that is, the number of fully correlated chain oscillators (red, blue, green circles for $N_p = 5$, 10 and 15, respectively). Black circles are benchmark results obtained with the bath in normal form. Lines serve as a guide to the eye.

The sticking problem was considered as a test case for both LCSA and for the linear-chain representation of the bath with MCTDH (**Figure 3**). Simulations with standard LCSA showed that the numerical instabilities were too severe to extract meaningful sticking coefficients, even if the energy dissipation was described quite accurately [35]. On the contrary, the energy-local variant eLCSA gave stable results but only in semi-quantitative agreement with the benchmark. A detailed analysis showed that this is due to an inadequate system–bath correlation in the adopted *ansatz*, which is crucial for the energy transfer and hence the sticking process. Excellent results, on the other hand, were obtained by applying MCTDH with a partially correlated linear chain of oscillators [27]. Importantly, the results were shown to steadily and quickly converge toward the benchmark when increasing the level of correlation.

5.2. Hydrogen atom dynamics on graphene

In the last decade, the *activated* dynamics of hydrogen sticking on graphitic/graphenic surfaces has been one of the most studied gas–surface scattering problems. Despite the apparent simplicity of the system, the presence of both dissipative and quantum features makes it a challenging dynamical problem.

Recently, we have devised a rather elaborate system–bath model to describe hydrogen chemisorption of graphene and used it in a fully quantum study of the sticking dynamics with the MCDTH method. The model consists of an accurate description of the hydrogen atom and its bonding carbon atom (a 4D system), which were then coupled to the graphene sheet described by a phonon bath. It rests on the following, reasonable assumptions: (i) The energy exchange that occurs between the system and the lattice for near equilibrium configurations is representative of energy dissipation; (ii) relaxation proceeds through sequential energy transfer from the hydrogen atom to the carbon atom; (iii) a mapping holds, at least approximately, which relates the classical Hamiltonian dynamics of the interesting C and H atoms to a GLE description. On this basis, the following form was adopted for the Hamiltonian

$$H = \frac{p_H^2}{2m_H} + \frac{p_C^2}{2m_C} + V_s(x_H, z_C) + \sum_{k=1}^{F} \left[\frac{p_k^2}{2} + \frac{\omega_k^2}{2} \left(q_k - \frac{c_k}{\omega_k^2} \left(z_C - z_C^{eq} \right) \right)^2 \right] \quad (39)$$

Here, x_H is the position of the H atom, z_C the height of the binding C atom above the surface, p_H and p_C the corresponding momenta, and $V_s(x_H, z_C)$ an appropriate 4D system potential. The frequencies ω_k and couplings c_k of the IO bath were chosen to sample the spectral density $J_c(\omega)$) that describes *the coupling of the C atom to the rest of the lattice*. The latter was obtained by applying twice the inversion procedure (Section 3.1), using as only input the position correlation function $C_H(t)$ describing the equilibrium dynamics of the *hydrogen atom* [40].

The thus-obtained SD $J_C(\omega)$ is shown in **Figure 5** and presents a clear separation between the low-frequency region (0–900 cm^{-1}) – associated with the "surface" stretching[19]—and the high-frequency region of the C–H stretching. Details on how it was extracted and thoroughly checked can be found in the original research paper [40].

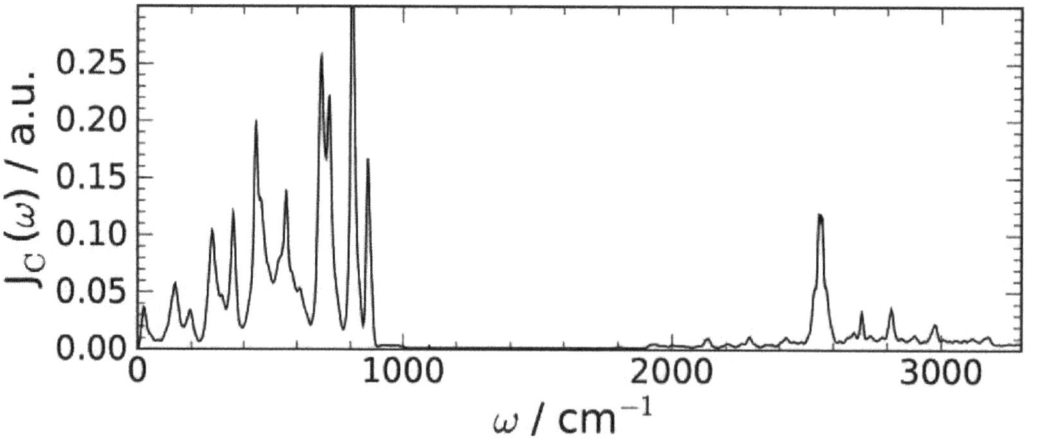

Figure 5. The function $J_c(\omega)$), that is, the spectral density "felt" by the carbon in the CH model of Eq. (39). The SD has been obtained from the inversion procedure described in the text and used for the high-dimensional quantum dynamics calculations of Refs. [40,41].

Once the coupling of the C atom with its environment was introduced, the Hamiltonian model of Eq. (39) could be used to investigate the hydrogen atom dynamics. We start here by considering the relaxation problem of the C–H bond. In this case, the system was prepared in an eigenstate of the V_s potential and allowed to relax because of the interaction with the bath (initially in its ground state, to mimic a $T = 0$ K situation).

[19] The "surface" stretching is one of the normal mode of the 4D system potential $V_s(x_H, z_C)$. This eigen-mode lying at ~460 cm^{-1} approximately corresponds to block oscillations of the C-H unit above the surface plane.

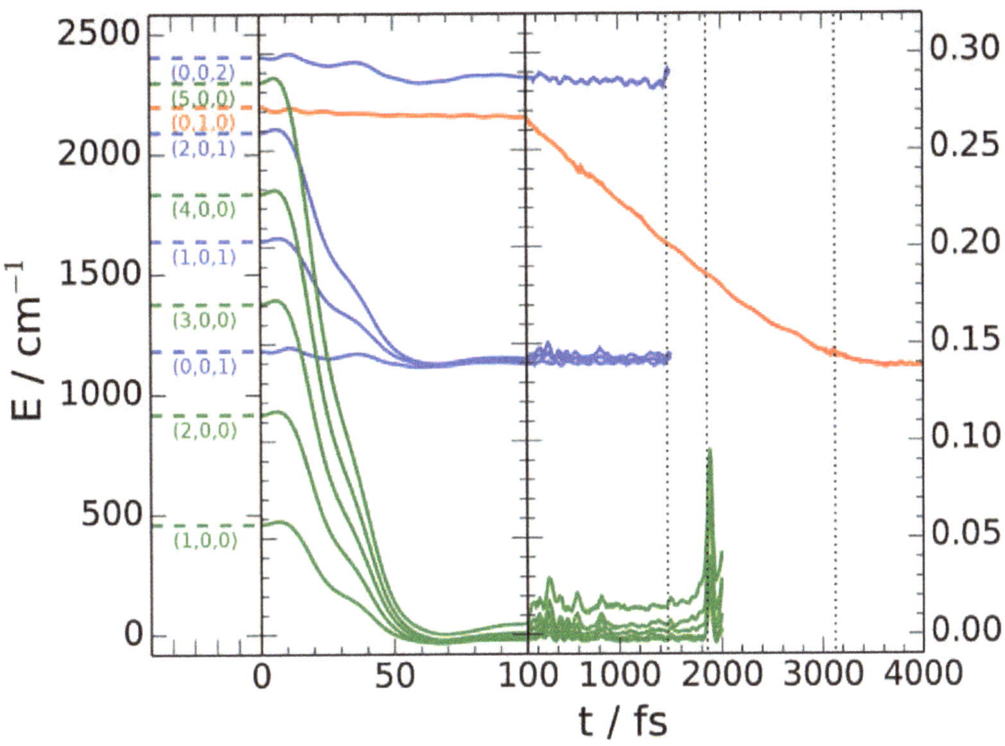

Figure 6. Time evolution of the system energy for the lowest lying vibrational states of the CH potential, computed with MCTDH and here labeled by the three appropriate quantum numbers ($\nu_{s,CH}$, $\nu_{s,surf}$, ν_b) where $\nu_{s,CH}$ is the C–H stretching, $\nu_{s,surf}$ is the surface stretching, and ν_b is the doubly degenerate C–H bending. The panel on the left gives the vibrational energy spectrum of the system. The dashed lines mark the recurrence time of the bath models adopted.

As is shown in **Figure 5**, relaxation from the surface stretching mode proceeds over a very short time scale and is complete in a few tens of fs. Despite the fast relaxation dynamics indicates a strong coupling between this coordinate and the bath, the energy decay shows essentially a Markovian behavior, except for the slippage at short time which extends for a considerable fraction of the relaxation window. This feature is related to the prepared initial states and to the switching on of the coupling term, which actually causes a slight *increase* of the system energy. The opposite behavior, on the other hand, is seen in the relaxation of the C–H stretching mode that takes place over a picosecond scale and seems to be complete on a time scale much larger than the 3.0 ps limit imposed by the recurrence time of our bath discretization. The resulting relaxation rate ($\tau \sim 5.0$ ps) is determined only by the background around the main peak in the SD. Its magnitude is (maybe incidentally) very close to the result obtained by Sakong and Kratzer [52], who applied perturbation theory from *first principles* and found $\tau = 5.2$ ps.

Next, we consider the quantum simulations of the collinear sticking dynamics [41], that is, the process in which a gas-phase hydrogen atom colliding at normal incidence above a carbon atom exchanges energy with the surface and gets trapped in the chemisorption well. We used the MCTDH method once again and, in addition, classical and quasi-classical methods to single out quantum effects in the results (**Figure 7**).

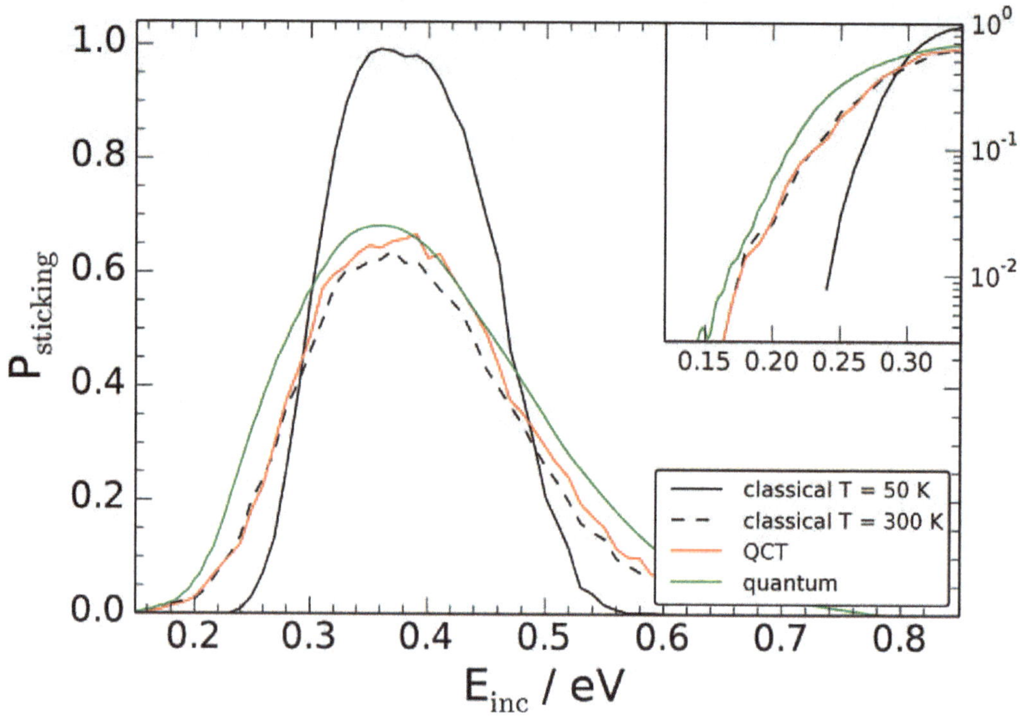

Figure 7. Sticking probability as a function of the collision energy as obtained from the MCTDH calculations with the CH system–bath model described in the main text (green curve). Also shown for comparison the classical sticking curves at T = 50 and 300 K (solid and dashed black lines, respectively) and the quasi-classical results (red line), obtained with the same IO model.

Before discussing the quantum dynamical results, it is instructive to first focus on some classical aspects of the sticking process. In **Figure 7**, the classical results are reported for two different surface temperatures, T=50 and 300 K. At low temperature, the sticking probability is negligibly small below the (static) barrier energy of ~ 0.24 eV, whereas above the barrier it reaches a saturation value $P_s \approx 1.0$ in a relatively narrow energy range, and decreases afterward. The detail analysis of the dynamical process [41] shows that in the below-barrier energy regime sticking is only determined by the probability that the projectile overcomes the barrier, since in the interaction region, the atom easily dissipates the (small) amount of energy required to get trapped in the chemisorption well. Above the barrier, energy transfer to the surface represents the limiting factor to sticking: Only if a large amount of energy can be transferred to the bath, the projectile is prevented to recross the barrier and to return to the gas phase. As a consequence, a simple, impulsive model of the dynamics describes the results rather well [41].

The quantum results differ from the classical ones in the whole energy range considered. While quantum effects (tunneling) can be invoked in the low-energy regime, above saturation the discrepancy is necessarily due to the quantum nature of the low-temperature surface which, in this $T = 0$ K limit, shows pronounced zero-point energy effects on the projectile dynamics. As a consequence, quasi-classical simulations show a rather good agreement with the quantum results, apart from the threshold region where tunneling through the barrier dominates. A logarithmic plot of the curves in this energy region (**Figure 7**) shows though that the effect of tunneling is moderate (less than one order of magnitude), in contrast with the effect of the lattice quantum fluctuations (up two orders of magnitudes). Furthermore, it has been shown that the quantum results are well described by the above-mentioned impulsive model of the dynamics, provided it is extended in order to account for the lattice quantum fluctuations and it is applied at energies not too small compared to the barrier height [41].

Acknowledgements

This work has arisen over the years through an intense and fruitful collaboration with several people. Among them, we are particularly in debt to Mathias Nest, Peter Saalfrank, Keith Hughes, and Irene Burghardt, who are sincerely acknowledged for their contribution. Support for the numerics from the Regione Lombardia and the CINECA High Performance Computing Center through the LISA initiative is also gratefully acknowledged.

Author details

Matteo Bonfanti[1] and Rocco Martinazzo[1,2*]

*Address all correspondence to: rocco.martinazzo@unimi.it

1 Department of Chemistry, University of Milan, Milano, Italy

2 Institute of Molecular Science and Technologies, Consiglio Nazionale delle Ricerche, Milano, Italy

References

[1] Nitzan A. Chemical dynamics in condensed phases—relaxation, transfer and reactions in condensed molecular systems. Oxford University Press; 2006.

[2] Lee H, Cheng Y-C, Fleming GR. Coherence dynamics in photosynthesis: protein protection of excitonic coherence. Science 2007;316:1462–5. doi:10.1126/science.1142188.

[3] Collini E, Scholes GD. Coherent intrachain energy migration in a conjugated polymer at room temperature. Science 2009;323:369–73. doi:10.1126/science.1164016.

[4] Hwang I, Scholes GD. Electronic energy transfer and quantum-coherence in π-conjugated polymers. Chem Mater 2011;23:610–20. doi:10.1021/cm102360x.

[5] Engel GS, Calhoun TR, Read EL, Ahn T-K, Mancal T, Cheng Y-C, et al. Evidence for wavelike energy transfer through quantum coherence in photosynthetic systems. Nature 2007;446:782–6. doi:10.1038/nature05678.

[6] Collini E, Wong CY, Wilk KE, Curmi PMG, Brumer P, Scholes GD. Coherently wired light-harvesting in photosynthetic marine algae at ambient temperature. Nature 2010;463:644–7. doi:10.1038/nature08811.

[7] Einstein A. Über die von der molekularkinetischen Theorie der Wärme geforderte Bewegung von in ruhenden Flüssigkeiten suspendierten Teilchen. Ann Phys 1905;322:549–60. doi:10.1002/andp.19053220806.

[8] Kubo R, Toda M, Hashitsume N. Statistical physics II—nonequilibrium statistical mechanics. Springer Verlag; 1991.

[9] Lemons DS, Gythiel A. Paul Langevin's 1908 paper "On the Theory of Brownian Motion" ["Sur la théorie du mouvement brownien," C. R. Acad. Sci. (Paris) 146, 530–533 (1908)]. Am J Phys 1997;65:1079–81. doi:10.1119/1.18725.

[10] Zwanzig R. Nonequilibrium statistical mechanics. USA: Oxford University Press; 2001.

[11] Tuckerman M. Statistical mechanics: theory and molecular simulation. OUP Oxford; 2010.

[12] Breuer HP, Petruccione F. The theory of open quantum systems. Oxford University Press; 2007.

[13] Caldeira AO, Leggett AJ. Influence of damping on quantum interference: an exactly soluble model. Phys Rev A 1985;31:1059–66. doi:10.1103/PhysRevA.31.1059.

[14] Ford GW, Lewis JT, O'Connell RF. Quantum Langevin equation. Phys Rev A 1988;37:4419–28. doi:10.1103/PhysRevA.37.4419.

[15] Zwanzig R. Nonlinear generalized Langevin equations. J Stat Phys 1973;9:215–20.

[16] Meyer H-D, Gatti F, Worth GA, editors. Multidimensional quantum dynamics: MCTDH theory and applications. Weinheim: Wiley-VCH Verlag GmbH & Co. KGaA; 2009.

[17] Caldeira AO, Leggett AJ. Influence of dissipation on quantum tunneling in macroscopic systems. Phys Rev Lett 1981;46:211–4. doi:10.1103/PhysRevLett.46.211.

[18] Weiss U. Quantum dissipative systems. 3rd ed. Singapore: World Scientific; 2008.

[19] Nonequilibrium statistical physics: linear irreversible processes. Oxford University Press; (2010). .

[20] Hughes KH, Christ CD, Burghardt I. Effective-mode representation of non-Markovian dynamics: a hierarchical approximation of the spectral density. I. Application to single surface dynamics. J Chem Phys 2009;131:024109. doi:10.1063/1.3159671.

[21] Chin AW, Rivas Á, Huelga SF, Plenio MB. Exact mapping between system-reservoir quantum models and semi-infinite discrete chains using orthogonal polynomials. J Math Phys 2010;51:092109. doi:10.1063/1.3490188.

[22] Prior J, Chin AW, Huelga SF, Plenio MB. Efficient simulation of strong system-environment interactions. Phys Rev Lett 2010;105:050404. doi:10.1103/PhysRevLett.105.050404.

[23] Martinazzo R, Vacchini B, Hughes KH, Burghardt I. Communication: Universal Markovian reduction of Brownian particle dynamics. J Chem Phys 2011;134:011101. doi:10.1063/1.3532408.

[24] Martinazzo R, Hughes KH, Burghardt I. Unraveling a Brownian particle's memory with effective mode chains. Phys Rev E 2011;84:030102(R). doi:10.1103/PhysRevE.84.030102.

[25] Burghardt I, Hughes KH, Martinazzo R, Tamura H, Gindensperger E, Köppel H, et al. Conical intersections coupled to an environment. In: Domcke W, Yarkony DR, Köppel H, editors. Conical Intersect., vol. 17, World Scientific; 2011, pp. 301–46.

[26] Martinazzo R, Hughes KH, Burghardt I. Hierarchical effective-mode approach for extended molecular systems. In: Hoggan EP, Brändas JE, Maruani J, Piecuch P, Delgado-Barrio G, editors. Adv. Theory Quantum Syst. Chem. Phys., Dordrecht, Netherlands: Springer; 2012, p. 269–83.

[27] Bonfanti M, Tantardini GF, Hughes KH, Martinazzo R, Burghardt I. Compact MCTDH wave functions for high-dimensional system–bath quantum dynamics. J Phys Chem A 2012;116:11406–13. doi:10.1021/jp3064504.

[28] Woods MP, Groux R, Chin AW, Huelga SF, Plenio MB. Mappings of open quantum systems onto chain representations and Markovian embeddings. J Math Phys 2014;55:032101. doi:10.1063/1.4866769.

[29] Bonfanti M, Hughes KH, Burghardt I, Martinazzo R. Vibrational relaxation and decoherence in structured environments: a numerical investigation. Ann Phys 2015;527:556–69. doi:10.1002/andp.201500144.

[30] Gottwald F, Ivanov SD, Kühn O. Applicability of the Caldeira–Leggett model to vibrational spectroscopy in solution. J Phys Chem Lett 2015;6:2722–7. doi:10.1021/acs.jpclett.5b00718.

[31] Gottwald F, Ivanov SD, Kühn O. Vibrational spectroscopy via the Caldeira–Leggett model with anharmonic system potentials. ArXiv E-Prints 2016.

[32] Meyer H-D, Manthe U, Cederbaum LS. The multi-configurational time-dependent Hartree approach. Chem Phys Lett 1990;165:73–8. doi:10.1016/0009-2614(90)87014-I.

[33] Beck MH, Jäckle A, Worth GA, Meyer H-D. The multiconfiguration time-dependent Hartree (MCTDH) method: a highly efficient algorithm for propagating wavepackets. Phys Rep 2000;324:1–105. doi:10.1016/S0370-1573(99)00047-2.

[34] Martinazzo R, Nest M, Saalfrank P, Tantardini GF. A local coherent-state approximation to system-bath quantum dynamics. J Chem Phys 2006;125:194102. doi:10.1063/1.2362821.

[35] Martinazzo R, Burghardt I, Martelli F, Nest M. Local coherent-state approximation to system-bath quantum dynamics. In: Shalashilin DV, De Miranda MP, editors. vol. Multidimensional quantum mechanics with trajectories, CCP6; 2009.

[36] López-López S, Nest M, Martinazzo R. Generalized CC-TDSCF and LCSA: The system-energy representation. J Chem Phys 2011;134:014102. doi:10.1063/1.3518418.

[37] Nest M, Meyer H-D. Dissipative quantum dynamics of anharmonic oscillators with the multiconfiguration time-dependent Hartree method. J Chem Phys 2003;119:24–33. doi: 10.1063/1.1576384.

[38] Burghardt I, Nest M, Worth GA. Multiconfigurational system-bath dynamics using Gaussian wave packets: energy relaxation and decoherence induced by a finite-dimensional bath. J Chem Phys 2003;119:5364–78. doi:10.1063/1.1599275.

[39] López-López S, Martinazzo R, Nest M. Benchmark calculations for dissipative dynamics of a system coupled to an anharmonic bath with the multiconfiguration time-dependent Hartree method. J Chem Phys 2011;134. doi:10.1063/1.3556940.

[40] Bonfanti M, Jackson B, Hughes KH, Burghardt I, Martinazzo R. Quantum dynamics of hydrogen atoms on graphene. I. System-bath modeling. J Chem Phys 2015;143. doi: 10.1063/1.4931116.

[41] Bonfanti M, Jackson B, Hughes KH, Burghardt I, Martinazzo R. Quantum dynamics of hydrogen atoms on graphene. II. Sticking. J Chem Phys 2015;143. doi:10.1063/1.4931117.

[42] Hughes KH, Christ CD, Burghardt I. Effective-mode representation of non-Markovian dynamics: a hierarchical approximation of the spectral density. II. Application to environment-induced nonadiabatic dynamics. J Chem Phys 2009;131:124108. doi: 10.1063/1.3226343.

[43] Mori H. Transport, collective motion, and Brownian motion. Prog Theor Phys 1965;33:423. doi:10.1143/PTP.33.423.

[44] Zwanzig R. Memory effects in irreversible thermodynamics. Phys Rev 1961;124:983–92. doi:10.1103/PhysRev.124.983.

[45] Kramer P, Saraceno M, editors. Geometry of the time-dependent variational principle in quantum mechanics. Lecture notes in physics. vol. 140. Berlin: Springer Verlag; 1981.

[46] Flanders H. Differential forms with applications to the physical sciences. Dover Publications; 1963.

[47] Worth GA, Beck MH, Jäckle A, Meyer H-D. The MCTDH Package, Version 8.2, (2000). H.-D. Meyer, Version 8.3 (2002), Version 8.4 (2007), Version 8.5 (2011). Current version: 8.5.3 (2013). See http://mctdh.uni-hd.de. University of Heidelberg, Germany; n.d.

[48] Burghardt I, Meyer H-D, Cederbaum LS. Approaches to the approximate treatment of complex molecular systems by the multiconfiguration time-dependent Hartree method. J Chem Phys 1999;111:2927–39. doi:http://dx.doi.org/10.1063/1.479574.

[49] Zhang DH, Bao W, Yang M. Continuous configuration time-dependent self-consistent field method for polyatomic quantum dynamical problems. J Chem Phys 2005;122:091101. doi:10.1063/1.1869496.

[50] Shalashilin DV, Child MS. Time dependent quantum propagation in phase space. J Chem Phys 2000;113:10028.

[51] Shalashilin DV, Child MS. The phase space CCS approach to quantum and semiclassical molecular dynamics for high-dimensional systems. Chem Phys 2004;304:103–20. doi: 10.1016/j.chemphys.2004.06.013.

[52] Sakong S, Kratzer P. Hydrogen vibrational modes on graphene and relaxation of the C–H stretch exmixed-citation from first-principles calculations. J Chem Phys 2010;133:054505. doi:10.1063/1.3474806.

4

Invariance in Quantum Walks

Miquel Montero

Abstract

In this Chapter, we present some interesting properties of quantum walks on the line. We concentrate our attention in the emergence of invariance and provide some insights into the ultimate origin of the observed behavior. In the first part of the Chapter, we review the building blocks of the quantum-mechanical version of the standard random walk in one dimension. The most distinctive difference between random and quantum walks is the replacement of the random coin in the former by the action of a unitary operator upon some internal property of the later. We provide explicit expressions for the solution to the problem when the most general form for the homogeneous unitary operator is considered, and we analyze several key features of the system as the presence of symmetries or stationary limits. After that, we analyze the consequences of letting the properties of the coin operator change from site to site, and from time step to time step. In spite of this lack of homogeneity, the probabilistic properties of the motion of the walker can remain unaltered if the coin variability is chosen adequately. Finally, we show how this invariance can be connected to the gauge freedom of electromagnetism.

Keywords: quantum walks, invariance, symmetry, Dirac equation, gauge transform

1. Introduction

In their origins [1–5], quantum walks (QWs) were thought as the quantum-mechanical generalization of the standard random walk in one dimension: the mathematical model describing the motion of a particle which follows a path that consists of a succession of jumps with fixed length whose direction depends on the random outcome of flipping a coin. In the quantum version, the coin toss is replaced by the action of a unitary operator upon some intrinsic degree of freedom of the system, a quantum observable with only two possible eigenvalues: for example, the spin of an electron, the polarization of a photon, or the chirality of a molecule.

After this preliminary analysis, it became clear that the similitude between these two processes was mainly formal and that random and QWs displayed divergent properties [6]. The most remarkable of these discrepancies is perhaps the ability of unbiased QWs to spread over the line, not as the square root of the elapsed time, the fingerprint of any diffusion process, but with constant speed [7]. This higher rate of percolation enables the formulation of quantum algorithms [8, 9] that can tackle some problems in a more efficient way than their classical analogs: For instance, QWs are very promising resources for optimal searching [10–12]. Today, QWs have exceeded the boundaries of quantum computation and attracted the attention of researchers from other fields as, for example, information theory or game theory [13–16].

As a consequence of this wide interest, diverse extensions of the discrete-time QW on the line have been considered in the past. Most of these variations are related with the properties of the unitary coin operator [17], backbone of the novel features of the process. Thus, one can find in the literature QWs whose evolution depends on more than one coin [18–20], QWs that suffer from decoherence [21, 22], or QWs driven by inhomogeneous, site-dependent coins [23–28]. There are also precedents where the temporal variability of the QW is explicit: in the form of a recursive rule for the coin selection, as in the so-called Fibonacci QWs [29, 30], through a given function that determines the value of the coin parameters [31–33], or by means of an auxiliary random process that modifies properties of the coin [34].

The main goal in most of these seminal papers is to find out new and exciting features that the considered modifications introduce in the behavior of the system, like the emergence of quasiperiodic patterns or the induction of dynamic localization. Recent works [35–37], however, have also regarded the issue from the opposite point of view, by exploring the conditions under which the evolution of the system results unchanged. In particular, Montero [37] considers the case of a discrete-time QW on the line with a time-dependent coin, a unitary operator with changing phase factors.

These phase factors are three parameters that appear in the definition of the coin operator whose relevance has been sometimes ignored in the past: When these phases are static magnitudes, they are superfluous [38], but if they are dynamic quantities, they can substantially modify the evolution of the system. This fact does not close the door to the possibility that a set of well-tuned variable phase factors can keep the process unchanged from a probabilistic perspective. This defines a control mechanism that can compensate externally induced decoherence and introduces a nontrivial invariance to be added to other well-known symmetries of QWs [39–41].

In this Chapter, we will review the approach taken in [37] and consider a generalization of it. Now, the evolution of the discrete-time quantum walker on the line will be subjected to the introduction of a fully inhomogeneous coin operator: The properties of the unitary operator will depend both on the location and on the present time through the action of the aforementioned phase factors. This extra variability leads to additional constraints to be satisfied by these magnitudes if one wants to guarantee that the properties of the motion of the walker remain unaltered. Finally, we will connect our results with those appearing in the study of Di Molfetta et al. [36], where the authors considered how the inclusion of time- and site-dependent phase factors in the coin operator of a quantum walk on the line may induce some dynamics which, in the continuous limit, can be linked with the propagation of a Dirac spinor coupled to some external electromagnetic field. We will also explore the implications of this mapping here.

2. Fundamentals of QWs

We begin this Chapter with a survey of the fundamental concepts required in the designing of discrete QWs on the line. In its simplest version, the particle represented by the walker can occupy detached and numerable locations on a one-dimensional space. This space of positions may be just a topological space (a graph or a chain, for instance) or can be endowed with a metric. In such a case, it is usual to consider that the sites are separated by a fixed distance , so that $X = n \cdot l$. Within this standard framework, time increases in discrete steps as well, [1] $T = t \cdot \tau$, τ being the sojourn time so that variable t becomes a non-negative integer index, $t \in \{0,1,2,\cdots\}$, and the evolution of the system is just a sequence of states, $|\psi\rangle_t$.

Up to this point, there is no significant difference between random and quantum walks. The major distinction is found in the nature of the random event that determines the progress of the particle. While in a world governed by the laws of classical mechanics, randomness is the way in which we describe the uncertain effect of multiple (and usually uncontrollable) external agents acting upon a system, in the realms of quantum mechanics randomness is not an exogenous ingredient. This means that we can use some internal degree of freedom in the quantum system with two possible eigenvalues (the spin, the polarization, or the chirality) as a proxy for the coin and understand that any change in this inner property is the result of the act of tossing. Therefore, to represent the state of the walker, we need two different Hilbert spaces: $\mathcal{H}_P$, the Hilbert space of particle positions spanned by the basis $\{|n\rangle : n \in \mathbb{Z}\}$, and the Hilbert space of the coin states, $\mathcal{H}_C$, which is spanned by the basis $\{|+\rangle, |-\rangle\}$. The expression of $|\psi\rangle_t$ in the resulting Hilbert space $\mathcal{H}$, $\mathcal{H} \equiv \mathcal{H}_C \otimes \mathcal{H}_P$, reads

$$|\psi\rangle_t = \sum_{n=-\infty}^{\infty} \left[\psi_+(n,t)\,|+\rangle \otimes |n\rangle + \psi_-(n,t)\,|-\rangle \otimes |n\rangle\right], \tag{1}$$

where we have introduced the wave-function components $\psi_\pm(n, t)$, the two-dimensional projection of the state of the walker into the elements of the basis:

$$\psi_+(n,t) \equiv \langle n| \otimes \langle +|\psi\rangle_t, \tag{2}$$

1 There is another kind of QW, called continuous quantum walk, in which the walker can modify its position at any time: this is the quantum counterpart of continuous-time random walk. The evolution of processes belonging to this category is ruled by a Hamiltonian and the corresponding Schrödinger equation. In spite they are different, discrete, and continuous QWs share common traits [42].

$$\psi_{-}(n,t) \equiv \langle n| \otimes \langle -|\psi\rangle_t. \tag{3}$$

Now, we have to consider the mechanism that connects these two properties, position and quirality, which eventually leads to a model for the dynamics of $\psi_{\pm}(n, t)$. Evolution in the discrete-time, discrete-space QW can be regarded as the result of the action of operator $\hat{\mathcal{T}}, \hat{\mathcal{T}} \equiv \hat{\mathcal{S}}\hat{\mathcal{U}}$, on the state of the system $|\psi\rangle_t$. As it can be observed, the practical implementation of operator $\hat{\mathcal{T}}$ has two stages: In the first one, the unitary operator $\hat{\mathcal{U}}$ modifies exclusively the internal degree of freedom of the quantum system, in what represents the throw of the coin as indicated earlier,

$$\begin{aligned}\hat{\mathcal{U}} \equiv \sum_{n=-\infty}^{\infty} e^{i\chi}[e^{i\alpha}\cos\theta|+\rangle\langle+| + e^{-i\beta}\sin\theta|+\rangle\langle-| \\ + e^{i\beta}\sin\theta|-\rangle\langle+| - e^{-i\alpha}\cos\theta|-\rangle\langle-|]\otimes|n\rangle\langle n|.\end{aligned} \tag{4}$$

In a second step, the shift operator $\hat{\mathrm{S}}$ *moves* the walker depending on the result obtained after the last toss: [2]

$$\hat{\mathcal{S}}\left(|\pm\rangle \otimes |n\rangle\right) = |\pm\rangle \otimes |n \pm 1\rangle. \tag{5}$$

Therefore, the state of the system at a later time $|\psi\rangle_{t+1}$ is recovered by application of $\hat{\mathrm{T}}$ to the preset state:

$$|\psi\rangle_{t+1} = \hat{\mathcal{T}}|\psi\rangle_t, \tag{6}$$

and the complete evolution of the system is determined once $|\psi\rangle_0 \equiv |\psi\rangle_{t=0}$ is selected. As in any quantum problem, one can consider for the initial state of the walker any combination of the elements in the basis of $\mathcal{H}$, a configuration that may lead to some degree of uncertainty in the position and/or the chirality of the system. However, the interest in establishing parallelisms between classical and quantum walkers encourages the choice in which, at the beginning, the particle position is known exactly, but its internal degree of freedom is aligned arbitrarily:

$$|\psi\rangle_0 = \left(\cos\eta|+\rangle + e^{i\gamma}\sin\eta|-\rangle\right)\otimes|0\rangle. \tag{7}$$

2 With the present definition, the problem is spatially homogeneous and the system displays translational invariance. Therefore, alternative shift rules may be considered with equivalent results, as in the case of directed quantum walks [43, 44], where the particle can either remain still in the place or proceed in a fixed direction but never move backward.

Needless to say that the linearity and the translational invariance of the problem ensure that the solution for a general initial state can be recovered by direct superposition of the evolution of Eq. (7), Eqs. (14) to (17) later.

The similarities and dissimilarities between classical and QWs must be grounded on the analysis of the probability mass function (PMF) of the process, $\rho(n, t)$, the probability that the walker can be found in a particular position n at a given time t. The PMF for a random walk is

$$\rho_{clas.}(n,t)=\begin{pmatrix} t \\ \frac{t+n}{2} \end{pmatrix} p^{\frac{t+n}{2}}(1-p)^{\frac{t-n}{2}}, \tag{8}$$

where p is the probability of obtaining a head as the result of flipping the coin. For the QW, $\rho(n, t)$ is the sum of the squared modulus of the wave-function components,

$$\rho(n,t)=|\psi_+(n,t)|^2+|\psi_-(n,t)|^2. \tag{9}$$

On the basis of the values of the moduli of $\psi_\pm(n, t)$ we can also express the probability of obtaining a head value or a tail value when measuring the global coin state of the walker:

$$P_\pm(t)\equiv\sum_{n=-\infty}^{\infty}|\psi_\pm(n,t)|^2, \tag{10}$$

or the value of $M(n, t)$,

$$M(n,t)\equiv|\psi_+(n,t)|^2-|\psi_-(n,t)|^2, \tag{11}$$

another interesting magnitude that can be connected with the local magnetization of the system if the internal degree of freedom has its origin in the spin of the particle [45].

2.1. General solution

The evolution operator $\widehat{T}$ induces the following set of recursive equations in the wave-function components,

$$\psi_+(n,t)=e^{i\chi}\,[e^{i\alpha}\cos\theta\,\psi_+(n-1,t-1)+e^{-i\beta}\sin\theta\,\psi_-(n-1,t-1)], \tag{12}$$

and

$$\psi_-(n,t) = e^{i\chi}[e^{i\beta}\sin\theta\ \psi_+(n+1,t-1) - e^{-i\alpha}\cos\theta\ \psi_-(n+1,t-1)], \tag{13}$$

whose general solution [38] can be written in a compact way by using $\psi_+(0,0)$ and $\psi_-(0,0)$,

$$\psi_+(0,0) = \cos\eta,$$
$$\psi_-(0,0) = e^{i\gamma}\sin\eta,$$

and the nonzero components of the wave function at time $t=1$,

$$\psi_+(+1,1) = e^{i\chi}\left[e^{i\alpha}\cos\eta\cos\theta + e^{i(\gamma-\beta)}\sin\eta\sin\theta\right],$$
$$\psi_-(-1,1) = e^{i\chi}\left[e^{i\beta}\cos\eta\sin\theta - e^{i(\gamma-\alpha)}\sin\eta\cos\theta\right],$$

since $\psi_+(-1,1)=\psi_-(+1,1)=0$, cf. Eqs. (12) and (13). In terms of the preceding quantities, and for $n\in\{-t,\ -t+2,\cdots,\ t-2,t\}$, one has

$$\psi_+(n,t) = e^{i(\chi\cdot t+\alpha\cdot n)}\left[\psi_+(0,0)\Lambda(n,t) + e^{-i(\chi+\alpha)}\psi_+(+1,1)\Lambda(n-1,t+1)\right], \tag{14}$$

and

$$\psi_-(n,t) = e^{i(\chi\cdot t-\alpha\cdot n)}\left[\psi_-(0,0)\Lambda(n,t) + e^{-i(\chi-\alpha)}\psi_-(-1,1)\Lambda(n+1,t+1)\right], \tag{15}$$

where

$$\Lambda(n,t) \equiv \frac{1}{t+1}\left\{\frac{1+(-1)^t}{2} + \sum_{r=1}^{t}\frac{1}{\cos\omega_{r,t}}\cos\left[(t-1)\cdot\omega_{r,t} - \frac{\pi r n}{t+1}\right]\right\}, \tag{16}$$

and

$$\omega_{r,t} \equiv \arcsin\left(\cos\theta\sin\frac{\pi r}{t+1}\right). \tag{17}$$

It is noted that in this picture the evolution of each component depends only on their own initial values. In fact, it can be shown [38] that $|\psi_+(+1,1)|^2$ can be understood as the "rightward initial velocity" of our quantum walker, whereas $|\psi_-(-1,1)|^2$ would play the role of the "leftward initial velocity."

Even though the expression for $\Lambda(n, t)$ is completely explicit, Eq. (16), it may be instructive to show how the set of equations that cross correlate the evolution of the two components of the wave function, Eqs. (12) and (13), turns now into a single, two-step recursive formula that governs the whole dynamics:

$$\Lambda(n,t) = \cos\theta\left[\Lambda(n-1,t-1) - \Lambda(n+1,t-1)\right] + \Lambda(n,t-2). \tag{18}$$

Equation (16) is recovered from the above relationship once one considers the initial condition $\Lambda(0,0)=1$, together with the boundary conditions $\Lambda(-n, t)=\Lambda(n, t)=0$, for $n \geq t \geq 1$.

Observe how $\Lambda(n, t)$ does not depend on χ, α, β, γ, or η. It is a function of θ through the value of $\cos\theta$, a property that can be also observed in Eq. (18). One could infer from this feature that $\cos^2\theta$ plays in QWs the same role of p in random walks and that the rest of parameters represent mathematical degrees of freedom without correspondence in the physical world. This impression can be strengthened by computing the value of the PMF in simple examples as, for instance, when n coincides with t: in this case, $\rho_{\text{clas.}}(t, t)=p^t$ while $\rho(t, t)=\cos^{2t}\theta$.

This conclusion is illusory, however. It is well known [19] that $\rho(n, t)$ does not depend on χ and that α, β and γ appear in the PMF only in the following combination $\varphi=\alpha+\beta-\gamma$. But it is true as well that one needs to specify θ, φ , and η to determine even the most basic aspects of the evolution of QWs. **Figure 1** illustrates this fact. In the upper panel, we observe how the probability is distributed unevenly for positive and negative values of n, although $\theta=\pi/4$. In the lower panel, we face the reversed situation, $\theta=\pi/8$ but $\rho(n, t)$ shows no clear asymmetry.

2.2. Stationary PMF

Figure 1 also shows us that the disparity in the bias is not the most striking aspect that distinguishes QWs from their classical analogues. These differences can be appreciated more easily when one considers the stationary limit [34]. It can be shown [38] that for $t \gg 1$, the probability mass function $\rho(n, t)$ is well described by $\bar{\rho}(n, t)$,

$$\bar{\rho}(n,t) = \frac{\sin\theta}{\pi}\frac{1}{t^2-n^2}\frac{1}{\sqrt{t^2\cos^2\theta - n^2}}[t + n(\cos 2\eta + \sin 2\eta \tan\theta\cos\varphi)], \tag{19}$$

in the range $-t\cos\theta < n < t\cos\theta$, $0<\theta<\pi/2$. As it can be seen in **Figure 1**, the agreement between $\rho(n, t)$ and $\bar{\rho}(n, t)$ is greater for small values of n, whereas when $|n|$ approaches to $t\cos\theta$, $\rho(n, t)$ displays an oscillatory behavior around $\bar{\rho}(n, t)$. Regardless of this, Eq. (19) captures the essence of $\rho(n, t)$: its U-shaped profile, with a central flat region and two local maxima in the vicinity of $\pm t\cos\theta$. These traits are in clear contrast to the bell-shaped contour of the classical PMF, centered around $(2p-1)\cdot t$, the mean value of the displacement of the random walker (**Figure 1**).

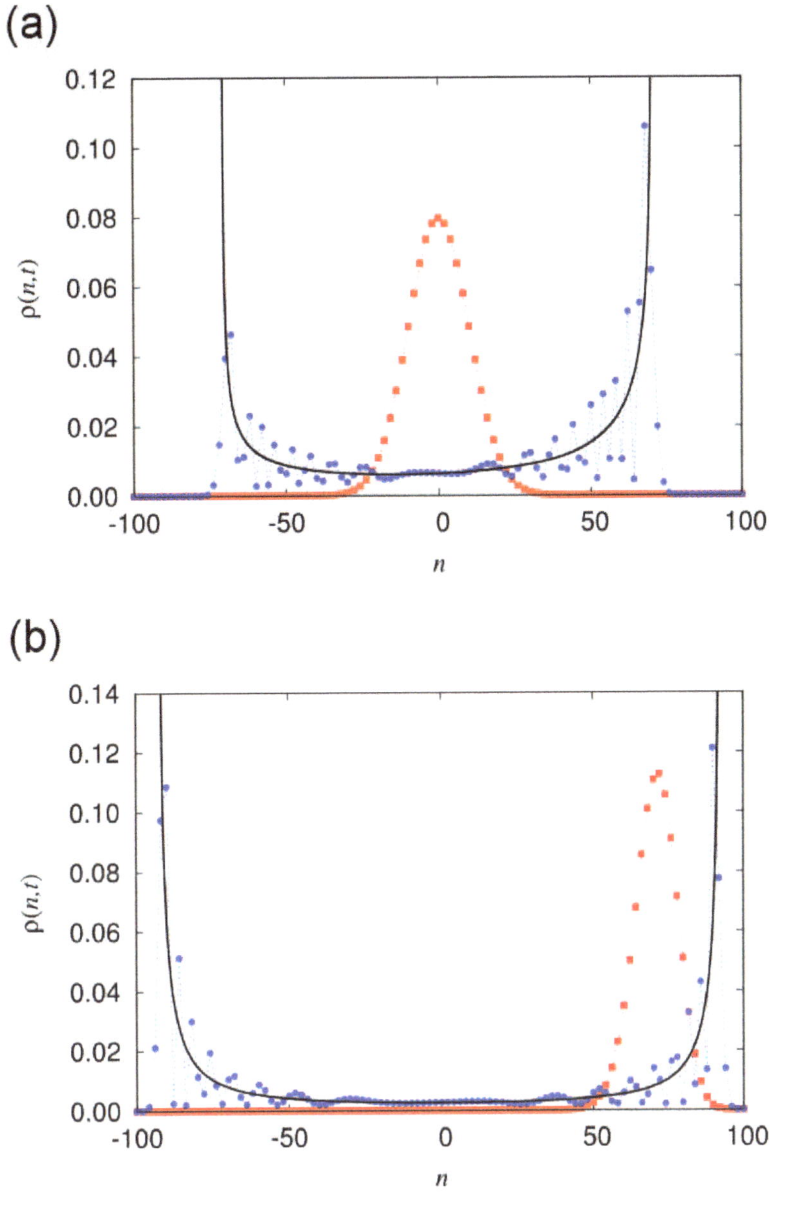

Figure 1. Probability mass function after t = 100 time steps. The dots correspond to the exact result for: (a) $\theta = \pi/4$, $\eta = \pi/16$, $\varphi = \pi$; (b) $\theta = \pi/8$, $\eta = 3\pi/16$, $\varphi = \pi$; the boxes represent classical probabilities with $p = \cos^2\theta$, whereas the black solid lines correspond to $\bar{\rho}(n, t)$, cf. Eq. (19). We have only depicted probabilities for even values of n, since in this case probabilities for odd values of n are identically null.

Regarding the expectation value of the position of the quantum walker, $\langle X \rangle_t$,

$$\langle X \rangle_t \equiv \ell \cdot \sum_{n=-\infty}^{\infty} n \rho(n,t), \tag{20}$$

its magnitude does not stem from the location of the largest maximum of $\rho(n, t)$, but has its origin in the skewness of the distribution. An elementary analysis of $\bar{\rho}(n, t)$ reveals that any bias in $\langle X \rangle_t$ is determined in the long run by the sign of the expression between parentheses in the right-hand side of Eq. (19). Therefore, as long as

$$\cos 2\eta + \sin 2\eta \tan\theta \cos\varphi \neq 0,$$

the expectation value of the position of the walker will increase linearly with time:

$$\langle X \rangle_t : \ \ell(1-\sin\theta)(\cos 2\eta + \sin 2\eta \tan\theta \cos\varphi)t, \tag{21}$$

as it can be checked in **Figure 2**. The converse is not true [40, 41]: in order to get quantum walkers that show an exact symmetry in the parity one has to demand that

$$\cos 2\eta + \sin 2\eta \tan\theta \cos\varphi = 0, \tag{22}$$

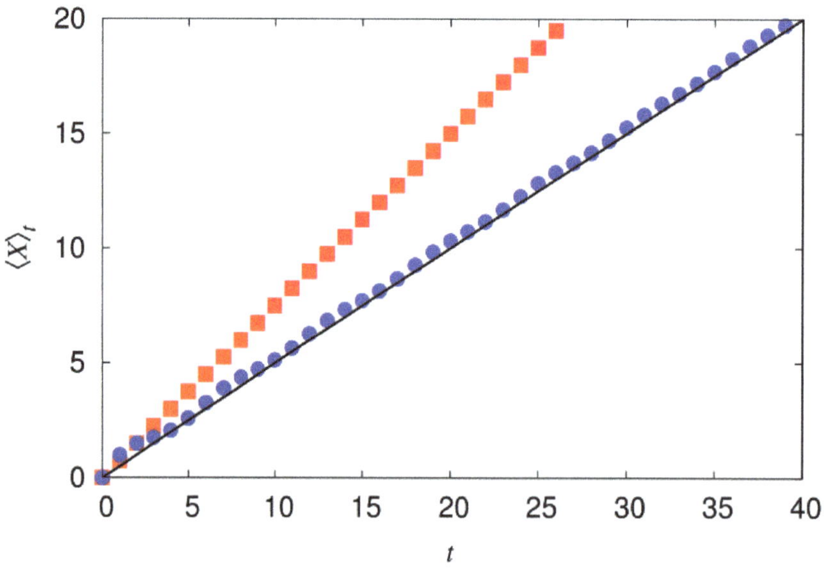

Figure 2. Expectation value of the position of the walker after $t = 40$ time steps. The dots correspond to the exact result for the QW with $\theta = \pi/6$, $\eta = \pi/6$, $\varphi = 0$, the boxes represent the classical mean position when $p = \cos^2\theta = 3/4$, whereas the black solid line corresponds to the approximate law, Eq. (21), which in this case reads $\langle X \rangle_t \sim t/2$ when l=1.

but also that [3]

$$\cos 2\eta + \sin 2\eta \tan 2\theta \cos\varphi = 0, \tag{23}$$

3 Eq. (23) implies $|\psi_+(+1, 1)|^2 = |\psi_-(-1, 1)|^2 = 1/2$, see Eqs. (26) and (27) below. In other words, this is the condition that ensures the absence of bias in the "initial velocities."

equations that have only three main families of solutions [38], being the most relevant of them the one corresponding to $\eta=\pi/4$, $\varphi=\pi/2$, for any choice of θ.

3. Inhomogeneous QWs

The fact that not only χ but even α and β (after a suitable choice of γ) can be completely ignored in the previous analysis can lead to the false conclusion that these phases can be disregarded in any other situation. We will devote the rest of this Chapter to the analysis of a framework where these magnitudes play a crucial role.

Consider a general inhomogeneous, time-dependent unitary operator $\hat{U}_t$:

$$\begin{aligned}\hat{\mathcal{U}}_t \equiv \sum_{n=-\infty}^{\infty} e^{i\chi_{n,t}}[e^{i\alpha_{n,t}} \cos\theta_{n,t} \,|+\rangle\langle+| + e^{-i\beta_{n,t}} \sin\theta_{n,t} \,|+\rangle\langle-| \\ + e^{i\beta_{n,t}} \sin\theta_{n,t} \,|-\rangle\langle+| - e^{-i\alpha_{n,t}} \cos\theta_{n,t} \,|-\rangle\langle-|]\otimes|n\rangle\langle n|,\end{aligned} \tag{24}$$

where $\alpha_{n,t}$, $\beta_{n,t}$, $\chi_{n,t}$, and $\theta_{n,t}$ are two-dimensional sets of real quantities. Now, we can define a new evolution operator $\hat{\mathcal{T}}_t$ based on $\hat{\mathcal{U}}_t$ and the standard shift operator $\hat{\mathcal{S}}$ Eq. (5), $\hat{\mathcal{T}}_t \equiv \hat{\mathcal{S}}\hat{\mathcal{U}}_t$ in such a way that the state of the particle at time $t+1$ is the result of the application of $\hat{T}_t$ to $|\psi\rangle_t$,

$$|\psi\rangle_{t+1} = \hat{\mathcal{T}}_t\,|\psi\rangle_t. \tag{25}$$

In this case, the information supplied by the initial state of the system is not so important: Assume that $|\psi\rangle_0$ is of the form depicted in Eq. (7). Then, one has that

$$|\psi_+(+1,1)|^2 = \frac{1}{2}\left[1+\cos 2\eta \cos 2\theta_{0,0} + \sin 2\eta \sin 2\theta_{0,0} \cos\varphi_{0,0}\right], \tag{26}$$

$$|\psi_-(-1,1)|^2 = \frac{1}{2}\left[1-\cos 2\eta \cos 2\theta_{0,0} - \sin 2\eta \sin 2\theta_{0,0} \cos\varphi_{0,0}\right], \tag{27}$$

with $\varphi_{0,0} \equiv \alpha_{0,0}+\beta_{0,0}-\gamma$. Note how this expression is invariant under the interchange

$$\begin{aligned}\eta &\leftrightarrow \theta_{0,0},\\ \gamma &\leftrightarrow \alpha_{0,0}+\beta_{0,0}.\end{aligned}$$

In practice, this means that we can modify $\theta_{0,0}$ and $\alpha_{0,0}+\beta_{0,0}$ in order to obtain any desired value for $|\psi_+(+1,1)|$ and $|\psi_-(-1,1)|$, irrespective of η and γ. The complex arguments of $\psi_+(+1,1)$ and $\psi_-(-1,1)$ can be recovered with a suitable choice of $\chi_{0,0}$ and $\alpha_{0,0}-\beta_{0,0}$.

The recursive equations of the wave-function components under the present dynamics induced by $\widehat{T}_t$ are straightforward variations of Eqs. (12) and (13):

$$\begin{aligned}\psi_+^\circ(n,t) = e^{i\chi_{n-1,t-1}^\circ}[e^{i\alpha_{n-1,t-1}^\circ}\cos\theta_{n-1,t-1}\psi_+^\circ(n-1,t-1)\\ +e^{-i\beta_{n-1,t-1}^\circ}\sin\theta_{n-1,t-1}\psi_-^\circ(n-1,t-1)],\end{aligned} \tag{28}$$

and

$$\begin{aligned}\psi_-^\circ(n,t) = e^{i\chi_{n+1,t-1}^\circ}[e^{i\beta_{n+1,t-1}^\circ}\sin\theta_{n+1,t-1}\psi_+^\circ(n+1,t-1)\\ -e^{-i\alpha_{n+1,t-1}^\circ}\cos\theta_{n+1,t-1}\psi_-^\circ(n+1,t-1)].\end{aligned} \tag{29}$$

Since we have a specific interest in revealing a new kind of invariance, we will introduce $\psi_\pm(n, t)$, the solution to a certain inhomogeneous, time-dependent appealing problem

$$\begin{aligned}\psi_+^\circ(n,t) = e^{i\chi_{n-1,t-1}^\circ}[e^{i\alpha_{n-1,t-1}^\circ}\cos\theta_{n-1,t-1}\psi_+^\circ(n-1,t-1)\\ +e^{-i\beta_{n-1,t-1}^\circ}\sin\theta_{n-1,t-1}\psi_-^\circ(n-1,t-1)],\end{aligned} \tag{30}$$

and

$$\begin{aligned}\psi_-^\circ(n,t) = e^{i\chi_{n+1,t-1}^\circ}[e^{i\beta_{n+1,t-1}^\circ}\sin\theta_{n+1,t-1}\psi_+^\circ(n+1,t-1)\\ -e^{-i\alpha_{n+1,t-1}^\circ}\cos\theta_{n+1,t-1}\psi_-^\circ(n+1,t-1)].\end{aligned} \tag{31}$$

Therefore, our task is to find out nontrivial relationships connecting both set of parameters. Regarding this, note that $\theta_{n,t}$ is the same in both cases: as we have seen in Section 2, there are some features of the process, which are exclusively encoded in these magnitudes, and therefore, we will exclude them from the present analysis.

4. Invariance

The properties of the system enumerated up to this point are based on the moduli of the components of the wave function. This means, in particular, that if one has that $\psi_{\pm}(n, t)$ and $\psi_{\pm}(n, t)$ are linked through the following identities:

$$\psi_{+}(n,t) = \psi_{+}^{\circ}(n,t)e^{i\xi_{n,t}}, \tag{32}$$

and

$$\psi_{-}(n,t) = \psi_{-}^{\circ}(n,t)e^{i\zeta_{n,t}}, \tag{33}$$

$\rho(n, t)$ or $M(n, t)$ will remain unchanged. The new magnitudes introduced in Eqs. (32) and (33), $\xi_{n,t}$ and $\zeta_{n,t}$, are two additional sets of arbitrary real constants, whose meaning will be discussed below.

If we assume the validity of Eqs. (32) and (33) and replace these expressions in Eqs. (28) and (29), the conditions to recover Eqs. (30) and (31) are

$$\begin{aligned}
\chi_{n,t}^{\circ} + \alpha_{n,t}^{\circ} &= \chi_{n,t} + \alpha_{n,t} + \xi_{n,t} - \xi_{n+1,t+1},\\
\chi_{n,t}^{\circ} - \alpha_{n,t}^{\circ} &= \chi_{n,t} - \alpha_{n,t} + \zeta_{n,t} - \zeta_{n-1,t+1},\\
\chi_{n,t}^{\circ} + \beta_{n,t}^{\circ} &= \chi_{n,t} + \beta_{n,t} + \xi_{n,t} - \zeta_{n-1,t+1},\\
\chi_{n,t}^{\circ} - \beta_{n,t}^{\circ} &= \chi_{n,t} - \beta_{n,t} + \zeta_{n,t} - \xi_{n+1,t+1}.
\end{aligned}$$

These equations lead to the following prescription to modify the phases leaving invariant the moduli of the components of the wave function:

$$\chi_{n,t} = \chi_{n,t}^{\circ} + \frac{\xi_{n+1,t+1} - \xi_{n,t} + \zeta_{n-1,t+1} - \zeta_{n,t}}{2}, \tag{34}$$

$$\alpha_{n,t} = \alpha_{n,t}^{\circ} + \frac{\xi_{n+1,t+1} - \xi_{n,t} - \zeta_{n-1,t+1} + \zeta_{n,t}}{2}, \tag{35}$$

$$\beta_{n,t} = \beta_{n,t}^{\circ} + \frac{\zeta_{n-1,t+1} + \zeta_{n,t} - \xi_{n+1,t+1} - \xi_{n,t}}{2}. \tag{36}$$

4.1. Invariance of global observables

The first conclusion that can be drawn from Eqs. (34)–(36) is that there is an infinite variety of choices for $\xi_{n,t}$ and $\zeta_{n,t}$ that does not modify the main properties of the quantum walker. The hard task is to identify those with a clear physical meaning or relevance. In a previous work [37], it has been considered one example that belongs to the following category:

$$\xi_{n+1,t+1} = \xi_{n,t}, \tag{37}$$

$$\zeta_{n-1,t+1} = \zeta_{n,t}. \tag{38}$$

This assumption simplifies enormously Eqs. (34)–(36):

$$\chi_{n,t} = \chi^{\circ}_{n,t}, \tag{39}$$

$$\alpha_{n,t} = \alpha^{\circ}_{n,t}, \tag{40}$$

$$\beta_{n,t} = \beta^{\circ}_{n,t} + \zeta_{n,t} - \xi_{n,t}. \tag{41}$$

One particular choice that satisfies the above requirements is $\beta_{n,t} = \beta_0$, a constant value for all n and t, and the following functional forms for $\xi_{n,t}$ and $\zeta_{n,t}$:

$$\xi_{n,t} = \frac{n-t}{2}(\beta_1 - \beta_0), \tag{42}$$

$$\zeta_{n,t} = \frac{n+t}{2}(\beta_1 - \beta_0), \tag{43}$$

a possible solution of Eqs. (37) and (38). The above expressions lead to the following homogeneous update rule for $\beta_{n,t}$, $t \geq 0$,

$$\beta_{n,t} = \beta_t = \beta_0 + (\beta_1 - \beta_0)t, \tag{44}$$

where β_1 is an arbitrary constant, whose value cannot be assessed on the basis of the knowledge of $\rho(t, n)$, $P_{\pm}(t)$, or $M(n, t)$: it can only be inferred from the relative phase of the spinor components.

We illustrate in **Figure 3** the invariance of $\rho(t, n)$ in spite of the time- and site-inhomogeneous phase shifts that Eq. (44) introduces in the wave-function components, cf. Eqs. (42) and (43). Here, we have set $\theta=\pi/3$, $\eta=\pi/3$, $\gamma=0$, $\chi=0$, $\alpha=0$, $\beta_0=0$, and $\beta_1=1/10$. With this choice, $\psi_{\pm}(n, t)$ are real functions that solve a stationary homogeneous problem, whereas $\psi_{\pm}(n, t)$ exhibit a complex, correlated behavior: for example, $\psi_{-}(n, t)$ is a symmetric function around $n=0$, while neither the real part nor the imaginary part of $\psi_{-}(n, t)$ shows this symmetry.

We can sketch a complementary picture that may help in the understanding the behavior of $\hat{U}_t$ when β_t follows Eq. (44), through a geometrical analogy. Let us introduce u_t, a time-dependent, unit-length vector in R^3. Let us denote by θ and β_t its polar and azimuthal spherical coordinates, respectively. Then, we can recover the coin operator $\hat{U}_t$ through the scalar projection of the Pauli vector of operators, $\hat{\boldsymbol{\sigma}}$, with Cartesian components

$$\hat{\sigma}_x \equiv |+\rangle\langle-| + |-\rangle\langle+|,$$
$$\hat{\sigma}_y \equiv -i|+\rangle\langle-| + i|-\rangle\langle+|, \text{ and}$$
$$\hat{\sigma}_z \equiv |+\rangle\langle+| - |-\rangle\langle-|,$$

onto the $\boldsymbol{u}_t$ direction, that is,

$$\hat{\mathcal{U}}_t \equiv (\boldsymbol{u}_t \cdot \hat{\boldsymbol{\sigma}}) \otimes \hat{I}_P, \tag{45}$$

where $\hat{I}_P$ is the identity operator defined in the position space $\mathcal{H}_P$. The evolution of $\boldsymbol{u}_t$ is a step-like precession around the North Pole. Observe how, as in the example shown in **Figure 2**, when $(\beta_1-\beta_0)/\pi$ is an irrational number, the precession of $\boldsymbol{u}_t$ is not a periodic phenomenon at all. The absence of periodicity implies that vector $\boldsymbol{u}_t$ defines an everywhere dense but enumerable subset of points in the ring associated with colatitude θ on the sphere, and thus, the unconditional probability of choosing a particular value for β_t is uniformly distributed in the stationary limit.

4.2. Exact invariance

Obviously, we can go further and demand exact invariance in the problem. This can be achieved by setting $\zeta_{n,t}=\xi_{n,t}$. Eqs. (34)–(36) read now [36]:

$$\chi_{n,t} = \mathring{\chi}_{n,t} + \frac{\xi_{n+1,t+1} + \xi_{n-1,t+1} - 2\xi_{n,t}}{2}, \tag{46}$$

$$\alpha_{n,t} = \alpha^{\circ}_{n,t} + \frac{\xi_{n+1,t+1} - \xi_{n-1,t+1}}{2}, \tag{47}$$

$$\beta_{n,t} = \beta^{\circ}_{n,t} - \frac{\xi_{n+1,t+1} - \xi_{n-1,t+1}}{2}. \tag{48}$$

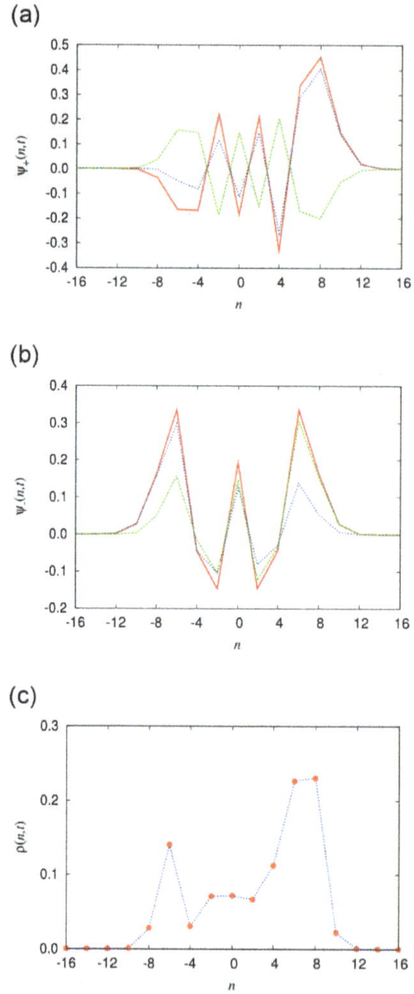

Figure 3. Comparison of the wave function after t = 16 time steps. The red solid lines and dots correspond to a timehomogeneous QW. The blue-dotted lines show the real parts of the magnitudes associated with a time-dependent QW, while the imaginary parts are depicted by green-dashed lines.

As is shown below, these equations can be expressed in terms of finite differences which in turn lead to partial derivatives. In fact, in the expression of $\chi_{n,t}$, it appears a time derivative, whereas the formulas for $\alpha_{n,t}$ and $\beta_{n,t}$ contain a spatial derivative. To illustrate these statements, consider the simple choice

$$\xi_{n,t} = a \cdot n \cdot t. \tag{49}$$

Equations (46) to (48) read, as we have anticipated,

$$\chi_{n,t} = \overset{\circ}{\chi}_{n,t} + a \cdot n, \tag{50}$$

$$\alpha_{n,t} = \overset{\circ}{\alpha}_{n,t} + a(t+1), \tag{51}$$

$$\beta_{n,t} = \overset{\circ}{\beta}_{n,t} - a(t+1). \tag{52}$$

This means, in particular, that we can transform an inhomogeneous coin into a time-dependent one

$$\overset{\circ}{\chi}_{n,t} = -a \cdot n \rightarrow \chi_{n,t} = 0,$$

$$\overset{\circ}{\alpha}_{n,t} = 0 \rightarrow \alpha_{n,t} = a(t+1),$$

$$\overset{\circ}{\beta}_{n,t} = 0 \rightarrow \beta_{n,t} = -a(t+1).$$

4.3. Continuous limit

Let us express Eqs. (34) to (36) in a slightly different way. Consider the discrete difference operators Δ_n and Δ_t defined as follows:

$$\Delta_n \xi_{n,t} \equiv \xi_{n+1,t} - \xi_{n,t}, \tag{53}$$

$$\Delta_t \xi_{n,t} \equiv \xi_{n,t+1} - \xi_{n,t}, \tag{54}$$

and similarly for $\Delta_n \zeta_{n,t}$ and $\Delta_t \zeta_{n,t}$. In terms of these operators, Eqs. (34) to (36) now read:

$$\chi_{n,t} = \overset{\circ}{\chi}_{n,t} + \frac{1}{2}\left[\Delta_n\left(\xi_{n,t+1} - \zeta_{n-1,t+1}\right) + \Delta_t\left(\xi_{n,t} + \zeta_{n,t}\right)\right], \tag{55}$$

$$\alpha_{n,t} = \overset{\circ}{\alpha}_{n,t} + \frac{1}{2}\left[\Delta_n\left(\xi_{n,t+1} + \zeta_{n-1,t+1}\right) + \Delta_t\left(\xi_{n,t} - \zeta_{n,t}\right)\right], \tag{56}$$

$$\beta_{n,t} = \beta^{\circ}_{n,t} + \zeta_{n,t} - \xi_{n,t} - \frac{1}{2}\left[\Delta_n\left(\xi_{n,t+1} + \zeta_{n-1,t+1}\right) + \Delta_t\left(\xi_{n,t} - \zeta_{n,t}\right)\right]. \tag{57}$$

Observe how the expression connecting $\beta_{n,t}$ and $\beta_{n,t}$ depends explicitly on $\xi_{n,t}$ and $\zeta_{n,t}$, in the sense that it is not merely a function of the increments, cf. Eq. (41) above. In fact, we can rearrange the previous expressions in order to emphasize the distinct effects of $\xi_{n,t}$ and $\zeta_{n,t}$:

$$\chi_{n,t} + \alpha_{n,t} = \chi^{\circ}_{n,t} + \alpha^{\circ}_{n,t} + \Delta_n\xi_{n,t+1} + \Delta_t\xi_{n,t}, \tag{58}$$

$$\chi_{n,t} - \alpha_{n,t} = \chi^{\circ}_{n,t} - \alpha^{\circ}_{n,t} - \Delta_n\zeta_{n-1,t+1} + \Delta_t\zeta_{n,t}, \tag{59}$$

$$\alpha_{n,t} + \beta_{n,t} = \alpha^{\circ}_{n,t} + \beta^{\circ}_{n,t} + \zeta_{n,t} - \xi_{n,t}. \tag{60}$$

At this point, it is appropriate to note that we are not taking into account the issue of the parity of indexes n and t: since the instances of $\xi_{n,t}$ and $\zeta_{n,t}$ that appear in Eqs. (34) to (36) are those whose subscripts have the same parity, only one of the two terms in the right-hand side of Eqs. (53) and (54) is relevant or even well defined.

However, our interest in this Section is to analyze the continuous limit, $\tau \to 0$, $l \to 0$. Up to the first order in τ and l, one has that discrete difference operators Δ_n and Δ_t become partial derivatives:

$$\Delta_n : \ \ell\frac{\partial}{\partial X},$$
$$\Delta_t : \ \tau\frac{\partial}{\partial T}.$$

We need to relate l and τ in order to obtain an unambiguous limit. We will assume that $l = c \cdot \tau$, where c is the characteristic speed associated with the action of the shift operator $\hat{S}$ upon the state of the walker. Therefore, depending on the physical nature of the system, c represents the velocity at which the information is transferred, and it may coincide with the speed of light in vacuum. With this prescription, one has that Eqs. (55)–(57) turn into

$$\chi : \ \chi^{\circ} + \frac{\ell}{2}\left[\partial_+\xi + \partial_-\zeta\right], \tag{61}$$

$$\alpha : \ \alpha^{\circ} + \frac{\ell}{2}\left[\partial_+\xi - \partial_-\zeta\right], \tag{62}$$

$$\beta: \ \beta^{\circ} + \zeta - \xi - \frac{\ell}{2}[\partial_{+}\xi - \partial_{-}\zeta], \tag{63}$$

where $\partial_{\pm}$ are defined as follows,

$$\partial_{\pm} \equiv \frac{\partial}{\partial X^{\pm}} = \frac{1}{c} \cdot \frac{\partial}{\partial T} \pm \frac{\partial}{\partial X}, \tag{64}$$

and $X^{\pm} = c \cdot T \pm X$ are the coordinates of the null geodesics in a flat (1+1) space and time. Observe how we have removed the subscripts: the dependency on X and T of all the magnitudes is implicitly assumed from now on.

The exact invariance, $\zeta = \xi$, was analyzed in detail in the study of Di Molfetta et al [36]. There, it is shown how the recurrence equations of the wave-function components of the walker, Eqs. (28) and (29), can be mapped into equations describing the propagation of a Dirac spinor with charge e and masses $m_{\pm}$ coupled to a two-dimensional Maxwell potential $\mathbf{A}$:

$$i\hbar\,\partial_{+}\psi_{+} + e(A_T + A_X)\psi_{+} - m_{+}c\psi_{-} = 0, \tag{65}$$

$$i\hbar\,\partial_{-}\psi_{-} + e(A_T - A_X)\psi_{-} - m_{-}c\psi_{+} = 0, \tag{66}$$

whose respective space-time components must change according to the formulas

$$A_T = A_T^{\circ} + \frac{\hbar}{ec}\lim_{\tau \to 0}\frac{\chi - \chi^{\circ}}{\tau}, \tag{67}$$

$$A_T = A_T^{\circ} + \frac{\hbar}{ec}\lim_{\tau \to 0}\frac{\chi - \chi^{\circ}}{\tau}, \tag{68}$$

Note that $\zeta = \xi$ implies that

$$\chi: \ \chi^{\circ} + \tau\frac{\partial \xi}{\partial T}, \tag{69}$$

$$\alpha: \ \alpha^{\circ} + \ell\frac{\partial \xi}{\partial X}, \tag{70}$$

$$\beta : \ \beta^{\circ} - \ell \frac{\partial \xi}{\partial X}, \tag{71}$$

and, when one introduces these relationships into Eqs. (67) and (68), one obtains the standard gauge transformations for the components of the potential **A**,

$$A_T = A_T^{\circ} + \frac{\hbar}{ec} \cdot \frac{\partial \xi}{\partial T}, \tag{72}$$

$$A_X = A_X^{\circ} + \frac{\hbar}{e} \cdot \frac{\partial \xi}{\partial X}, \tag{73}$$

a transform that keeps invariant the electric field E_X acting upon the system,

$$E_X \equiv \frac{\partial A_X}{\partial T} - c \frac{\partial A_T}{\partial X} = \frac{\partial A_X^{\circ}}{\partial T} - c \frac{\partial A_T^{\circ}}{\partial X} = E_X^{\circ}. \tag{74}$$

If we reconsider the example introduced at the end of Section 4.2,

$$\xi = \frac{eE_X^{\circ}}{\hbar} \cdot X \cdot T,$$

we can conclude that it corresponds to a case in which the electric field E_X is constant, where we are replacing the electric potential ϕ , $\phi = -c \cdot A_T$, by a time-dependent magnetic potential A_X,

$$A_T^{\circ} = -\frac{E_X^{\circ}}{c} \cdot X \to A_T = 0,$$
$$A_X^{\circ} = 0 \to A_X = E_X^{\circ} \cdot T.$$

In the most general case, when $\zeta \neq \xi$, the transformation rule for **A** is

$$A_T = A_T^{\circ} + \frac{\hbar}{2e}\left[\partial_{+}\xi + \partial_{-}\zeta\right], \tag{75}$$

$$A_X = A_X^{\circ} + \frac{\hbar}{2e}\left[\partial_{+}\xi - \partial_{-}\zeta\right], \tag{76}$$

which departs from the gauge invariance of potential **A**. However, if we investigate the change in the electric field induced by Eqs. (75) and (76) we find

$$E_X - E_X^{\circ} = \frac{\partial\left[A_X - A_X^{\circ}\right]}{\partial T} - c\frac{\partial\left[A_T - A_T^{\circ}\right]}{\partial X} = \frac{\hbar c}{2e}\left[\partial_-\partial_+\xi - \partial_+\partial_-\zeta\right]. \tag{77}$$

Clearly, $\zeta = \xi$ is not the only solution to the constraint

$$\partial_-\partial_+\xi - \partial_+\partial_-\zeta = 0, \tag{78}$$

that results in the invariance of the electric field. A possible choice is to demand that both ξ and ζ satisfy the 2-dimensional wave equation by their own

$$\frac{1}{c^2}\frac{\partial^2\xi}{\partial T^2} - \frac{\partial^2\xi}{\partial X^2} = \frac{1}{c^2}\frac{\partial^2\zeta}{\partial T^2} - \frac{\partial^2\zeta}{\partial X^2} = 0. \tag{79}$$

Another alternative solution to Eq. (78) has appeared above, in Section 4.1. The equivalent expressions for Eqs. (37) and (38) in the continuous limit read:

$$\partial_+\xi = \partial_-\zeta = 0,$$

what provides another solution to Eq. (78). Note that in this case Eqs. (65) and (66) show not merely covariance but perfect invariance in the mass-less case, $m_+ = m_- = 0$, since

$$A_T + A_X = A_T^{\circ} + A_X^{\circ} + \frac{\hbar}{e}\partial_+\xi, \tag{80}$$

$$A_T - A_X = A_T^{\circ} - A_X^{\circ} + \frac{\hbar}{e}\partial_-\zeta. \tag{81}$$

5. Conclusion

Along this Chapter, we have analyzed some interesting aspects of discrete-time QWs on the line, specifically those related with the emergence of invariance. In the first part, we have elaborated a succinct but comprehensive review covering the main features of the most elementary version of this process, when the unitary operator which assumes the function of the coin in the classical analog is kept fixed. We have described the dynamics that determines the evolution of the walker, supplied explicit formulas for assessing the precise state of the system at any time and approximate expressions that capture the main traits of the process in the stationary limit. These equations have been very useful to pinpoint the role played by the different parameters on the solution to the problem and put into context the generalization considered afterward.

The second part of the Chapter contemplates the situation in which the coin is time and site dependent. In particular, we have focused our interest on the phase parameters that define the unitary operator and determined the constraints that must be imposed in these changing phases if one wants to obtain invariance. This invariance can be demanded at two different levels: one can require that the invariance connects states belonging to the same ray of the Hilbert space or a milder condition, that the transformation modifies unevenly the two wave-function components. In this latter case, global properties (e.g., the probability that the particle is in a particular place or in a given spin state) remain unaltered but some other local quantum properties depending on the relative phase of these components can become modified.

The Chapter ends by analyzing the introduced invariance in the continuous limit. This approach unveils that the evolution of a time- and site-inhomogeneous quantum walk can be understood in terms of the dynamics of a particle coupled to an electromagnetic field and that the new symmetry shown by the walker can be interpreted as a manifestation of the well-known gauge invariance of electromagnetism.

Acknowledgements

The author acknowledges partial support from the Spanish Ministerio de Economa y Competitividad (MINECO) under Contract No. FIS2013-47532-C3-2-P and from Generalitat de Catalunya under Contract No. 2014SGR608.

Author details

Miquel Montero

Address all correspondence to: miquel.montero@ub.edu

Department of Fundamental Physics, University of Barcelona (UB), Barcelona, Spain

References

[1] Aharonov Y, Davidovich L, Zagury N: Quantum random walks, Physical Review A. 1993;48:1687–1690. DOI: 10.1103/PhysRevA.48.1690

[2] Travaglione B C, Milburn G J: Implementing the quantum random walk, Physical Review A. 2002;65:032310. DOI: 10.1103/PhysRevA.65.032310

[3] Konno N: Quantum random walks in one dimension, Quantum Information Processing. 2003;1:345–354. DOI: 10.1023/A:1023413713008

[4] Kempe J: Quantum random walks: an introductory overview, Contemporary Physics. 2003;44:307–327. DOI: 10.1080/00107151031000110776

[5] Venegas-Andraca S E: Quantum walks: a comprehensive review, Quantum Information Processing. 2012;11:1015–1106. DOI: 10.1007/s11128-012-0432-5

[6] Childs A, Farhi E, Gutmann S: An example of the difference between quantum and classical random walks, Quantum Information Processing. 2003;1:35–43. DOI: 10.1023/A:1019609420309

[7] Ambainis A, Bach E, Nayak A, Vishwanath A, Watrous J. One-dimensional quantum walks. In: Proceedings of the thirty-third annual ACM symposium on Theory of computing (STOC '01); 06–08 July 2001; Heraklion. New York: ACM; 2001. p. 37–49. DOI: 10.1145/380752.380757

[8] Shor P W: Polynomial-time algorithms for prime factorization and discrete logarithms on a quantum computer, SIAM Journal on Computing. 1997;26:1484–1509. DOI: 10.1137/S0097539795293172

[9] Farhi E, Gutmann S: Quantum computation and decision trees, Physical Review A. 1998;58:915–928. DOI: 10.1103/PhysRevA.58.915

[10] Shenvi N, Kempe J, Whaley K B: Quantum random-walk search algorithm, Physical Review A. 2003;67:052307. DOI: 10.1103/PhysRevA.67.052307

[11] Agliari E, Blumen A, Nülken O: Quantum-walk approach to searching on fractal structures, Physical Review A. 2010;82:012305. DOI: 10.1103/PhysRevA.82.012305

[12] Magniez F, Nayak A, Roland J, Santha M: Search via quantum walk, SIAM Journal on Computing. 2011;40:142–164. DOI: 10.1137/090745854

[13] Flitney A P, Abbott D, Johnson N F: Quantum walks with history dependence, Journal of Physics A. 2004;37:7581–7591. DOI: 10.1088/0305-4470/37/30/013

[14] Bulger D, Freckleton J, Twamley J: Position-dependent and cooperative quantum Parrondo walks, New Journal of Physics. 2008;10:093014. DOI: 10.1088/1367-2630/10/9/093014

[15] Chandrashekar C M, Banerjee S: Parrondo's game using a discrete-time quantum walk, Physics Letters A. 2011;375:1553–1558. DOI: 10.1016/j.physleta.2011.02.071

[16] Romanelli A, Hernández G: Quantum walks: Decoherence and coin-flipping games, Physica A. 2011;390:1209–1220. DOI: 10.1016/j.physa.2010.12.006

[17] Chandrashekar C M, Srikanth R, Laflamme R: Optimizing the discrete time quantum walk using a SU(2) coin, Physical Review A. 2008;77:032326. DOI: 10.1103/PhysRevA. 77.032326

[18] Brun T A, Carteret H A, Ambianis A: Quantum walks driven by many coins, Physical Review A. 2003;67:052317. DOI: 10.1103/PhysRevA.67.052317

[19] Tregenna B, Flanagan W, Maile R, Kendon V: Controlling discrete quantum walks: coins and initial states, New Journal of Physics. 2003;5:83. DOI: 10.1088/1367-2630/5/1/383

[20] Venegas-Andraca S E, Ball J L, Burnett K, Bose S: Quantum walks with entangled coins, New Journal of Physics. 2005;7:221. DOI: 10.1088/1367-2630/7/1/221

[21] Brun T A, Carteret H A, Ambianis A: Quantum random walks with decoherent coins, Phyical Review A. 2003;67:032304. DOI: 10.1103/PhysRevA.67.032304

[22] Kendon V, Tregenna B: Decoherence can be useful in quantum walks, Physical Review A. 2003;67:042315. DOI: 10.1103/PhysRevA.67.042315

[23] Wójcik A, Łuczak T, Kurzyn'ski P, Grudka A, Bednarska M: Quasiperiodic dynamics of a quantum walk on the line, Physical Review Letters. 2004;93:180601. DOI: 10.1103/ PhysRevLett.93.180601

[24] Romanelli A, Auyuanet A, Siri R, Abal G, Donangelo R: Generalized quantum walk in momentum space, Physica A. 2005;352:409–418. DOI: 10.1016/j.physa.2005.01.026

[25] Shikano Y, Katsura H: Localization and fractality in inhomogeneous quantum walks with self-duality, Physical Review E. 2010;82:031122. DOI: 10.1103/PhysRevE.82.031122

[26] Konno N, Łuczak T, Segawa E: Limit measures of inhomogeneous discrete-time quantum walks in one dimension, Quantum Information Processing. 2013;12:33–53. DOI: 10.1007/s11128-011-0353-8

[27] Zhang R, Xue P, Twamley J: One-dimensional quantum walks with single-point phase defects, Physical Review A. 2014;89:042317. DOI: 10.1103/PhysRevA.89.042317

[28] Xue P, Qin H, Tang B, Sanders C: Observation of quasiperiodic dynamics in a one-dimensional quantum walk of single photons in space, New Journal of Physics. 2014;16:053009. DOI: 10.1088/1367-2630/16/5/053009

[29] Ribeiro P, Milman P, Mosseri R: Aperiodic quantum random walks, Physical Review Letters. 2004;93:190503. DOI: 10.1103/PhysRevLett.93.190503

[30] Romanelli A: The Fibonacci quantum walk and its classical trace map, Physica A. 2009;388:3985–3990. DOI: 10.1016/j.physa.2009.06.022

[31] Romanelli A: Driving quantum-walk spreading with the coin operator, Physical Review A. 2009;80:042332. DOI: 10.1103/PhysRevA.80.042332

[32] Romanelli A, Segundo G: The entanglement temperature of the generalized quantum walk, Physica A. 2014;393:646–654. DOI: 10.1016/j.physa.2013.08.050

[33] Bañuls M C, Navarrete C, Pérez A, Roldán E, Soriano J C: Quantum walk with a time-dependent coin, Physical Review A. 2006;73:062304. DOI: 10.1103/PhysRevA.73.062304

[34] Ahlbrecht A, Vogts H, Werner A H, Werner R F: Asymptotic evolution of quantum walks with random coin, Journal of Mathematical Physics. 2011;52:042201. DOI: 10.1063/1.3575568

[35] Di Molfetta G, Brachet M, Debbasch F: Quantum walks as massless Dirac fermions in curved space-time, Physical Review A. 2013;88:042301. DOI: 10.1103/PhysRevA.88.042301

[36] Di Molfetta G, Brachet M, Debbasch F: Quantum walks in artificial electric and gravitational fields, Physica A. 2014;397:157–168. DOI: 10.1016/j.physa.2013.11.036

[37] Montero M: Invariance in quantum walks with time-dependent coin operators, Physical Review A. 2014;90:062312. DOI: 10.1103/PhysRevA.90.062312

[38] Montero M: Quantum walk with a general coin: exact solution and asymptotic properties, Quantum Information Processing. 2015;14:839–866. DOI: 10.1007/s11128-014-0908-6

[39] Chandrashekar C M, Srikanth R, Banerjee S: Symmetries and noise in quantum walk, Physical Review A. 2007;76:022316. DOI: 10.1103/PhysRevA.76.022316

[40] Asbóth J K: Symmetries, topological phases, and bound states in the one-dimensional quantum walk, Physical Review B. 2012;86:195414. DOI: 10.1103/PhysRevB.86.195414

[41] Kitagawa T: Topological phenomena in quantum walks: Elementary introduction to the physics of topological phases, Quantum Information Processing. 2012;11:1107–1148. DOI: 10.1007/s11128-012-0425-4

[42] Konno N: Limit theorem for continuous-time quantum walk on the line, Physical Review E. 2005;72:026113. DOI: 10.1103/PhysRevE.72.026113

[43] Hoyer S, Meyer D A: Faster transport with a directed quantum walk, Physical Review A. 2009;79:024307. DOI: 10.1103/PhysRevA.79.024307

[44] Montero M: Unidirectional quantum walks: evolution and exit times, Physical Review A. 2014;88:012333. DOI: 10.1103/PhysRevA.88.012333

[45] Souza A M C, Andrade R F S: Coin state properties in quantum walks, Scientific Reports. 2013;3:1976. DOI: 10.1038/srep01976

5

Quantum Dynamics, Entropy and Quantum Versions of Maxwell's Demon

Paul Bracken

Abstract

Several subjects which reside in the overlap area of quantum mechanics, statistical physics and thermodynamics are investigated in depth. This collection of subjects shares a common domain which is referred to as Maxwell's demon. The classical version of this idea is introduced, and then, the contribution made by Szilard to the subject is presented. Several demons are considered, and it is shown that to best understand this area, quantum mechanics and the role information plays in it must be appreciated deeply.

Keywords: quantum dynamics, entropy, wave function, information, density matrix, quantum measurement

1. Introduction

This is an introduction to the paradox referred to as Maxwell's demon from both the perspective of the second law of thermodynamics and its consequences for quantum dynamics and recent progress in resolving it [1–3]. The classical paradox of Maxwell's demon has been around for a century and a half and has had a strong influence on the study of the second law of thermodynamics and statistical mechanics in general in addition to the area of quantum dynamics and information theory. Many new ideas and paradigms have been introduced as a result of this paradox. Maxwell first introduced the idea in one of his books in 1871 as a way of discussing limitations of the second law of thermodynamics [4, 5]. Clausius's version of the second law states that, "It is impossible to devise an engine which, working in a cycle, will produce no effect other than the transfer of heat from a colder to a hotter body."

In brief, the classical Maxwell's demon has the capacity to separate hot particles, atoms or molecules from cold particles and therefore obtains work from a single heat bath, which seemed to violate the second law of thermodynamics [6, 7]. The classical paradox which arose well before quantum mechanics was developed and it will be introduced from the point of view of a physical model. The demon operates a tiny door in a partition that divides a box into two parts of equal volumes and contains a gas in thermal equilibrium, and hence, temperature is uniform over the box. The demon observes the molecules in the left side, and if a molecule is seen approaching the door with speed less than the average speed of the molecules, he opens the door and lets the molecules go right; if he sees one approaching with a speed greater than average, he opens the door to let it move into the left. Once a small temperature difference has been induced between the right and left sides, his action continues to transfer heat from a colder to a hotter region without exerting any work. This violates Clausius's form of the second law of thermodynamics. This process is referred to as temperature demon.

Another type of demon can be imagined which produces a difference in pressure. This is referred to as pressure demon. This demon runs a cycle by making the gas interact with a heat bath at constant temperature after generating a pressure inequality. This cycle converts heat transferred from the bath to work. This violates the second law in terms of the Kelvin form of the second law. Thus, in either case, for either type of demon, based on temperature or pressure, the role of the demon is to decrease the entropy of the whole system in a cyclic process.

It was Szilard who first made some progress subduing the demon [8]. Szilard had the idea of treating the demon's memory or intelligence, as a form of information that could be linked to physics, to thermodynamics in fact. Szilard took a gas consisting of a single molecule, and as a first step, a thin massless adiabatic partition is inserted into the chamber quickly which divides it into two parts of equal volume. The demon measures the position coordinate of the molecule on the left or right sides and records the result for the next step. A mass is attached to the partition on the side where the molecule is found. Using a heat bath to stabilize the temperature T, the demon lets the gas do work W by quasistatic isothermal expansion. The gas returns to its initial state and occupies the whole volume of the chamber. While this takes place, heat Q is extracted from the bath and so $W = Q$, as it is an isothermal process. The cycle is completed as the extracted heat Q is turned into an equal amount of mechanical work.

During isothermal expansion of the gas then, if k_B is Boltzmann's constant, the amount of extracted work is given by the following equation:

$$W = k_B T \int_{V/2}^{V} V^{-1}\, dV = k_B T \ln(2). \tag{1.1}$$

It may be asked whether it is reasonable to say the one-molecule gas is a normal ideal gas. Consider an ensemble of one-molecule gases, as it is done in statistical mechanics. By taking averages over the ensemble, calculations will be made as if it were an ideal gas composed of a large number of molecules. By considering the position of the molecule, one is led to its dual interpretation in terms of both thermodynamics and information theory. As a result of the perfect conversion of heat Q into work W, the entropy of the heat bath is reduced by the following:

$$\frac{Q}{T} = \frac{W}{T} = k_B \ln(2). \tag{1.2}$$

The second law demands an entropy increase of at least the same amount somewhere in order to compensate for this decrease. Szilard attributed this increase to measurement. Although Szilard thought that the demon's memory was important in analyzing the engine, he failed to uncover its role in terms of the second law.

2. A classical Szilard engine

The first to state clearly that the paradox of Maxwell's demon could be solved by considering the increase in entropy due to memory erasure was Penrose in his well-known book on statistical mechanics [4]. It was later on that Landauer, and independently Bennett, arrived at a similar conclusion [9, 10]. Bennett realized that a major result from the thermodynamics of computation by Rolf Landauer could be used to show that a Maxwell's demon could not violate the second law. Information processing must be carried out by some sort of physical system, and thus, it follows that there should be a one-to-one correspondence between logical and physical states. Let us state that logical states can be described as an abstract set of variables on which the task of information processing will be carried out. A reversible process, which means an injective or one-to-one mapping for logical states, corresponds to a reversible physical process. If a correspondence between logical and physical entropies is assumed, it implies that a reversible logical process can be realized physically by a process which is isentropic. For our purposes, an isentropic process is an entropy-preserving process.

On the other hand, a logically irreversible process is a many-to-one mapping and thus noninjective. The mapping cannot be inverted and many initial states correspond to a single resulting state. Memory erasure is a logically irreversible process because many possible states of memory should be in one specific state in order not to carry any information. The specific state after erasure is called standard state. In terms of physical states, a logically irreversible process reduces degrees of freedom of a system, implying an entropy decrease. It was Landauer who realized that logical irreversibility must involve dissipation, and thus, erasure of information in a memory implies entropy increase in the environment. This result has been formulated into a statement that is now called Landauer's erasure principle. Alternatively, Landauer's principle states that although logically reversible computational processes can in principle be performed with arbitrarily little dissipation, erasure is a logically irreversible act that has a threshold entropy production. Landauer then states that for each erased bit, the entropy sent to the environment is at least $k_B \ln(2)$. A complete thermodynamic analysis of a demon's cyclic operations requires that its memory be brought back to its initial state. The entropy returned to the environment is just large enough to save the second law.

Bennett made another important contribution regarding the physics of information. Measurement can be carried out reversibly without any change in entropy, provided the measuring apparatus is in a standard state. This means that storing information in the memory does not involve the erasure of information previously stored in the same memory. Measurement will be regarded as a process that correlates the memory with the system, which can be achieved reversibly in principle. This process will be considered as one that copies the memory state to another system in a standard state [11].

Since measurement will be taken without energy use, it is dissipation due to erasure that compensates the entropy decrease induced by the demon in Szilard's model. The demon commits the result to memory by establishing the position in the box. The molecule is either on the left (L) or right (R) side, depending on the information it stores. The prescription for erasing the stored information is to remove the partition, insert a piston on the right when the standard memory state is (L) and move it left isothermally at temperature T until the compressed volume is $V/2$. The resulting state (L) is for both initial states and the information is erased. The erasing process should not depend on the initial state of the memory, and thus, erasure should be independent of the initial memory state. The work invested to compress the volume from V to $V/2$ is $W_{\text{erasure}} = k_B T \ln(2)$. This is dissipated as heat into the environment, thereby increasing its entropy by $k_B \ln(2)$, in agreement with Landauer's principle.

It is shown that the erasure work is proportional to the amount of information stored, that is, $W_{\text{erasure}} = k_B T \ln(2)\, H(p)$, where p is the probability for the molecule to be in state L and $H(p)$ is the binary Shannon entropy. In other words, suppose a tendency in the frequency of appearance of a particular memory state exists, can we ask what the erasure work is. An unbalanced tendency between L and R can be expressed by the numbers of molecules in each region. Consider only an ideal gas, so removing the partition at the beginning allows the gas an undesired irreversible adiabatic expansion or compression. Let the gases in both parts expand or contract isothermally by making the partition free to move without friction. The gases generate work toward the outside. Let P_L, P_R and V_L denote the pressures in the left and right sides and volume on the left of the partition, respectively. The work done by the gases is given as follows:

$$\begin{aligned} W &= \int_{V/2}^{pV} (P_L - P_R)\, dV_L = Nk_B T \int_{V/2}^{pV} \left(\frac{p}{V_L} - \frac{1-p}{V - V_L}\right) dV_L \\ &= Nk_B T[\ln(2) + p\ln(p) + (1-p)\ln(1-p)]. \end{aligned} \tag{2.1}$$

This work can be expressed in terms of the Shannon entropy which is defined to be

$$H(p) = -p\log_2 p - (1-p)\log_2(1-p). \tag{2.2}$$

To do so, transform $\ln(x)$ to $\log_2(x)$ by means of the base change formula $\log_2(x) = \ln(x)/\ln(2)$to obtain

$$W = Nk_BT\ln(2)[1+p\log_2 p+(1-p)\log_2(1-p)] = Nk_BT\cdot\ln(2)\cdot[1-H(p)]. \tag{2.3}$$

Since the pressures in the left and right are equal, this returns us to the configuration with the partition in the middle. Hence, at least $Nk_BT\ln(2)$ of work needs to be consumed to set the memory to the standard state. In total, after dividing out N, an amount

$$W_{erasure} = k_BT\ln(2) - W = k_BT\ln(2)H(p) \tag{2.4}$$

of work per molecule is invested. Supposing the memory is in the standard state to begin with, W_{erasure} gives a measure of the additional energy required to erase the memory due to its information content. Without violating the second law, the state of the whole system consisting of the heat engine and demon is restored after completing a thermodynamic cycle.

3. The von Neumann entropy as the quantum entropy

Entropy is not an observable property, and thus, there does not exist an operator with the property that its expectation value in some state would give the entropy. Entropy is a state function. In order to study entropy in the context of quantum systems, it is necessary to give a precise definition of the concept before proceeding [7].

To provide some motivation for this, it may be supposed that the ground state of a quantum system is described by its density matrix ρ. If $\mathbf{A}$ is an observable which pertains to the system characterized by ρ, the spectral theorem permits the spectral decomposition of $\mathbf{A}$ as $\mathbf{A}=\sum_i a_i\mathbf{P}_i$, where $\mathbf{P}_i$ is a projection operator onto the state with eigenvalue a_i. The probability of obtaining a_j in a measurement is given by $p_j=\text{Tr}(\rho\mathbf{P}_j)=\text{Tr}(P_j\rho)$. The uncertainty in a given observable can be expressed by means of the P_i through the Shannon entropy, $S(\mathbf{p})$, not to confuse it with von Neumann entropy $S_N(\mathbf{A})$.

Definition 3.1. The uncertainty in a collection of possible classical states $\{a_i\}$ with corresponding probability distribution $\mathbf{p} = p(a_i)$ is given by its entropy, the Shannon entropy

$$S(\mathbf{p}) = -\sum_i p(a_i)\log_2(p(a_i)), \tag{3.1}$$

The Shannon information is a good indicator of how much two given observables are correlated. This quantity is inherently classical, as it describes the correlations between single observables. The quantity that is related to the correlations in the overall state as a whole is the von Neumann entropy or mutual information. It is assigned to a state as a whole, so naturally it depends on the density matrix.

Definition 3.2. The von Neumann entropy of a quantum system described by the density matrix ρ is defined to be

$$S_N(\rho) = -\ \mathrm{Tr}(\rho \ln(\rho)). \tag{3.2}$$

The von Neumann entropy can also be defined in terms of the base two logarithms. Defined in this way, Eq. (3.2) implies that if λ_i are the eigenvalues of ρ. the von Neumann entropy can be expressed in Shannon form as $S_N(\rho) = -\sum_i \lambda_i log_2(\lambda_i) = S(\lambda(\rho))$. There have been several studies recently of quantum operations that preserve the von Neumann entropy of quantum states. The Shannon entropy $S(\mathbf{p})$ is equal to the von Neumann entropy only when it describes the uncertainties in the values of observables that commute with the density matrix and $S(\mathbf{p}) \geq S_{(N)}$ $(\mathbf{A})$ otherwise. Here, **A** is any observable of a system described by the density matrix ϱ. There are two important properties of the entropy in Eq. (3.2) which should be noted.

(a) additivity

$$S_N(\rho_A \otimes \rho_B) = S_N(\rho_A) + S_N(\rho_B). \tag{3.3}$$

(b) concavity

$$S_N(\sum_i \lambda_i \rho_i) \geq \sum_i \lambda_i S_N(\rho_i). \tag{3.4}$$

As in the classical instance, property (a) states that entropies add up. The concavity property (b) simply reflects the fact that mixing increases uncertainty. The von Neumann mutual information refers to the correlation between whole subsystems rather than that relating only two variables and it is introduced as follows.

Definition 3.3. The von Neumann mutual information between two subsystems ρ_U and ρ_V of a joint state ρ_{UV} is defined as follows:

$$I_N(\rho_U : \rho_V : \rho_{UV}) = S_N(\rho_U) + S_N(\rho_V) - S_N(\rho_{UV}). \tag{3.5}$$

This quantity can be interpreted as a distance between two quantum states.

Definition 3.4. The von Neumann relative entropy between two states σ and ρ is defined as follows:

$$S_N(\sigma \| \rho) = Tr\sigma(\ln\sigma - \ln\rho). \tag{3.6}$$

The relative entropy expresses how difficult it is to distinguish the state σ from the state ρ.

4. A quasi-classical erasure process

An example of an erasure process that can be studied in detail is a way to erase classical information encoded in quantum states [6]. It straddles the classical-quantum interface and can be accomplished by thermal randomization. Thermal randomization makes use of the randomness of states in a heat bath, which is in thermal equilibrium. If a heat bath at temperature T and a state, which could represent a message, are brought into contact, the state will approach equilibrium with the heat bath. A message state ρ_i changes gradually after an interaction with the heat bath. A sufficient number of collisions make the state indistinguishable from that of the heat bath. In this way, the information that was carried by ρ_i is lost irreversibly. The entropy of the whole system necessarily increases.

Suppose each external 'message' state is in a pure state. Before erasure, the whole message consists of an ensemble $\{p_i, |\phi_i\rangle\}$. Thus, its average state is described by a density operator of the form $\rho = \sum_i p_i |\phi_i\rangle\langle\phi_i|$. The process of thermalization brings all states $|\phi_i\rangle$ to the same state, φ in thermal equilibrium at temperature T. Consequently, the density matrix is as follows:

$$\varphi = \frac{1}{Z} e^{-\beta \hat{H}} = \sum_j q_j |e_j\rangle\langle e_j|, \tag{4.1}$$

where $\hat{H} = \sum_i e_i |e_i\rangle\langle e_i|$ is the Hamiltonian operator of the message state in terms of energy eigenstates $|e_i\rangle$, and $Z = \mathrm{Tr}(e^{-\beta \hat{H}})$ is the partition function. It follows from Eq. (4.1) that $\mathrm{Tr}\varphi = 1$.

The total entropy $\Delta S_{\mathrm{erasure}}$ is the sum of the entropy change of the message or reference system and of the heat bath: $\Delta S_{\mathrm{erasure}} = \Delta S_{\mathrm{system}} + \Delta S_{\mathrm{bath}}$. Since the state before erasure is pure and its state after erasure is the same as the heat bath, the minimum entropy change in the reference state is given by the following equation:

$$\Delta S_{system} = k_B \ln(2) S(\varphi), \tag{4.2}$$

where $S(\varphi) = -\mathrm{Tr}(\varphi \ln(\varphi))$ is the von Neumann entropy of state φ.

The entropy change in the heat bath is equal to the average heat transfer from the bath to the reference system divided by temperature T,

$$\Delta S_{bath} = \frac{1}{T} \Delta Q_{bath}. \tag{4.3}$$

The heat change in the heat bath must agree with that of the system, but with an opposite sign, $\Delta Q_{\mathrm{bath}} = -\Delta Q_{\mathrm{system}}$. When heat transfer is done quasistatically, the mechanical work required for the state change is arbitrarily close to zero. Moreover, energy conservation requires ΔQ_{system}

be equal to the change of internal energy of the system, ΔU_{system}. This can be computed as the change of average values of the Hamiltonian $\hat{H}$ before and after the erasure process,

$$\Delta S_{bath} = -\frac{1}{T}\Delta Q_{system} = -\frac{1}{T}\Delta U_{system} = -\frac{1}{T}[Tr(\varphi\hat{H}) - Tr(\rho\hat{H})] = -\frac{1}{T}Tr[(\varphi-\rho)\hat{H}]. \tag{4.4}$$

The Hamiltonian $\hat{H}$ can be expressed using Eq. (4.1) as $\hat{H} = -k_B T\ln(Z\varphi)$. As Z is a trace, it follows from the properties of the trace, converting the natural logarithm to base two that the following result holds

$$\Delta S_{bath} = k_B Tr[(\varphi-\rho)\ln(Z\varphi)] = k_B Tr[(\varphi-\rho)\ln(\varphi)] = -k_B \ln(2)[S(\varphi) + Tr(\rho\log_2\varphi)]. \tag{4.5}$$

Therefore, the required total energy change is given by the following equation:

$$\Delta S_{erasure} = \Delta S_{system} + \Delta S_{bath} = -k_B \ln(2)\; Tr(\rho\log_2\varphi). \tag{4.6}$$

Since the quantum relative entropy satisfies the following inequality

$$S(\rho\,\|\,\varphi) = -S_N(\rho) - \mathrm{Tr}(\rho\log_2\varphi) \geq 0, \tag{4.7}$$

the minimum of the entropy change $\Delta S_{\mathrm{erasure}}$ can be obtained,

$$\Delta S_{\mathrm{erasure}} = -k_B \ln(2)\; \mathrm{Tr}(\rho\log_2\varphi) \geq S_N(\rho). \tag{4.8}$$

The minimum corresponding to equality in Eq. (4.8) is achieved by choosing the temperature of the bath and $\{p_i, |\phi_i\rangle\}$ such that $\rho = \sum_i p_i |\phi_i\rangle\langle\phi_i|$ is the same as the thermal equilibrium state.

Consequently, the minimum entropy increase required for erasure of the classical information which is encoded in quantum states is given by the von Neumann entropy $S_N(\rho)$, where ρ is the average state of the system, in place of the Shannon entropy, $H(\rho)$, in the case of information erasure of classical states.

5. Quantum entanglement and Maxwell's demon

Formally, entanglement is defined as a form of quantum correlation that is not present in any separable states and represents a true type of quantum behavior [12, 13]. Entanglement can be approached from the point of view of Maxwell's demon. Let $\mathcal{H}^A$ and $\mathcal{H}^P$ be the Hilbert spaces for two spatially separated or noninteracting subsystems, called A and P, and define the entire

Hilbert space $\mathcal{H}^{AP} = \mathcal{H}^{A} \otimes \mathcal{H}^{P}$. Let $S(\mathcal{H})$ denote the state space that consists of a set of density operators acting in $\mathcal{H}$. The case studied here is that of bipartite entanglement.

A state of a bipartite system is said to be separable or classically correlated if its density operator can be expressed as a convex sum of products of density operators

$$\rho = \sum_{i=1}^{n} p_i \rho_i^A \otimes \rho_i^P, \tag{5.1}$$

where all the p_i are nonnegative and $\sum_i p_i = 1$. Any state that cannot be written in the form of Eq. (5.1) is called entangled. Let S_{sep} denote the subspace that contains all separable states.

It is natural to ask whether or not a given state $\rho \in S(\mathcal{H}^{AP})$ is separable considering how important the subject of entanglement is in quantum mechanics. This can usually be expressed in terms of an operator or function. In general, it is very hard to obtain a good separability criterion, that is, something that is efficient and singles out as many entangled states as possible. The separable subspace formed by all separable states is convex and is the main reason for the difficulty of the problem.

Another question pertains to the amount of entanglement a pair or set of quantum objects contains. The amount plays a major role when it comes to characterization or manipulation of entanglement. This topic might be approached by trying to quantify entanglement by means of a thermodynamic quantity.

In a chamber such as the one encountered in Szilard's engine, it can be thought of as a general information-storage apparatus, that is, physical states are distinguishable by measurement, and therefore, stored information will be extracted. Information could be transferred from a different system to the memory of a Szilard engine, if the initial state of the engine is in a standard state.

Now that the memory has been identified with the one molecule gas of Szilard's engine, consider the following picture. From an ensemble of memories, each of which stores the value of an n-bit random variable η, mechanical work can be extracted whose average amount per single memory register is

$$W_c = n - H(\eta), \tag{5.2}$$

where $H(\eta)$ is the Shannon entropy of η, in units such that $k_B \ln 2 = 1$. The extractable work is the work done by the gas, so it is $W = N k_B T ln(2)[1 - H(\eta)]$ when $n = 1$. To understand Eq. (5.2) more completely, suppose there are N memory registers. If all N registers are measured, the remaining uncertainty in the memory is zero, hence Nn bits of work can be obtained. However, the information due to the measurement on memory is kept, and this must be erased in order to consider the amount of extractable work. The least energy consumption needed to erase information is $NH(\eta)$ bits, according to the erasure principle. The maximum total extractable work is $N[\eta - H(\eta)]$. Also, from the thermodynamic point of view Eq. (5.2) follows, as the work done by the gas in an isothermal process is equal to the entropy change multiplied by the temperature.

This argument can be applied to the case of work extraction from quantum bits. Let ρ be the density operator for the state in a given ensemble. The qubits are in a known pure state after measurements, which is essentially classical in terms of information. Information stored in this set of pure states can be copied to the Szilard memory, and each register gives one bit of work. After erasing the information acquired by measurement, the net maximum amount of work we obtain becomes $1 - S_N(\rho)$ bits of work.

The work deficit is a difference between the globally and locally extractable work within the context of local operations, or at least when ρ is a system with spatially separated subsystems. Suppose there is an n-qubit state ρ_{AP} shared by A and P, then the optimal extractable work is given by

$$W_{global} = n - S_N(\rho^{AP}), \tag{5.3}$$

If the entire system can be accessed globally. On the other hand, letting W_{local} be the largest amount of work that A and P can locally extract from the same system under local operations and classical communication. Define the deficit to be $\Delta = W_{global} - W_{local}$. In order to understand this, the deficits for a classically correlated state

$$\rho_{cl}^{AP} = \frac{1}{2}(|00\rangle\langle 00| + |11\rangle\langle 11|), \tag{5.4}$$

and a maximally entangled state

$$|\Phi^{AP}\rangle = \frac{1}{\sqrt{2}}(|00\rangle + |11\rangle) \tag{5.5}$$

will be calculated.

The globally extractable work W_{global}^{cl} from ρ_{cl}^{AP} is simply one bit. The locally extractable work W_{local}^{cl} is also one bit. The strategy is as follows: observer A will measure its bit in the basis $\{|0\rangle, |1\rangle\}$ and will send the result to observer A, who obtains one bit of work from it. Although A can extract one bit of work from A's own bit, using A's measurement result, A needs to consume all this energy to erase the information stored in the memory used to communicate with observer P. Thus, the deficit for state ρ_{cl}^{AP} is $\Delta_{cl} = 1 - 1 = 0$. The locally extractable work is the same, one bit, even if the state is maximally entangled as in Eq. (5.5). However, as this state is globally pure, it must be $W_{global} = 2 - S_N(|\Phi^{AP}\rangle\langle\Phi^{AP}|) = 2$, and therefore, $\Delta_{entangled} = 2 - 1 = 1$.

It can be shown that the deficit is bounded from below as follows: $\Delta \geq max\{S_N(\rho^A), S_N(\rho^B)\} - S_N(\rho)$, where ρ^A and ρ^B are given by $\rho^A = Tr_P(\rho)$ and $\rho^P = Tr_A\rho$. The bound can be achieved when the state is pure and turns out to be equal to the entanglement measure for pure states. This is simply due to the fact that a pure state can be expressed as $|\psi\rangle = \sum_i \alpha_i |e_i\rangle |f_i\rangle$ in the Schmidt decomposition, then $\Delta = S_N(\rho^A) = E(\psi)$, where $\rho^A = Tr_P |\psi\rangle\langle\psi|$ and E. is the entanglement measure for pure states.

6. A nuclear magnetic resonance demon

A very sophisticated model of a demon which is based on a type of nuclear magnetic resonance experiment to introduced. The model studied here is close to the one originally proposed by Lloyd [14] in 1997. It is worth discussing the model though as it may be possible to realize it experimentally. It permits a detailed discussion on several levels and unites several subjects. This gives a nontrivial, fully quantum mechanical model that unlike semiclassical models allows the thermodynamics of the demon's entire cycle of operation to be treated within a unified quantum picture. In this section, the abbreviation $\beta_i = 1/k_B T_i$ will be used.

A spin is immersed in a magnetic field B. If the spin has the same direction as B, it has energy $-\mu B$, where μ is the spin's magnetic dipole moment, B is the magnitude of the field. In the opposite direction, the spin has energy $+\mu B$. The spin can be flipped from one energy to another by applying a π pulse at the spin's precession frequency, $\omega = 2\mu B/\hbar$. When the spin flips, it exchanges energy with the oscillatory field. When it absorbs one photon of energy $\hbar\omega$ from the field, it goes from lower to higher energy, and proceeding the other way, it coherently emits a photon of energy $\hbar\omega$ to the field. When the field is in a coherent state, as with fields which are normally produced by lasers or masers, the energy exchange involves no information exchange, entropy increase or loss of quantum coherence and so the oscillating field may be treated as if it were classical.

A device that acquires information about such a spin could use the information to make the spin do work. Suppose a device can measure whether the spin is in the low-energy quantum state $|\downarrow\rangle$ or the high-energy state $|\uparrow\rangle$. If it is in the high-energy state, a π pulse is sent to extract its energy. This device waits for the spin to come to equilibrium at $T_1 >> 2\mu B/k_B$ and repeats the operation. Each time, it converts an average of μB of heat into work. It is Landauer's principle which prevents such a device from violating the second law of thermodynamics. To proceed in a cyclic fashion, the device must erase the information that it has gained about the state of the spin. At erasure, entropy $S_{out} \geq k_B ln(2)$ is forced into the environment, which compensates for the entropy $S_{spin} = k_B ln(2)$ in the spin originally. If the environment is a heat bath at T_2 different from the temperature of the spin heat bath T_1, heat $k_B T_2 ln(2)$ flows to the heat bath along with entropy, decreasing the energy available to convert into work. To account for energy and entropy in the cycle, the heat in is $Q_{in} = T_1 S_{in}$, and the heat out is $Q_{out} = T_2 S_{out}$. Thus, the work out is $W_{out} = Q_{in} - Q_{out}$, and the efficiency ϵ

satisfies $\epsilon = W_{out}/Q_{in} \leq 1 - T_2/T_1 = \epsilon_C$, where ϵ_C is the Carnot efficiency. Landauer's principle implies that instead of violating the second law, the device operates as a heat engine, pumping heat from a high-temperature reservoir to one at low-temperature and doing work in the process.

It can be shown that such a device operates as a heat engine that undergoes a cycle analogous to a Carnot cycle. Thus, a quantum device that interacts with a thermal environment can get information and use it to do useful work, but not by violating the second law of thermodynamics. A detailed picture of the erasure model agrees with Landauer's principle. Two sets of modes of the electromagnetic field constitute the environment for the spins. The first will be a set of modes at T_1 with average frequency ω_1 and frequency spread greater than the coupling constant $|\kappa|$ between the spins but less than $\omega_1-\omega_2$. The second will be a set of modes at temperature T_2 with average frequency ω_1 and the same frequency spread. This can be achieved by immersing the spins in incoherent radiation with the given frequencies and temperatures. This should provide separate heat reservoirs for spin 1 and 2. This means spin 1 interacts strongly with the on-resonance radiation at frequency ω_1, and weakly with the off-resonance radiation at ω_2, with the reverse for spin 2. Spin 1 can be regarded as interacting only with mode 1, and spin 2 as interacting only with mode 2.

With respect to this approximation, the initial probabilities for the state of the j-th spin are

$$p_{\uparrow,j} = \frac{1}{Z_j} e^{-\mu_j B\beta}, \qquad p_{\downarrow,j} = \frac{1}{Z_j} e^{\mu_j B\beta}, \qquad \beta = \frac{1}{k_B T}. \tag{6.1}$$

Using Eq. (6.1), the energy can be calculated to be the following:

$$E_j = -\mu_j B p_{\uparrow,j} + \mu_j B p_{\downarrow,j} = -\mu_j B \tanh(\mu_j \beta B). \tag{6.2}$$

The entropy is given by

$$S_j = -k_B \sum_{i=\uparrow,\downarrow} p_{(i),j} \ln(p_{(i),j}) = \frac{1}{T} E_j + k_B \ln(Z_j), \tag{6.3}$$

where $Z_j = e^{-\mu_j \beta B} + e^{\mu_j \beta B} = 2cosh(\mu_j \beta B)$. Spin 2 can acquire information about spin 1, and this information can be exploited to perform work. Therefore, the spins can function as a heat engine by passing through the following cycle:

First, using spin coherence double resonance, flip spin 2 if spin 1 is in state $|\uparrow\rangle_1$. This causes spin 2 to acquire information $(\tilde{S}_2 - S_2)/k_B ln(2)$ about spin 1 at the cost of $W_1 = p_{\uparrow,1} 2\mu_2 Btanh(\mu_2 B\beta)$ of work supplied by the oscillating field, where $\tilde{S}_2 = -k_B \sum_{i=\uparrow,\downarrow} \tilde{p}_{(i),2} ln(p_{(i),2})$, such that

$\tilde{p}_{\uparrow,2} = p_{\uparrow,1}p_{\downarrow,2} + p_{\downarrow,1}p_{\uparrow,2}$ and $\tilde{p}_{\downarrow,2} = p_{\downarrow,1}p_{\downarrow,2} + p_{\uparrow,1}p_{\uparrow,2}$ are the two probabilities for the states of spin 2 after the conditional spin flip.

Second, flip spin 1 if spin 2 is in the state $|\uparrow\rangle_1$. This step permits spin 2 to take the amount $(S_2 - S_1)/k_B ln(2)$ of the information it has acquired, and in the process carry out work $-\mu_1 B[tanh(\mu_1 B\beta_1) - tanh(\mu_2 B\beta_2)]$ back on the field.

Third, spin 2 still has information in the amount $(\tilde{S}_2 - S_1)/k_B ln(2)$ about spin 1, which can be converted into work by flipping spin 2 if spin 1 is in state $|\uparrow\rangle_1$, thereby carrying out work in the amount $p_{\uparrow,2} 2\mu_2 B tanh(\mu_1 B\beta_2)$ on the field.

The set of pulses has finally exchanged the information associated in 1 with the information in 2. Next, after the three conditional spin flips, spin 1 has probabilities $p'_{(i),1} = p_{(i),2}$, while 2 has probabilities $p'_{(i),2} = p_{(i),1}$. Consequently, $S'_1 = S_2$ and $S'_2 = S_1$ and the new energies of the spins are given by the following equation:

$$E_1' = -\mu_1 B \tanh(\mu_2 B\beta_2), \qquad E_2' = -\mu_2 B \tanh(\mu_1 B\beta_1). \tag{6.4}$$

Thus, the total amount of work done by the spins on the field is given as follows:

$$W = -(E_1' + E_2' - E_1 - E_2) = -(\mu_1 - \mu_2)B[\tanh(\mu_1 B\beta_1) - \tanh(\mu_2 B\beta_2)] \tag{6.5}$$

When the temperatures satisfy the inequalities $T_i >> \mu_i B / k_B$, then work W simplifies to the form

$$W = -(\mu_1 - \mu_2)\left(\frac{\mu_1}{T_1} - \frac{\mu_2}{T_2}\right)\frac{B^2}{k_B}. \tag{6.6}$$

These results for work done are a function of only conservation of energy and not the pulse. If $T_1 = T_2$, Eq. (6.6) implies W is zero or negative, no work can be extracted from the spins at equilibrium. The cycle can be completed by allowing the spins to re-equilibrate with their reservoirs. The following two steps then can be included to allow the spins to re-equilibrate isentropically.

Return spin 1 to its original state: take the spin out of contact with its reservoir by varying the frequency of the reservoir modes; next vary the field according to $B \to \hat{B} = BT_1/T_2$ adiabatically, with no heat flowing between spin and reservoir; thirdly, slowly change $\hat{B} \to B$, keeping the spin in contact with the reservoir at temperature T_1, so that heat flows isentropically between the spin and reservoir. Entropy $S_1 - S_2$ moves from the spin to the reservoir while the spin does work of $E_1 - E_{1'} - T_1(S_2 - S_1)$ on the field.

Spin 2 returns to its original state by the same steps. The total work done by the spins on the electromagnetic field throughout the cycle is $W_C = (T_1 - T_2)(S_1 - S_2)$.

Consider a simple model in which spin 1 is initially in the state

$$|\rightarrow\rangle_1 = \frac{1}{\sqrt{2}}(|\uparrow\rangle_1 + |\downarrow\rangle_1). \tag{6.7}$$

This state has nonminimum free energy available for conversion into work. Apply a $\pi/2$ pulse to rotate spin 1 into state $|\downarrow\rangle_1$, thereby adding energy $\mu_1 B$ to the field. Suppose the demon operates in a mode whereby, instead of taking the energy directly, it uses magnetic resonance to correlate the state of 2 with the state of 1. Suppose 2 is in state $|\downarrow\rangle_2$ initially; coherently flipping 2 if 1 is in state $|\uparrow\rangle_1$ results in the state

$$\frac{1}{\sqrt{2}}[|\uparrow\rangle_1|\uparrow\rangle_2 + |\downarrow\rangle_1|\downarrow\rangle_2]. \tag{6.8}$$

This represents an entangled state in which the state 2 is perfectly correlated with the state of 1. Energy extraction can be continued by flipping spin 1 if spin 2 is in state $|\uparrow\rangle_2$ which allows energy of $(\mu_1 - \mu_2)B$ to be taken from the spin. The resulting state of the spins is $\frac{1}{\sqrt{2}}|\downarrow\rangle_1|\rightarrow\rangle_2$. No extra thermodynamic cost has been incurred up till now. Since the conditional spin flipping occurs coherently, the process may be reversed by repeating the steps in reverse order to return to the original state $|\rightarrow\rangle_1$, with a total energy and entropy change of zero.

In the original cycle, decoherence occurs when 2 is placed in contact with the reservoir to erase it. The energy exchange between spin and reservoir is an incoherent process such that the pure state $|\rightarrow\rangle_2 = 1/\sqrt{2}(|\uparrow\rangle_2 + |\downarrow\rangle_2)$ transforms into the mixed state which has density matrix $\tilde{\rho} = 1/\sqrt{2}(|\uparrow\rangle_2\langle\uparrow| + |\downarrow\rangle_2\langle\downarrow|)$. This is significant since interaction of the spin with the reservoir turns the process by which 2 coherently picks up quantum information about 1 into a decoherent measurement process and creates one bit of information. The bit corresponds to an entropy increase in $k_B\ln(2)$. In agreement with Landauer's principle, erasure results in the transfer of entropy from spin 2 to the low-temperature reservoir.

The amount of inefficiency generated by decohering 2 to measure spin 1 increasing the entropy may be measured by means of a Carnot cycle model. A general state for the 1 spin would have density matrix

$$\rho = \rho_1(\uparrow_\vartheta)|\uparrow_\vartheta\rangle_1\langle\uparrow_\vartheta| + p_1(\downarrow_\vartheta)|\downarrow_\vartheta\rangle_1\langle\downarrow_\vartheta| \tag{6.9}$$

where the basis $\{|\uparrow_\vartheta\rangle, |\downarrow_\vartheta\rangle\}$ is made up of spin states along an axis making an angle ϑ with the z-axis. Let temperature T_1 and field B be chosen so that

$$p_1(\uparrow_\vartheta)=\frac{1}{Z_1}e^{-\mu_1 B\beta_1}, \qquad p_1(\downarrow_\vartheta)=\frac{1}{Z_1}e^{\mu_1 B\beta_1}. \tag{6.10}$$

Eq. (6.9) is not an equilibrium state and has free energy that may be extracted by applying a tipping pulse that rotates the spin by ϑ and carries out the mapping $|\uparrow_\vartheta\rangle\rightarrow|\uparrow\rangle$ and $|\downarrow_\vartheta\rangle\rightarrow|\downarrow\rangle$. The amount of extracted work is

$$W^* = E_1^* - E_1 = \mu_1 B[p_1^*(\uparrow)-p_1^*(\downarrow)]-\mu_1 B[p_1(\uparrow)-p_1(\downarrow)], \tag{6.11}$$

with $p_1^*(\uparrow)=p_1(\uparrow)cos^2\vartheta + p_1(\downarrow)sin^2\vartheta$ and $p_1^*(\downarrow)=p_1(\downarrow)cos^2\vartheta + p_1(\uparrow)sin^2\vartheta$. Undergoing a Carnot cycle through the five steps just presented above extracts work of $(T_1 - T_2)(S_1 - S_2)$ and energy is extracted isentropically with no entropy increase. By performing measurements with respect to which the density matrix is diagonal, the upper limit $\epsilon_Q = \epsilon_C$ can be attained. If the tipping pulse is used at first to remove the free energy from 1, the demon is carrying out measurements such that the density matrix possesses off-diagonal elements. As a result, the measurement introduces information and the efficiency satisfies the inequality $\epsilon_Q < \epsilon_C$.

The steps can be summarized in this way: three conditional spin flips swap the states of 1 and 2 so that spin 1 is in state ρ_2 and 2 is in state $\rho'_2=\rho'_1$. Interaction with the heat reservoir decohers 2 and destroys the off-diagonal elements in the density matrix so that

$$\rho_2' \rightarrow p_1^*(\uparrow)|\uparrow\rangle\langle\uparrow| + p_1^*(\downarrow)|\downarrow\rangle\langle\downarrow|, \tag{6.12}$$

which has entropy

$$S_1^* = -k_B \sum_{i=\uparrow,\downarrow} p_1^*(i)\ln(p_1^*(i)), \tag{6.13}$$

with $\Delta S_0=S_1^*-S_1$ as the extra entropy introduced by quantum decoherence. The entropy $S_1^*-S_2$ that flows out to reservoir 2 is then greater than the entropy $S_{in} = S_1 - S_2$ that flowed in from 1. The total amount of work done is then $T_1(S_1-S_2)-T_2(S_1^*-S_2)+W^*$ and less than $(T_1 - T_2)(S_1 - S_2) + W^*$, which is done by simply undoing the tipping pulse and running the engine as before.

7. Conceptual systems for future speculation and conclusions: quantum Szilard engines

The preceding model showed how a quantum system that obtains information about another quantum system can function in the capacity of a Maxwell's demon using information to perform work. It would be interesting if progress could be made on the experimental front in this area. At some point, nucleon-nucleon double resonance methods could be used to construct a demon in the form of a maser that functions as a demon and carries out a net amplification of pulses that causes spins to flip. Two different species of nucleons must start the cycle at different temperatures, which could be achieved by preparing one of them in a low temperature state using electron-nucleon double resonance as in the Pound-Overhauser effect.

Three final systems are presented to provide a summary and conclusion, both illustrate what has been achieved and provide work for the future. These are more conceptual in nature and are given in summary form. The processes in the first two examples are likely to be difficult to realize in practice. The first shows that there is a close relationship between dynamical evolutions, which violate some fundamental principle of quantum theory and those forbidden by the second law of thermodynamics. Thermodynamics does impose severe constraints on the choice of the fundamental axioms of quantum theory. This relies on the equivalence of the von Neumann entropy to ordinary entropy appearing in thermodynamics. Perhaps some of these physical situations would end up providing a test of this. The von Neumann entropy is essential in the second model, which represents a true quantum version of a Szilard engine, and it appears in the analysis of two cycles. The last system that involves a molecule in a variable double well may be closer to realization experimentally. A solvable model is proposed and used to describe it here. The conclusion to be drawn is that if the integrity of the axiomatic structure of quantum theory is not strictly respected, then every aspect of the theory must be examined.

(1) The first model to be studied is due to Peres and is a conceptual experiment based on the distinguishability of quantum states [12, 13]. Peres showed that if it were possible to distinguish nonorthogonal quantum states perfectly, then the second law of thermodynamics would necessarily be violated. Consider an elementary work extraction process that uses a collection of pure orthogonal states. A chamber is partitioned into two sections with volumes $p_1{}^V$ and $p_2{}^V$ such that $p_1 + p_2 = 1$. The chamber contains a gas of molecules whose quantum internal degree of freedom is a spin; as a first example before the more novel case of Peres, consider a gas with spin up $|\uparrow\rangle$ on the left and spin down $|\downarrow\rangle$ on the right. Similar experiments could also be imagined using polarized photons as well. The existence of semipermeable membranes is essential to all of this formalism. In this event, introduce two membranes $M_\uparrow$ and $M_\downarrow$ which distinguish the orthogonal states. The convention is that the membrane $M_\uparrow$ is completely transparent to the $|\downarrow\rangle$ spin gas and opaque to the $|\uparrow\rangle$ spin gas. The membrane $M_\uparrow$ has exactly the opposite properties. If the membranes replace the partition so that $M_\uparrow$ and $M_\downarrow$ face the $|\downarrow\rangle$ and $|\uparrow\rangle$ gases, respectively, the gases give work expanding isothermally by contact with a heat bath at temperature T. The total extractable work is then $W = -p_1 \log_2 p_1 - p_2 \log_2 p_2$.

The cycle imagined by Peres is related to this example; however, the key to its impact is the use of nonorthogonal states. The volume of the chamber is $2V$ and in the initial state, the gas of volume V is divided into two equal volumes $V/2$ and separated by an impenetrable wall. On the left side, the gas molecules are in the state $|\uparrow\rangle$ but on the right side, they are in the linear combination state

$$|\rightarrow\rangle = \frac{1}{\sqrt{2}}(|\uparrow\rangle + |\downarrow\rangle). \tag{7.1}$$

Both sections contain the same number of gas molecules $N/2$ and hence the same pressure.

The first step of the cycle is to let the gas expand isothermally at temperature T, so the entire chamber is finally occupied. During the expansion, the gases exert work equal to $Nk_BT\ln(2)$ toward the outside and absorbing the same amount of heat from the bath.

In the second step, conceptual membranes are introduced, which have the ability to distinguish nonorthogonal states. The partition at the center has to be replaced by these membranes. Next, insert an impenetrable piston on the right side of the vessel. The membrane $M_\uparrow$ transparent to $|\rightarrow$, but opaque to $|\uparrow>$, is fixed at the center, while the other membrane $M_\rightarrow$ of opposite transparency to $M_\uparrow$ can move in the left-hand region. As the piston is inserted on the right, $M_\rightarrow$ is forced to the left at the same speed so that the volume and pressure of the spin $|\rightarrow$ gas in between the piston and membrane $M_\rightarrow$ will remain constant. Due to the nature of the membranes, this process can be done without friction or resistance, and consequently, there is no work consumption or heat transfer required.

At this point, the vessel is a mixture of two spin states. In terms of the basis $\{|\uparrow>, |\downarrow>\}$, the density matrix for the mixture using Eq. (7.1) is

$$\rho = \frac{1}{2}|\uparrow\rangle\langle\uparrow| + \frac{1}{4}(|\uparrow\rangle + |\downarrow\rangle)(\langle\uparrow| + \langle\downarrow|). \tag{7.2}$$

This ρ can be mapped to a matrix representation given as follows:

$$\rho = \begin{pmatrix} \frac{3}{4} & \frac{1}{4} \\ \frac{1}{4} & \frac{1}{4} \end{pmatrix} \tag{7.3}$$

The eigenvalues of matrix Eq. (7.3) are found to be

$$\lambda_1 = \frac{1}{2}(1 + \frac{1}{\sqrt{2}}), \qquad \lambda_2 = \frac{1}{2}(1 - \frac{1}{\sqrt{2}}), \tag{7.4}$$

and the corresponding eigenvectors in terms of the $\{|\uparrow>, |\downarrow>\}$ bases have the form,

$$|\phi_1\rangle = \cos(\frac{\pi}{8})|\downarrow\rangle + \sin(\frac{\pi}{8})|\uparrow\rangle, \qquad |\phi_2\rangle = \cos(\frac{3\pi}{8})|\downarrow\rangle - \sin(\frac{3\pi}{8})|\uparrow\rangle. \tag{7.5}$$

If the membranes $M_\uparrow$, $M_\rightarrow$ are replaced by two new membranes that distinguish between the two orthogonal states $|\varphi_i\rangle$ in Eq. (7.5), which we call $M_{|\phi_i\rangle}$, the reverse process separates the states $|\varphi_1\rangle$ and $|\varphi_2\rangle$. After replacing the semipermeable membrane by an impenetrable wall, the gases on the left and right segments are compressed isothermally until the total volume and pressure of the gases become equal to the initial ones. This compression requires work of $-\lambda_1 log_2\lambda_1 - \lambda_2 log_2\lambda_2 = 0.6$ upon using Eq. (7.4), which is dissipated into the heat bath.

Finally, an opaque wall is inserted into the vessel which divides the volume V occupied by the gases in two. Next rotate the state $|\phi_1\rangle \rightarrow |\uparrow\rangle$ in the first region on the left $|\phi_1\rangle \rightarrow |\rightarrow\rangle$ in the center and $|\phi_2\rangle \rightarrow |\rightarrow\rangle$ on the right, along with a trivial spatial shift restores the initial state. Rotations are unitary transformations and an isentropic process, so any energy that has to be supplied can be reversibly recaptured. Hence work expenditure need not be considered in principle when the process is isentropic.

Throughout this cycle, the network gained is 1.0–0.6 = 0.4 bits. Therefore, it is concluded that Peres's system can complete a cycle that can withdraw heat from a heat bath and converts it into mechanical work without leaving any other effect on the environment. This model actually implies that the second law itself sets a strict barrier to quantum state discrimination.

(2) Quantum variants of the basic Szilard engine have seen renewed interest recently [15–17], partly because of their overlap with statistical mechanics and also due to links with quantum information theory and computation. Zurek [18, 19] examined the Szilard one-particle gas obeying Boltzmann statistics quantum mechanically. He begins by noting that a one-molecule gas is a microscopic system and it may be wondered whether conclusions of Szilard's classical analysis remain valid in the quantum domain. Following Jauch and Baron [20], it may be argued that Szilard's analysis is inconsistent because it employs two different, incompatible classical idealizations of the one-molecule gas, dynamical and thermodynamical. Zurek shows that the apparent inconsistency pointed out by Jauch and Baron is removed by a quantum treatment. Thermodynamic entropy is incompatible with classical mechanics, as it becomes infinite in the limit $\hbar \rightarrow 0$. Thus, he views partitioning as a slow, reversible process that creates a potential barrier of some height which is large relative to $k_B T$. The system's energy levels are then modified.

In the quantum Szilard engine, the appearance of a wall is signaled by the increase in height of a potential barrier, which becomes infinite when it is impermeable. This is crucial since the energy levels in the box vary with potential height and boundary conditions. The energy levels contribute to the quantum thermodynamic work and internal energy of the device. The position of the barrier and the rate of its appearance influence the level shifts. The faster the height of the barrier increases, the greater the change of internal energy of the system. The energy becomes infinite when the height of the barrier approaches infinity instantaneously. If the system is initially in the ground state and the barrier appears in an adiabatic fashion with the barrier off center, the particle will end up in the larger region of the box, and different from the classical situation. For isothermal insertion, the effect of energy level shifts is concealed by heat exchange. In order to show quantum effects of the quantum Szilard engine completely, it is necessary to consider adiabatic insertion. In this instance, a cycle of the quantum Szilard engine can be treated with fully quantum considerations.

Recently, such a model has been worked out in detail under the hypothesis that the von Neumann entropy determines the entropy of the quantum state. The model is based on a one-dimensional infinite square well, and the device is allowed to pass through two different cyclic processes. Here, the model will simply be introduced and then two distinct cycles will be outlined. It is too long to look at entirely. Consider a single particle of mass m which is confined to a one-dimensional infinite well of width a. The eigenvalues E_n and eigenstates $| E_n$ depend on the dimension of the box and are given as follows:

$$E_n(a) = \frac{\hbar^2 \pi^2}{2ma^2} n^2, \quad |E_n(a)\rangle = \begin{cases} \sqrt{\frac{2}{a}} \sin\left(\frac{n\pi(x - \frac{a}{2})}{a}\right), & n{=}2k, \\ \sqrt{\frac{2}{a}} \cos\left(\frac{n\pi(x - \frac{a}{2})}{a}\right), & n{=}2k-1, \end{cases} \tag{7.6}$$

where k is a positive integer and $0 \le x \le a$.

Since the particle's state is determined by a wave function, crucial properties such as the number of nodes in the wave function are determined by the dimension of the box. These may vary during an expansion or contraction phase.

Assume that the system is initially in thermal equilibrium with a bath at temperature T. The density matrix $\rho_0(a)$ takes the form

$$\rho_0(a) = \sum_n p_n(a) | E_n(a)\rangle\langle E_n(a)|, \qquad p_n(a) = \frac{e^{-\beta E_n}}{Z(a)}, \tag{7.7}$$

and $p_n(a)$ is the probability of the particle residing in the eigenstate $| E_n >$ and is normalized to unity. Also, $Z(a)$ is the associated partition function given by the following:

$$Z(a) = \sum_{n=1}^{\infty} e^{-\beta E_n} \tag{7.8}$$

The internal energy of the internal energy $U_0(a)$ and von Neumann entropy S_0 are given in terms of Eqs. (7.6) and (7.7) as

$$U_0(a) = \sum_{n=0}^{\infty} p_n(a) E_n(a), \qquad S_0 = -k_B Tr(\rho_0 \ln \rho_0) = -k_B \sum_{n=1}^{\infty} p_n(a) \ln(p_n(a)). \tag{7.9}$$

The model is capable of working over at least two different cycles. There is a cycle where the expansion is isothermal and a different cycle in which the expansion phase is adiabatic. Each of these cycles is composed of four segments. First, there is adiabatic insertion, measurement, expansion, and finally, there is extraction. The first two steps can usually be performed simultaneously and so can be regarded as one. Based on these cycles, the physics can be revealed by calculating the physical quantities, namely the internal energy, work, heat and entropy changes over each of the segments of the cycle being studied using the formulas for the basic physical quantities such as those in Eq. (7.9).

(3) Recently, Landauer's thought experiment has been realized by using a colloidal particle, which is trapped in a double-well potential that has been produced by two strongly focused laser beams. This could be regarded as an extension of the previous model above, but it is closer to realization experimentally. Such a system has two distinct states, that is, the particle may be in the right or left well of the double-well system. The particle is confined with equal probability to one of two optical potential wells and constitutes one bit of information. It may thus be thought to store one bit of information. The bit can be erased by means of the following procedure. First, the potential barrier between the two wells is lowered by varying or modulating the laser intensity. Next, the particle is pushed to the right by, in effect, inclining or tilting the trapping potential. Finally, the potential is restored to its original shape. The barrier places the particle in the right well, regardless of which well it started off in. Moreover, it will end up in the right well with probability close to one irrespective of the particle's initial position. The final configuration corresponds to zero bits of information. For a full erasure cycle, the average heat dissipated into the environment is equal to the average work needed to modulate the form of the double well potential. In the limit of long erasure cycles, the heat dissipated during the erasure process approaches, but does not drop below $k_b T \ln(2)$, in accord with Landauer's principle.

It is worth mentioning that such a system could be modeled by a potential well model. The inclining of the potential could be modeled at a more sophisticated level by raising the level of the potential on the left half. A primitive version could be modeled as follows. The Schrödinger equation can be solved in each of the three potential regions of the well. The corresponding solutions are as follows:

$$\psi_1(x) = A\sin(kx), \quad \psi_2(x) = B_1 e^{\omega x} + B_2 e^{-\omega x}, \quad a < x < a + b,$$

$$\psi_3(x) = C\sin(k(2a + b - x)), \quad a + b < x < 2a + b, \tag{7.10}$$

$$k = \sqrt{\frac{2mE}{\hbar}}, \qquad \omega = \sqrt{\frac{2m(V_0 - E)}{\hbar}}. \tag{7.11}$$

It is required that the wave function and derivative remain continuous at $x = a$ and $x = a + b$. Manipulating this system of equations, it can be realized in the following form:

$$A(\frac{\omega}{k}\tan(ka)+1)e^{\omega b} - C(\frac{\omega}{k}\tan(ka)-1) = 0, \quad A(\frac{\omega}{k}\tan(ka)-1)e^{-\omega b} - C(\frac{\omega}{k}\tan(ka)+1) = 0. \tag{7.12}$$

From the condition that the determinant of the coefficients of system Eq. (7.12) vanishes, it is found that

$$\tan(ka)(1 \mp e^{-\omega b}) = -\frac{k}{\omega} \mp \frac{k}{\omega} e^{-\omega b}. \tag{7.13}$$

By solving Eq. (7.13) in the form

$$\tan(ka) = \frac{k}{\omega}(\frac{-1 \mp e^{-\omega b}}{1 \mp e^{-\omega b}}), \tag{7.14}$$

from which the corresponding energies can be obtained at least numerically as in Eq. (7.6).

It has been observed that much progress has been made in this area. Similar to setup (3), Piechocinska verified Landauer's principle within the domains of both classical and quantum mechanics [21]. It is assumed that the particle is in a bistable potential well. Piechocinska assumes that the bit to be erased is in contact with a constant temperature reservoir. It is also assumed the reservoir begins in energy eigenstate $|n\rangle_{res}$. The external field is turned on and splits the degeneracy until the probability of the higher energy state being occupied is very small, so the lower energy state $|1\rangle$ is occupied with high probability. This accomplishes erasure. The external field is removed, with the final reservoir and bit states being $|m\rangle_{res}$ and $|1\rangle$.

Further work remains to be done, as the story is likely not finished. Armen Allahverdyan and Theo Nieuwenhuizen recently reported [22, 23] violations of Landauer's principle for two model systems. It concerns a Brownian particle in contact with a constant-temperature reservoir. Landauer's principle seems to break down in the extreme quantum domain. This is when the particle and reservoir are in an entangled quantum state. The total entropy cannot be written as a sum of system and reservoir entropies. The Clausius inequality and Landauer's principle both seem to be violated. Consequently, their work suggests that Landauer's principle is not a universal law. But this would be a subject for future work [24–27].

Author details

Paul Bracken

Address all correspondence to: paul.bracken@utrgv.edu

Department of Mathematics, University of Texas, Edinburg, TX, USA

References

[1] Leff H S and Rex A F, editors, Maxwell's Demon 2; Entropy, Classical and Quantum Information, Computing, Bristol: Adam Hilger, 2003.

[2] Wheeler J A and Zurek W H, editors, Quantum Theory and Measurement, Princeton: Princeton University Press, 1983.

[3] Sheehan D P, editor, Quantum Limits to the Second Law, AIP Conference Proceedings, p. 643, New York: AIP Publishers, 2002.

[4] Penrose O, Foundations of Statistical Mechanics, Oxford: Pergamon Press, 1970.

[5] Planck M, Treatise on Thermodynamics, London: Longmanns, Green and Co, 1997.

[6] Marayama K, Nori F and Vedral V, The physics of Maxwell's demon and information, Rev. Mod. Phys., 2009; 81, 1-23

[7] Vedral V, The role of relative entropy in quantum information theory, Rev. Mod. Phys., 2002; 74, 197–234.

[8] Szilard L, Thermodynamic system by the intervention, Z. f. Phys., 1929; 53, 840–856.

[9] Bennett C H, The thermodynamics of computation-a review, Int. J. Theor. Phys., 1982; 21, 905–940.

[10] Landauer, Irreversibility and heat generation in the computing process, IBM J. Res. Dev., 1961; 5, 183–191.

[11] Sherchenko S N, Ashhab S and Nori F, Landau-Zener-Stückelberg interferometry, Phys. Rep., 2010; 492, 1–30.

[12] Raimond J M, Brune M and Haroche S, Manipulating quantum entanglement with atoms and photons in a cavity, Rev. Mod. Phys., 2001; 73, 565–582.

[13] Peres A, Quantum Theory: Concepts and Methods, Dordrecht: Kluwer Academic Publishers, 1995.

[14] Lloyd S, Quantum mechanical Maxwell's demon, Phys. Rev. A, 1997; 56, 3374–3382.

[15] Li H, Zou J, Li J-G, Shao B and Wu L-A, Revisiting the quantum Szilard engine with fully quantum considerations, Ann. Phys. 2012; 327, 2955–2971.

[16] Bracken P, A quantum version of the classical Szilard engine, Central Eur J. Phys., 2014; 12, 1–8.

[17] Bracken P, A quantum Carnot engine in three-dimensions, Adv. Stud. Theor. Phys., 2014; 8, 627–633.

[18] Zurek W H, Algorithmic randomness, physical entropy, measurements, and the demon of choice. In: Hey J, editor. Feynman and Computation: Exploring the Limits of Computers. Reading: Perseus; 1999.

[19] Zurek W H, Algorithmic randomness and physical entropy, Phys. Rev. A, 1989; 40, 4731–4751.

[20] Jauch J M and Baron J G, Entropy, information and Szilard's paradox, Helv. Phys. Acta, 1972; 45, 220–232.

[21] Piechocinska B, Information erasure, Phys. Rev. A, 2000; 61, 062314.

[22] Allahverdyan A E and Nieuwenhuizen T M, Breakdown of the Landauer bound for information erasure in the quantum regime, Phys. Rev. E, 2001; 64, 056117.

[23] Allahverdyan A E, Balian R and Nieuwenhuisen T M, Understanding quantum measurement from the solution of dynamical models, Phys. Rep., 2013; 525, 1–166.

[24] Landauer R, Information is a physical entity, Physica A, 1999; 263, 63.

[25] Park J, Kim K-H, Sagawa T and Kim S W, Heat engine driven by purely quantum information, Phys. Rev. Lett., 2011; 111, 230402.

[26] Plenio M and Vitelli V, The physics of forgetting: Landauer's erasure principle and information theory, Contemp. Phys., 2001; 42, 25–60.

[27] Sagawa T and Ueda M, Minimal energy cost for thermodynamic information processing: measurement and information erasure, Phys. Rev. lett. 2009; 102, 250602.

6

Electronic and Molecular Dynamics by the Quantum Wave Packet Method

Zhigang Sun

Abstract

A solution to the time-dependent Schrödinger equation is required in a variety of problems in physics and chemistry. In this chapter, recent developments of numerical and theoretical techniques for quantum wave packet methods efficiently describe the dynamics of molecular dynamics, and electronic dynamics induced by ultrashort laser pulses in atoms and molecules will be reviewed, particularly on the development of grid methods and time-propagation or pseudo-time evolution methods developed recently. Applications of the quantum wave packet for studying the reactive resonances in $F + H_2/HD$ and $O + O_2$ reaction, dissociative chemisorption of water on transition-metal surfaces, state-to-state reaction dynamics, state-to-state tetra-atomic reaction dynamics using transition wave packet method and reactant coordinate method, and electronic dynamics in H_2^+ and H_2 molecules will be presented.

Keywords: wave packet method, molecular and electronic dynamics, grid methods, time propagators, molecular dynamics on surface

1. Introduction

Accurate and efficient solution to the Schrödinger equation is important in the fields of atomic physics, molecular physics, and chemical dynamics, which include the dynamics of atoms and molecules in time-dependent electromagnetic fields, a variety of atomic and/or molecular collision problems, molecular photodissociation, molecular dynamics on a surface, and in describing the behavior of materials subjected to internal and external forces, and the production of solitons and quantum vortices in Bose-Einstein condensates. In many situations, especially at the nanoscale level, the evolution of the physical system occurs over multiple length and timescales, and a computationally efficient approach is a necessity to make progress. Theoretical modeling of all of these phenomena is important to obtain a fundamental understanding of atomic processes at the nanoscale level and to provide a scientific base for the design and development of nanostructured materials and for the optimal control over a chemical reaction.

The numerical solution of the time-dependent (TD) Schrödinger equation relies heavily on the discretization of the variables (r, t) in coordinate space, or named as grid method [1–4]. The usage of numerical grid methods, which have numerically spectral convergence, has become the most important tools to solve the Schrödinger equation and has been well developed in the past decades because the grid methods are convenient for evolving the TD Schrödinger equation with an initial value with iterative approaches. The early attempts to demonstrate the computational viability of various grid methods for molecular dynamics have been largely superseded by applications to specific problems and deeper research into more sophisticated molecular systems. The grid methods, however, for efficiently solving the electronic Schrödinger equation are under active development currently, mainly due to the Coulomb-attractive and repulsive singularities, which are quite different from the potential energy surface (PES) experienced by the nuclei in molecular dynamics calculations.

Once a grid method has been chosen, a suitable iterative method, or an evolving method for the initial wave function, must be determined accordingly. The efficiency of the evolving methods of the Schrödinger equation with time-independent (TID) or TD Hamiltonian is crucial for computer modeling of the quantum dynamics. They have been well developed before 1990; among them, the most famous two are the Chebyshev polynomial expansion with a TID Hamiltonian and split operator with a TD or TID Hamiltonian [5,6]. Recently, there is particular interest in developing efficient and accurate propagators with a TD Hamiltonian.

An efficient quantum dynamics code comes from an optimal combination of the time propagator and the spatial grid method. Especially, recent rapid developments in hardware capabilities, as exemplified by the Intel's many integrated cores, Nvidia graphics processing unit cards, have added another dimension to increasing the number of floating-point operations per second that can be executed.

In the final of this part, we would like to note that, in the past, usually we would like to state that the solutions of the Schrödinger equation with the TID method, that is, matrix diagonalization methods, are much more computer consuming than the solutions with the TD method, that is, iterative methods for an initial value problem. Along with the development of the numerical method, such as the filter method or spectral transform method, this is true only conditionally [4,7,8]. Especially, with the intention of the real Chebyshev wave packet method, there is no distinction between the TID and TD method anymore [4].

In this work, the spatial grid methods and time propagators to solve the Schrödinger equation with TD and TID Hamiltonian will be reviewed, and some of their applications, that is, describing the resonances and isotope effects in molecular reaction dynamics, molecular dynamics on surface, and electronic dynamics in atoms and molecules induced by ultrashort laser pulses, will be presented.

2. The Schrödinger equation

The time-dependent Schrödinger equation usually can be written as

$$i\frac{\partial}{\partial t}\Psi = \hat{H}\Psi = (\hat{T} + \hat{V})\Psi$$

and for the dynamics of a molecule, the Hamiltonian and wave function are functions of **R** and *r*, where ***r*** is a collection of the coordinates of electrons and **R** is a collection of the coordinates of nuclei. In the case where we are interested in the dynamics of nuclei, usually, first the electronic structure theory is applied to construct the potential energy surface (PES) using the Born-Oppenheimer approximation, then the coordinate ***r*** disappears in the equation and we can conveniently carry out molecular dynamics simulations. In the case where we are interested in the electronic dynamics processes, especially which are induced by ultrashort laser pulses, the coordinate ***r*** has to be considered explicitly, and usually the coordinate **R** is fixed since the movement of the nuclei is much more slow.

In any case, in a modern solution to the TD Schrödinger equation, the spatial coordinates of the Hamiltonian have to be discretized to facilitate the wave function propagation, where iterative evaluations of the action of Hamiltonian operator on the wave function are required. Sometimes, the time variable also requires discretization; thus, the wave function propagation is accomplished by a series of advancing the wave function from t to $t + \Delta t$, where Δt is the time interval or time step. In the following, recent developments concerning these two aspects will be briefly reviewed. The author notes that the selected contents purely come from personal interest and expresses his apologies to the authors whose intelligent works are not included here.

3. The spatial discretization methods

Discretization of the spatial coordinates is very important in an efficient and accurate quantum dynamics calculation. For a special problem, a careful designed grid method would lead to substantial decrease of the computational effort. Basically, there are three approaches to the discretization of the spatial coordinates of the Hamiltonian: (1) local methods, such as the finite difference methods; (2) global methods, which includes all of the spectral methods and the corresponding grid methods; and (3) spectral element methods. Each has its advantages and limitations.

The local method usually applies with low-order finite difference method and is simple to use, thus leading to very sparse Hamiltonian matrix. However, they usually have limited accuracy since they converge with both the number of the grid points and the spacing of the grid points. When boundary region is important in the model, we need to impose the boundary conditions carefully, which results in an asymmetric Hamiltonian matrix, and unphysical states may arise. The grid points in finite difference calculations connect with each other only "locally," that is, connect with the nearby grid points; thus, parallel algorithm can be efficiently applied with MPI.

On the contrary, the global methods usually are derived based on the classical orthogonal polynomial. The grid point connects with all of the other points, necessarily or unnecessarily, which leads to the dense Hamiltonian matrix. However, the global method usually has spectral convergence, and results of machinery accuracy can be obtained. Due to limited accuracy of finite element method, spectral element method is much more popular in a quantum molecular and atomic dynamics calculation [9–11]. The grid points in an element of the spectral element are determined by the Lobatto-Chebyshev or Lobatto-Legendre quadrature. To set up the connection between different elements, many proposals have been put forward. The spectral element usually leads to numerical accuracy between the spectral method and low-order finite difference method, and results in Hamiltonian matrix sparser than that using the global method. Currently, the spectral element method has popular applications in solving electronic Schrödinger equation induced by ultrashort laser pulses. However, due to the denser grid points around both ends of each element, the spectral range of the Hamiltonian matrix usually is necessarily large; especially, high-order Lagrange polynomials are applied in each element for obtaining results of high accuracy. This unnecessary huge spectral range results in much difficulty in searching for an efficient matched time propagator.

Besides these three spatial discretization methods, we also have higher order finite difference method, spectral difference method, and the distributed approximation functional (DAF) method [12–16]. They usually exhibit spectral convergence, exactly same as the global spectral methods. However, they connect only with the nearby "necessary" grid points, thus leading to denser Hamiltonian matrix than that with low-order finite difference method, but sparser Hamiltonian matrix than that with the global methods. The grid points distribute evenly in the spectral difference method, but the grid points distribute exactly the same as their precedent Gauss quadrature of the kernel in the DAF method. They avoid the discontinuities between elements in the spectral method, thus an optimal choice in a calculation with smooth interaction potential functions. Since the grid points connect with each other semilocally, methods of this kind are suitable in a parallel calculation also, similar to the spectral element method. The higher order difference or spectral difference method and DAF method have been applied successfully in the molecular dynamics field. Recently, Sun also puts forward DAF method, which is capable of accurately describing the Coulomb singularity.

In the following, brief introductions to several spatial discretization methods, which currently are popular in a quantum dynamics calculation, will be given.

3.1. Discrete variable representation (DVR)

The DVR has been used quite successfully for a variety of problems, including the calculation of the ro-vibrational spectra of complex molecules and the reactive scattering cross sections of atoms and molecules [17,18]. In a usual variational or spectral calculation, one chooses some, in principle, complete, orthonormal basis set, φ, with truncation to a finite number of terms, N, and then evaluates all of the matrix elements of the Hamiltonian analytically, or by a quadrature technique. If the spectral basis is chosen to be one of the classical orthogonal functions, satisfying the three-term recursion relationship,

$$\beta_n \phi_n = (x - \alpha_{n-1})\phi_{n-1} - \beta_{n-1}\phi_{n-2} \tag{1}$$

then there exists an associated Gauss quadrature rule with the same number of points and weights, which integrates exactly any integrand of degree (2N - 1) or less. By diagonalizing the tridiagonal matrix built from the recursion coefficients α and β, the points and weights may be obtained. Under these conditions, it is possible to transform from the original spectral basis of N functions, φ, to a new basis of DVR or coordinate functions, called O, as follows:

$$O_i(x) = \sum_{n=0}^{N-1} c_n \phi_n(x) \tag{2a}$$

$$c_n = \langle \phi_n | O_i \rangle = \int_a^b \omega(x)\phi_n(x) O_i(x) dx = \sqrt{\omega_i}\phi_n(x_i) \tag{2b}$$

$$O_i(x) = \sum_{n=0}^{N-1} \sqrt{\omega_i}\phi_n(x_i)\phi_n(x) \tag{2c}$$

$$\langle O_i | x | O_j \rangle = \delta_{ij} x_i \tag{2d}$$

Here, all of the integrals are performed using the Gauss quadrature rule; there are no approximations as a consequence of the fact that the quadrature rule is exact for all integrands (2N - 1) or less. The Fourier functions or particle in a box, which are not even polynomials, can be verified to exactly satisfy Eq. (2) with a suitably chosen quadrature rule. In all of these cases, there exists a unitary transformation between the original spectral and coordinate basis. In the DVR, the potential $V(x)$ can be written as

$$\langle O_i | V | O_j \rangle = V(x_i)\delta_{ij}$$

since $V(x)$ can be written as

$$V(x) = \sum_{k=0}^{\infty} a_k x^k$$

Thus, the matrix elements of the potential operator are equal to their values at the Gauss quadrature points and diagonal. In a wide class of problems, this approximation has been proved to be accurate. As a result, there are significant advantages in transforming to the DVR basis over the original spectral basis in that the complex matrix elements of the potential are diagonal and are simply equal to their values as the quadrature points. This is very convenient in a numerical calculation since it leads to a sparse form of the Hamiltonian matrix.

As the other part of the Hamiltonian, the matrix elements of the kinetic energy operator, while not diagonal in the DVR basis, usually can be evaluated simply and exactly or analytically. Since the kinetic operator part of the Hamiltonian matrix is a separable sum over the particle and coordinate variables, a direct product DVR basis results in a very sparse matrix representation in n-body problems. This has a number of advantages which can be well exploited in the iterative stage, where the action of the Hamiltonian matrix on the wave function needs to be evaluated many times, such as the time propagation step or the iterative computation in Lanczos method for calculating the eigenstates. In a practical calculation, the basis functions are not necessary to be the classical polynomials or the trigonometric functions. Usually, the eigenstates calculated in a reduced dimensional model are taken as the basis function, which often leads to efficient potential optimized DVR. In the past decades, complicated but efficient sequential truncation techniques and phase optimization methods have been developed for solving ro-vibrational states of polyatomic molecules [8,19]. Calculations of reactive scattering and ro-vibrational state of molecules with dimensionality of nine using the DVR technique in a direct product form have been reported.

Usually, the DVR invented in the field of molecular dynamics does not work well with electronic dynamics; especially, the Coulomb singularity was rigorously included. The exception is the Lobatto-DVR method, which based on the Legendre polynomials but with Gauss-Lobatto quadrature [20]. However, with suitable phase optimization technique, the trigonometric functions can work well with the Coulomb singularities, even for describing the electrons of diatomic molecule in cylindrical coordinates.

The Lagrange meshes (LM) method is popular in the field of atomic physics [21]. Same as the DVR method, the LM method is a global method, using information from the whole domain of definition of the studied problem, an approximate spectral method taking the form of a collocation method. The LM method is a special case of DVR method, and when classical orthogonal polynomials are adopted as the basis, the LM and the DVR methods are identical. However, the LM method is derived from the Lagrange functions, which make it to be convenient for regularization of the singularity, starting from the basis function. However, usually, the quadrature in a DVR calculation is obtained by diagonalizing the coordinate matrix $X_{ij}=\langle\phi_i \mid x \mid \phi_j\rangle$, and the kinetic matrix in DVR is given by

$$T_{ij}^{FBR} = -\frac{\hbar^2}{2m}\int_a^b \phi_i(x)\frac{d^2}{dx^2}\phi_j(x)$$

$$T^{\mathrm{DVR}} = OT^{\mathrm{FBR}}O^{-1}$$

where O is an orthogonal matrix, which diagonalizes the coordinate matrix, X, defined as

$$O_{ij} = \sqrt{\omega_j}\phi_i(x_j)$$

This fact makes DVR be convenient for various generalizations and for the optimization of the DVR. The basic idea is to get rid of the constraints on the choice of the mesh points and associated weights. They are close to the collocation spirit, but they cannot be so easily related to the LM method. It has been proven that these generalizations are capable of giving very accurate results.

The quadrature discretization (QD) method is efficient for calculating low-bound states of molecules [22]. Same as the DVR and the LM methods, the QD method is a global method. The QD method is very similar to DVR method, but using nonclassical polynomials. In the QD method, a differential operator is replaced by a matrix. The mth derivative d^m/dx^m is represented by the mth power of a matrix representing d/dx in a basis of nonclassical orthogonal polynomials. Like in the DVR method, a matrix transformation leads to a basis where the potential is represented by a diagonal matrix. In a basis of N polynomials, $\phi_j(x)$, orthogonal with respect to a weight function, $\omega(x)$, a Gauss quadrature is obtained with the zeros x_i of $\phi_N(x_i)$ and weights ω_i, and the Hamiltonian matrix can be written as

$$H_{ij} = \sum_{k=1}^{N} D_{ki}D_{kj} + [V(x_i) - \tilde{V}(x_i)]\delta_{ij}$$

where matrix D is obtained by an orthogonal transformation based on O from the matrix elements of d/dx in the polynomial basis,

$$D_{ij} = \sum_{l=1}^{N}\sum_{k=1}^{N} O_{ik}O_{jl}\int_a^b \omega(x)\phi_k(x)\phi_l'(x)dx$$

with the auxiliary potential that reads as

$$\tilde{V}(x) = \frac{\omega''(x)}{2\omega(x)} - \frac{1}{4}\left[\frac{\omega'(x)}{\omega(x)}\right]^2$$

When $\omega(x)$ is chosen as $V(x)$, it means

$$\omega(x) = [\psi_0(x)]^2$$

where $\psi_0(x)$ is the ground-state wave function of the potential $V(x)$. Thus, this method is particularly useful for calculating low-lying bound states.

3.2. Multidomain spectral element method

The Coulomb-attractive singularities between nuclei and electrons are very difficult to describe accurately. At the same time, usually the range of the radial coordinate in a calculation has to be large, which leads to large Hamiltonian matrix using a global spectral method and results in heavy computational effort. The Lobatto-DVR, using the Lagrange interpolation polynomials on the Gauss-Legendre-Lobatto quadrature, is particularly suitable for describing the Coulomb-attractive singularity, whose quadrature, however, clusters densely around the two grid ends. Thus, average efficiency of the quadrature is not high. To overcome these two difficulties, the multidomain spectral method was purposed [3,4]. In this method, the whole range of the radial coordinate is divided into M segmentations or finite elements with boundaries,

$$x^0 \le x^1 \le \cdots x^{M-1} \le x^M$$

The wave functions are represented in a basis of functions which are local to each finite element. In this spectral element method (which is also named as finite element [FE]-DVR), a few of the Lobatto-DVR functions,

$$L_i(x) = \prod_{j \ne i} \frac{x - x_j}{x_i - x_j}$$

which fulfill

$$L_i(x_j) = \delta_{ij}$$

are applied for each of the element. The basis function is then defined explicitly with N quadrature as

$$\phi_j^i(x) = \begin{cases} \dfrac{L_N^i(x) + L_1^i(x)}{\sqrt{\omega_N^{i-1} + \omega_1^i}}, m = 1(bridge) \\ \dfrac{L_j^i(x)}{\sqrt{\omega_j^i}}, else(element) \end{cases}$$

where the "bridge" function is used to ensure the communication between two adjacent elements and guarantees the continuity of the wave function. Correspondingly, the kinetic matrix is given by

$$T_{mn}^{ij}=\begin{cases}\delta_{ij}\dfrac{t_{mn}^{i}}{\sqrt{\omega_m^i\omega_n^i}},2\le m\le N-1\\ \delta_{i,j-1}\dfrac{t_{mn}^{i}}{\sqrt{\omega_m^i(\omega_1^j+\omega_N^{j-1})}}+\delta_{i,j}\dfrac{t_{mn}^{i}}{\sqrt{\omega_m^i(\omega_1^j+\omega_N^{j-1})}},2\le m\le N-1,n=1\\ \delta_{i-1,j}\dfrac{t_{mn}^{i}}{\sqrt{\omega_n^j(\omega_1^i+\omega_N^{i-1})}}+\delta_{i,j}\dfrac{t_{mn}^{i}}{\sqrt{\omega_n^j(\omega_1^i+\omega_N^{i-1})}},m=1,2\le n\le N-1\\ \dfrac{\delta_{i,j}(t_{11}^i+t_{NN}^{i-1})+\delta_{i,j-1}t_{11}^i+\delta_{i-1,j}t_{N1}^{i-1}}{\sqrt{(\omega_1^i+\omega_N^{i-1})(\omega_1^j+\omega_N^{j-1})}},m=1,n=1\end{cases}$$

assuming each element has same N Lagrange functions. Further, t_{mn}^{i} is given by

$$t_{mn}^{i}=\sum_{k=1}^{N}\frac{d\phi_m^i(x_k^i)}{dx}\frac{d\phi_n^i(x_k^i)}{dx}\omega_k^i=$$

$$\sum_{k=1}^{N}\frac{L_{N-1}(s_k)}{L_{N-1}(s_m)}\frac{1}{x_m^i-x_k^i}\frac{L_{N-1}(s_k)}{L_{N-1}(s_n)}\frac{1}{x_n^i-x_k^i}$$

where $s_k=[2x_k^i-(x^{i-1}+x^i)/(x^i-x^{i-1})]$ and $L_{N-1}(s)$ are the N-order Legendre polynomials. For $m=k=2, 3, \cdots, N-1$,

$$\frac{d\phi_m^i(x_k^i)}{dx}=0$$

thus, the first and last diagonal values can be found to be

$$\frac{d\phi_1^i(x_1^i)}{dx}=-\frac{d\phi_N^i(x_N^i)}{dx}=-\frac{1}{2}\frac{N(N-1)}{x^i-x^{i-1}}$$

Figure 1 illustrates the structure for the kinetic energy matrix of one dimension. Since the many-particle kinetic energy operators are the sum of one-particle operators, the generalization to tensor product basis sets is straightforward. **Figure 2** shows the gird points of a typical FE-DVR grid with 11 basis functions per element and the resulting grid spacing.

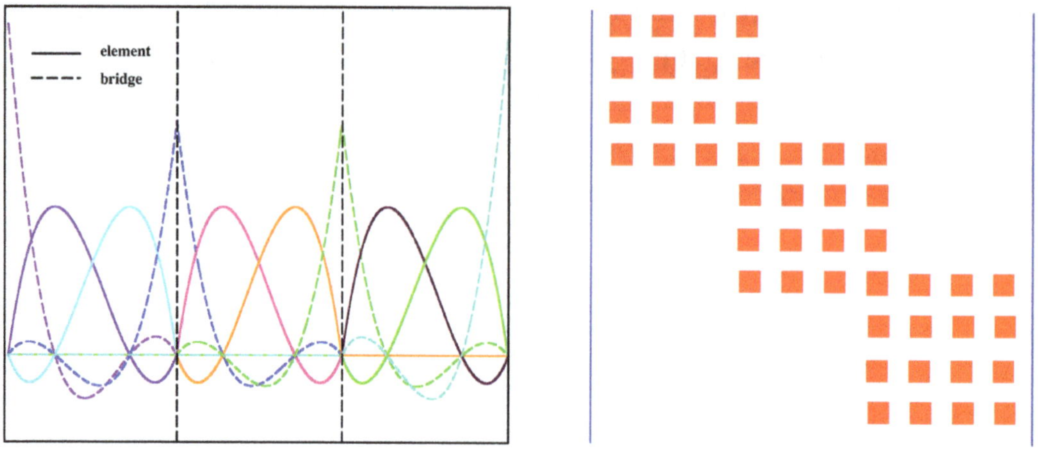

Figure 1. Left: FE-DVR basis functions for three elements, where each element has four basis functions. The *solid* and *dashed lines* correspond to the element and bridge functions, respectively. Right: The structure of the 1D Hamiltonian matrix, corresponding to the basis in the left panel.

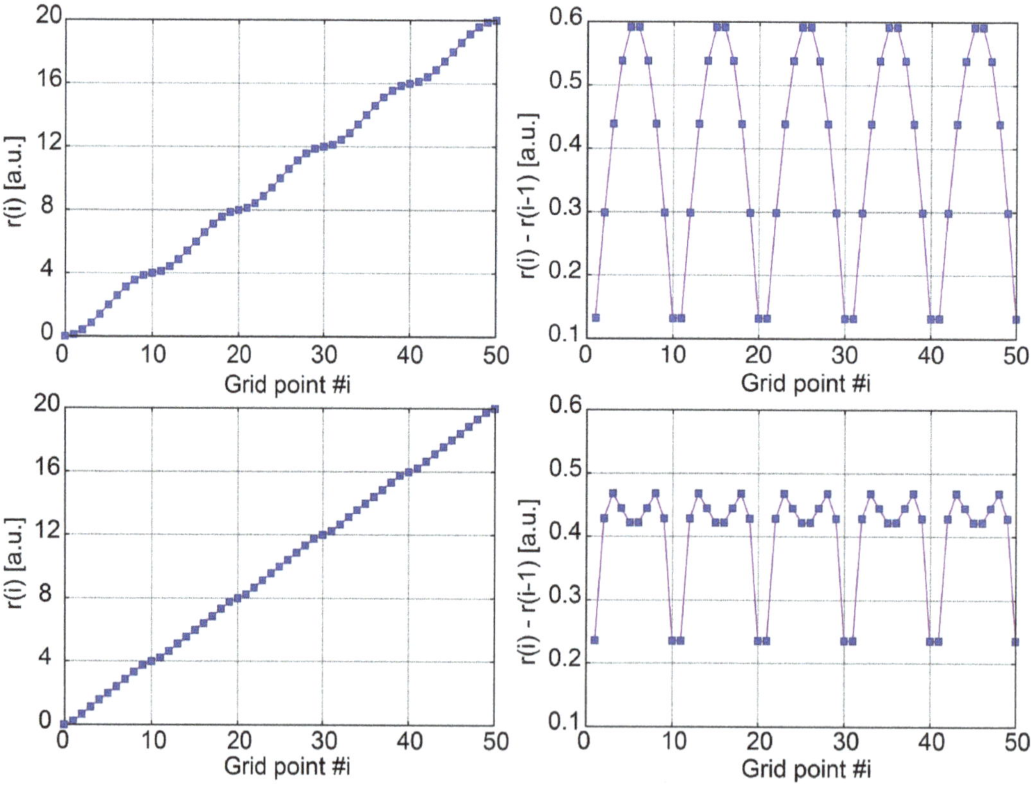

Figure 2. Grid points (left panels) and their spacing (right panels) of a typical FE-DVR grid (upper panels) and optimized FE-DVR grid (lower panels) with 11 basis functions per finite element. The oscillatory structure in the grid spacing of upper panels is a consequence of the Gauss-Lobatto quadrature. However, after optimization, the oscillation is much more gentle.

Even the finite-element DVR has been proven to work well in many cases; the oscillatory distribution of the grid and the bridge function leads to unphysically large spectral range of the Hamiltonian matrix, as shown in the upper panels of **Figure 3**. This fact not only leads to low grid efficiency but also makes the iterative propagator, such as short-time iterative Lanczos (SIL) method, converge slowly [23]. To overcome this difficulty, one can use the variable mapping method or prolate spheroidal wave functions to get a more uniform spatial grid far from the origin and dense spatial grid around the origin, to increase the numerical efficiency [24,25]. In the bottom panels of **Figure 2**, the gird distribution becomes nearly even after optimization, which improves the efficiency of the grid. At the same time, the spectral range of the Hamiltonian matrix reduces several times against that of the original FE-DVR.

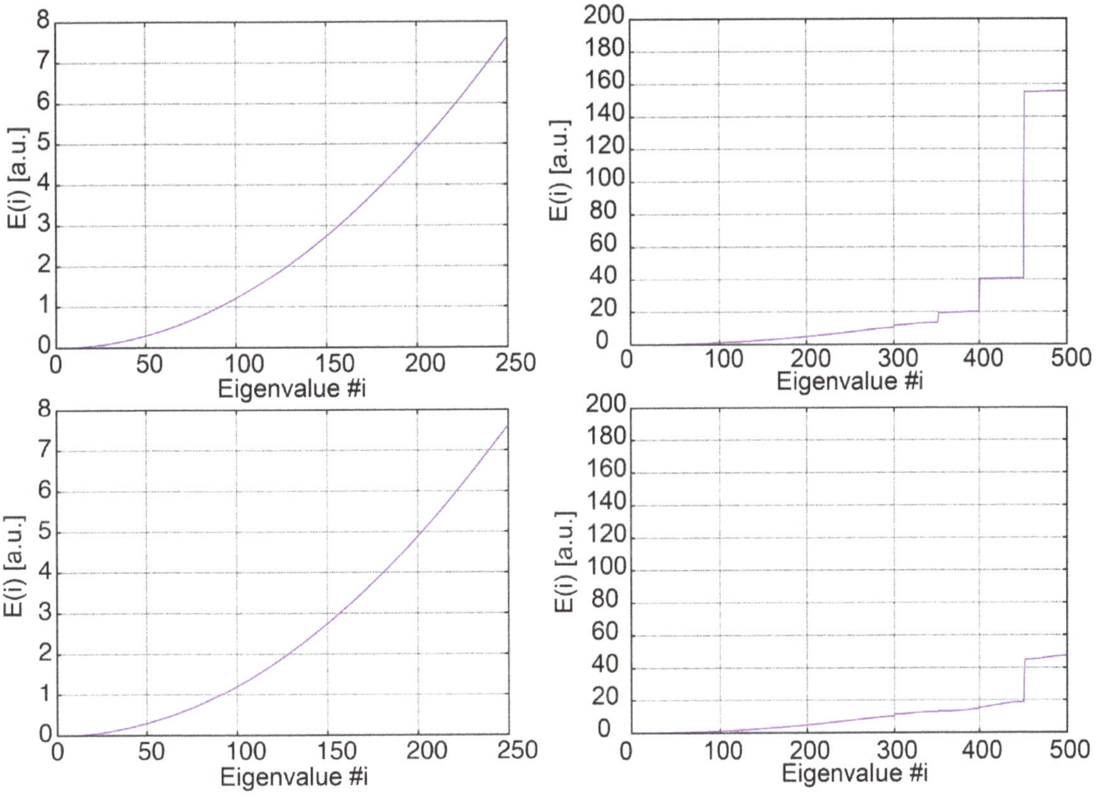

Figure 3. Left panels: The low eigen energies of the kinetic matrix grow quadratically of a usual FE-DVR (upper) and optimized FE-DVR (bottom). Right panels: For higher energies, the spectrum contains "unphysical" steps due to the division into finite elements, which reduces much of optimized FE-DVR (bottom right panel).

On the other hand, in order to exploit the sparseness of the Hamiltonian matrix in finite element DVR, the powerful split operator propagator cannot be applied, which even is insensitive to the spectral range of the Hamiltonian. The common choice is the short-time iterative Lanczos method, which requires a significant amount of computer memory. And, due to the fact that the continuity between elements only is guaranteed with the wave function other than the derivatives, which introduce numerical complications, results with spectral accuracy cannot be obtained.

3.3. Distributed approximating functionals (DAF) method

Even though it is known that for well-conditioned problems, that is, when infinitely differentiable solutions exist almost everywhere, global methods provide a better accuracy than local methods using information from pieces of the domain, such that the Lobatto-DVR method is capable of giving results of higher accuracy than finite element DVR method, the drawback is that the Hamiltonian matrix is dense and sometimes requires more computational effort. The higher-order finite difference (FD) method is a good choice for obtaining results of high accuracy, but possibly with sparsest Hamiltonian matrix, especially with spectral difference method where accelerating weights are applied.

However, in the case of Coulomb potential with singularity, the boundary conditions are very important to the convergence of the FD method. Usually with well-defined boundary conditions, the resulted Hamiltonian matrix is asymmetric, which leads to unphysical states that may be numerically unstable.

The distributed approximating functional (DAF) method presents a good solution to these problems, which reduce the correlated region around the grid point by a controllable Gaussian weight $\varpi(x, \sigma)$, but retaining the numerical accuracy by monitoring the convergence [12,26–30]. This kind of method uses the fact that the grid is unnecessary to correlate with all of the remaining grids, especially in the case where large spatial grid range has to be applied. The basis functions with specified DVR functions $\phi_i(x)$ as the kernel of the DAF may be defined as

$$\Phi_i(x,\sigma) = \phi_i(x)\varpi(x,\sigma)$$

where $\varpi(x, \sigma)$ is defined as

$$\varpi(x,\sigma) = \exp(-\frac{(x - x_i)^2}{2\sigma^2})$$

with a controllable parameter σ. In the DAF, the momentum and kinetic operator can be derived from those of $P_{i,j}$ and $T_{i,j}$ in the corresponding DVR method directly by the following equations:

$$\bar{P}_{i,j} = P_{i,j}\varpi_j(x_i,\sigma) + \delta_{i,j}\varpi'_j(x_i,\sigma) \tag{3a}$$

and

$$\bar{T}_{i,j} = T_{i,j}\varpi_j(x_i,\sigma) + 2P_{i,j}\varpi'_j(x_i,\sigma) + \delta_{i,j}\varpi''_j(x_i,\sigma) \tag{3b}$$

The potential operator has exactly the same form as that in the precedent DVR method. Thus, the Hamiltonian matrix for one dimension has similar form as that using higher-order FD method, as shown in **Figure 4**.

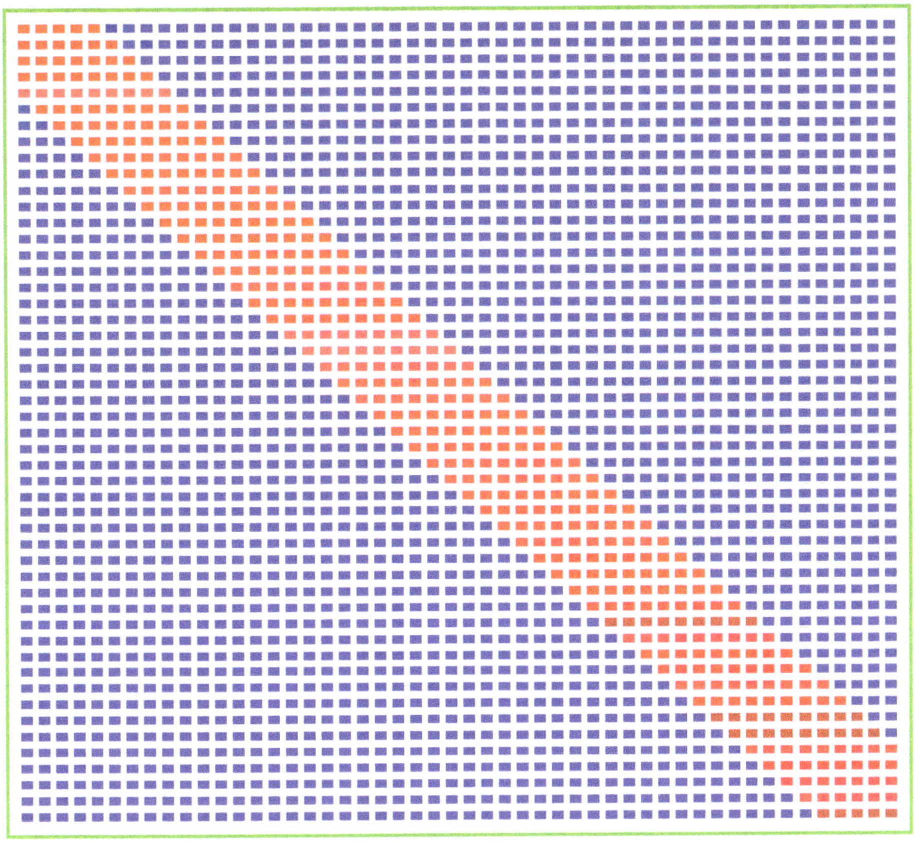

Figure 4. The matrix in the DVR method (*red* and *blue elements*) and the corresponding DAF method (*red elements*). The matrix using the DAF is much sparser.

As observed from Eq. (3), the DAF concept can be conveniently applied with many DVR kernels. The usage of DAF and the corresponding DVR method can be switched back and forth conveniently for checking convergence and saving computational effort, since the introduction of the DAF only reduces the computational effort, without affecting the numerical efficiency of its precedent DVR. The DAF method has been successfully applied for solving vibrational bound state, reactive scattering process, and calculating electronic states with explicit Coulomb singularity.

3.4. Sparse grid (SG)

With increasing dimensionality, the computational effort for solving Schrödinger equation increases exponentially using the basis space by a direct product expansion of a one-dimensional (1D) basis, regardless of how efficient the grid is for one dimension. As we understand, the physically required phase space does not increase so fast with increasing dimensionality in a specified energy range. Thus, many grid points (combinations) from different degrees of freedom in a direct product form are actually useless.

In 1963, Smolyak first introduced a multivariable integral method to overcome this difficulty [31]. This approach constructs a multidimensional multilevel basis by a special truncation of the tensor product expansion of a one-dimensional multilevel basis, which holds the promise of alleviating the restriction of excessively large numbers of unknowns and solves a set of differential equations using significantly fewer unknown. In the limit of high accuracy, the number of unknowns required by the sparse grid combination technique is independent of the spatial dimensionality of the problem. For example, asymptotically, a spatial 3D problem requires the same order of unknowns as a spatial 1D problem.

Sparse grids for solving a discretized partial differential equation (PDE) were introduced in 1990 by Zenger [32], in order to significantly reduce the number of degrees of freedom, while causing only a marginal increase in the representation error relative to the standard discretization. In 1992, Griebel et al. [33] showed that, for two and three dimensions, the sparse grid complexity and representation error can also be achieved by the so-called combination technique which is suitable for a parallel computation.

The sparse grid combination technique can be understood as a multivariate extrapolation technique. Instead of solving a set of differential equations on a single grid, solutions are obtained on a number of semi-coarsened grids. After solving these semi-coarsened problems, the solutions are combined to obtain a single, more accurate solution. Currently, there is focused interest for solving various PDE using sparse grid technique with different quadratures and in combination with multidomain techniques.

There are many studies on the introduction of the sparse grid method in the literature, and their theoretical details are not presented here. With multidimensional sparse grid and hierarchical bases, vibrational states of a molecule with dimensionality of 64 have been obtained [34], electronic structures of small molecules have been solved [35–40], and solution of the time-dependent Schrödinger equation of model potentials with split operator method, with hierarchical Fourier basis, have been reported [41].

In passing, we note that the coordinate choice is important in an efficient quantum dynamics calculation, especially for a calculation using sparse grid technique where the combination technique can be effectively realized. In spirit, the usage of hierarchical basis is very similar to the high-dimensional representations (HDMR) and multiconfiguration time-dependent Hartree (MCTDH) method [42], and general applications of the multidimensional sparse grid and hierarchical basis in the high-dimensional atomic and molecular dynamics are expected in the near future [43]. Inspecting the prospective development of multidomain technique for solving PDE and the coordinate problem in a state-to-state reactive scattering calculation, we might also expect that an exploration of multidomain technique in the field of molecular reaction dynamics would be prospective.

4. Time discretization methods

Besides the importance of the coordinate discretization methods for solving the TD Schrödinger equation, the efficiency of a time propagator is also vital. Especially, it is not surprised that the efficiency of a time propagator is only high for special spatial discretization methods. During the past decades, various time propagators have been developed, which roughly can be classified into two classes: One is short-time propagator which often is of low order and the other one is long-time propagator by a polynomial expansion which often is accurate. In the following, only the most useful ones are briefly reviewed.

4.1. Split operator method (second and higher order)

The second-order split operator method is the most popular one in a quantum dynamics calculation, especially in a molecular quantum dynamics calculation in combination with the DVR method [1]. The second-order split operator method is realized as

$$\exp(-i\Delta t\hat{H}) = \exp(-i\Delta t\hat{T} / 2)\exp(-i\Delta t\hat{V})\exp(-i\Delta t\hat{T} / 2)$$

or

$$\exp(-i\Delta t\hat{H}) = \exp(-i\Delta t\hat{V} / 2)\exp(-i\Delta t\hat{T})\exp(-i\Delta t\hat{V} / 2)$$

for the TID Hamiltonian or TD Hamiltonian, ignoring the time ordering. With the DVR method or the Fourier basis, where fast Fourier transform (FFT) can be applied, the split operator can be conveniently implemented. Higher order split operator can be numerically much more efficient than the second-order split operator for evolving the wave function of a diatomic molecule or triatomic reactive scattering [44,45].

4.2. Crank-Nicolson and spectral transformed Crank-Nicolson propagator

The Crank-Nicolson is particularly popular for solving a TD Schrödinger equation in the field of atomic physics. One reason, which is well known, is that the Coulomb singularity is difficult for a spectral method, but the finite difference is good for describing the Coulomb singularity thus often adopted. The finite difference in combination with the Crank-Nicolson propagator is very efficient as a propagator of TD Schrödinger equation by solving a linear equation. The other one reason is not so explicit and needs a bit more explanation, which comes from the concept of spectral transformed Hamiltonian [46]. As we know, with a TID Hamiltonian $\hat{H}$,

$$\exp(-i\Delta t\hat{H}) = \cos(\Delta t\hat{H}) - i\sin(\Delta t\hat{H})$$

and

$$\exp(i\Delta t\hat{H}) = \cos(\Delta t\hat{H}) + i\sin(\Delta t\hat{H})$$

Thus

$$\exp(-i2\Delta t\hat{H}) = \frac{\exp(-i\Delta t\hat{H})}{\exp(i\Delta t\hat{H})} = \frac{1 - i\tan(\Delta t\hat{H})}{1 + i\tan(\Delta t\hat{H})}$$

By retaining the first Taylor expansion term of *tan*(x), we obtain the usual Crank-Nicolson propagator

$$\exp(-i\Delta t\hat{H}) \approx \frac{1 - i\Delta t\hat{H}/2}{1 + i\Delta t\hat{H}/2}$$

On the other hand, we can define a spectral transformed Hamiltonian $\hat{\Theta}$ as

$$\hat{\Theta} = 2\arctan\left(\frac{\Delta t\hat{H}}{2}\right) \tag{4}$$

Hamiltonian $\hat{\Theta}$ has the exactly same eigenfunctions as the original Hamiltonian $\hat{H}$, but with different eigenvalues. The TD Schrödinger equation with Hamiltonian $\hat{\Theta}$ can be written as

$$i\frac{\partial \Psi}{\partial t} = \hat{\Theta}\Psi$$

For this equation, we conveniently have an exact short-time propagator

$$\exp(-i\Delta t\hat{\Theta}) = \frac{1 - i\tan(\Delta t\hat{\Theta}/2)}{1 + i\tan(\Delta t\hat{\Theta}/2)} = \frac{1 - i\Delta t\hat{H}/2}{1 + i\Delta t\hat{H}/2}$$

Therefore, in the traditional usage of the Crank-Nicolson propagator, the wave function actually is propagated with the spectral transformed Hamiltonian $\hat{\Theta}$. In the case where the time step is small, the energy of a wave function of the Hamiltonian $\hat{H}$ can be effectively approximated by the one of $\hat{\Theta}$.

In the calculation involving Coulomb singularity, the spectral range of the Hamiltonian usually is quite huge, which imposes the upper limit for the time step by

$$\Delta t \leq \frac{2\pi}{E_{\max} - E_{\min}}$$

As seen from Eq. (4), this spectral transformation reduces the original huge spectral range of Hamiltonian $\hat{H}$ into a very limited one, $[-\pi/2, \pi/2]$, which alleviates the upper limit of the time step much. This fact explains why the split operator in the exponential form is less accurate than that in the Crank-Nicolson form, contrary to intuition.

4.3. Short-time iterative Lanczos method

Another popular short-time propagator is the short-time iterative Lanczos (SIL) method [47]. The Lanczos method initially was proposed to calculate the eigenvalues and eigenfunctions of a symmetric Hamiltonian matrix by solving a tridiagonal matrix, which was given by a similar transformation of the original Hamiltonian matrix through the Krylov vectors. Therefore, the fundamental idea of the Lanczos method is to iteratively build a small, orthornormal set of vectors using the Krylov subspace generated from the solution to the time-dependent Schrödinger equation at the previous time step as the initial vector. This small set of vectors are used to diagonalize the Hamiltonian, which enables the exponentiation of the matrix to be carried out. The SIL method is realized by initializing

$$q_0 = \Psi(0)$$

$$\hat{H}q_0 = \alpha_0 q_0 + \beta_0 q_1$$

and by expressing generally

$$\hat{H}q_j = \beta_{j-1}q_{j-1} + \alpha_j q_j + \beta_j q_{j+1}$$

where the coefficients are given by

$$\alpha_j = \langle q_j | \hat{H} | q_j \rangle$$

$$\beta_{j-1} = \langle q_{j-1} | \hat{H} | q_j \rangle$$

The projected subspace representation of the Hamiltonian operator has the tridiagonal form

$$Z = \begin{Bmatrix} \alpha_0 & \beta_0 & & & & \\ \beta_0 & \alpha_1 & \beta_1 & & & \\ & \beta_1 & \alpha_2 & \beta_2 & & \\ & & \ddots & \ddots & \ddots & \\ & & & \beta_{n-3} & \alpha_{n-2} & \beta_{n-2} \\ & & & & \beta_{n-2} & \alpha_{n-1} \end{Bmatrix}$$

Then, the propagation can be accomplished by diagonalizing a small matrix Z:

$$\begin{aligned}\Psi(\Delta t) &= \exp(-i\Delta t\hat{H})\Psi(0) \\ &\approx \exp(-i\Delta t Z)\Psi(0)\end{aligned}$$

The SIL method requires memory usually one order larger than that required in a calculation using the split operator or Crank-Nicolson method. However, since SIL does not need to switch between different representations, unlike that using the split operator method, the SIL method can explore the block sparseness of the kinetic operator matrix for one dimension. Several works have been reported on the combination usage of the SIL method and the FE-DVR method. On the other hand, the iteration number for a convergent result using the SIL has direct ratio to the spectral range of the Hamiltonian. Therefore, even the SIL method takes well the advantage of the sparseness of the Hamiltonian matrix using the FE-DVR method; the unphysically large spectral range resulted from the finite element method reduces the convergence of the SIL method, which requires more iterations, and thus more memory and more computational effort.

4.4. Symplectic propagator

Symplectic time propagators for time-dependent Schrödinger equation are of particular interest, especially higher order ones. They can be very accurate and implemented simply [48]. For a typical symplectic propagator, the implementation can be done by separately expressing the wave function into real and imaginary parts,

$$\Psi = \varphi + i\gamma$$

where both φ and γ are real. Then, the time-dependent Schrödinger equation can be written as

$$\frac{\partial}{\partial t}\varphi(t) = \hat{H}\gamma(t)$$

$$\frac{\partial}{\partial t}\gamma(t) = -\hat{H}\varphi(t)$$

which can be rewritten as

$$\frac{\partial}{\partial t}\begin{Bmatrix}\varphi \\ \gamma\end{Bmatrix} = \begin{pmatrix} 0 & \hat{H} \\ -\hat{H} & 0 \end{pmatrix}\begin{Bmatrix}\varphi \\ \gamma\end{Bmatrix} = (A+B)\begin{Bmatrix}\varphi \\ \gamma\end{Bmatrix} \tag{5}$$

with

$$A = \begin{pmatrix} 0 & \hat{H} \\ 0 & 0 \end{pmatrix}, \text{and } B = \begin{pmatrix} 0 & 0 \\ -\hat{H} & 0 \end{pmatrix}$$

The evolution operator of Eq. (5) can be written as

$$O(\Delta t) = \begin{pmatrix} \cos(\Delta t\hat{H}) & \sin(\Delta t\hat{H}) \\ -\sin(\Delta t\hat{H}) & \cos(\Delta t\hat{H}) \end{pmatrix} \tag{6}$$

which is an orthogonal and symplectic matrix. Direct evaluation of Eq. (6) is expensive. By using composition methods, evaluation of Eq. (6) can be written as

$$O(\Delta t) = e^{\Delta t b_k B} e^{\Delta t a_k A} \cdots e^{\Delta t b_1 B} e^{\Delta t a_1 A}$$

where

$$e^{\Delta t A} = \Delta t \begin{pmatrix} I & \hat{H} \\ 0 & I \end{pmatrix}, \text{and}$$

$$e^{\Delta t B} = \Delta t \begin{pmatrix} I & 0 \\ -\hat{H} & I \end{pmatrix}$$

Various coefficients determined by the composition methods have been reported, which have been proven highly accurate for solving a time-dependent Schrödinger equation to molecular dynamics.

Even symplectic propagator implemented in this form is accurate and straightforward; it is unstable and gives diverged results when the adopted time step is larger than an upper limit. Usually, this upper limit is not large, which is determined by the spectral range of the Hamiltonian. Thus, the numerical efficiency of symplectic propagator, usually, is not high. This is quite different from the split operator in exponential form whose time step is only determined by the numerical error. This fact explains its much less popularities in a practical quantum dynamics calculation.

4.5. Chebyshev and real Chebyshev polynomial expansion with a TID Hamiltonian

The most popular long-time propagator is Chebyshev polynomial propagator, introduced by Taz and Kosloff [2]. It is accomplished as follows:

$$\Psi(t) = \exp(-it\hat{H})\Psi(0) = \exp\left[-\frac{i(E_{\max} - E_{\min})t}{2}\right]\sum_{n=1}^{N} a_n P_n\left(\frac{E_{\max} + E_{\min} - 2H}{E_{\max} - E_{\min}}\right)\Psi(0)$$

where P_n is the n-order of Chebyshev polynomial. E_{max} and E_{min} are the maximal and minimal energies of the Hamiltonian. The expansion coefficient a_n is given by

$$a_n = i^n (2 - \delta_{n0}) J_n[(E_{\max} - E_{\min})t / 2]$$

where J_n is the n-order Bessel function of first class. Since the Chebyshev polynomials can be obtained by the following iteration,

$$P_n(x) = 2P_{n-1}(x) + 2P_{n-2}(x)$$

$$P_0(x) = 1, \text{ and } P_1(x) = x$$

the propagation of the wave function by the Chebyshev polynomial expansion can be implemented by an iterative procedure.

For the initial wave function which has only the real part, the time evolution of the wave function can be achieved in the order domain of Chebyshev polynomial, instead of the time domain [49,50]. This can be understood as follows:

$$\Psi(\Delta t) = \exp(-i\Delta t\hat{H})\Psi(0) = \cos(\Delta t\hat{H})\Psi(0) - i\sin(\Delta t\hat{H})\Psi(0)$$

$$\Psi(-\Delta t) = \exp(i\Delta t\hat{H})\Psi(0) = \cos(\Delta t\hat{H})\Psi(0) + i\sin(\Delta t\hat{H})\Psi(0)$$

With $\Psi = \varphi + i\gamma$, we have

$$\varphi(\Delta t) = -\varphi(-\Delta t) + 2\cos(\Delta t\hat{H})\varphi(0)$$

and

$$\gamma(\Delta t) = -\gamma(-\Delta t) + 2\cos(\Delta t\hat{H})\gamma(0)$$

Therefore, the real and imaginary parts of the wave function can be propagated separately. By introducing the spectral transformed Hamiltonian,

$$\Xi = \arccos(\Delta t \hat{H})$$

Then, in the space of Ξ, the evolution of the real-wave function φ can be written as

$$\varphi(\Delta t) = -\varphi(-\Delta t) + 2\Delta t \hat{H} \varphi(0) \tag{7}$$

Here, the time step Δt is only an arbitrary parameter; thus, we simply set it as 1.0. Equation (7) then becomes

$$\varphi(k+1) = -\varphi(k-1) + 2\Delta t \hat{H} \varphi(k) \tag{8}$$

with $\varphi(1)=\hat{H}\varphi(0)$; the evolution of the wave function can then be carried out simply by the iterative action of the Hamiltonian on the wave function, according to Eq. (8). It has been proven that the real Chebyshev wave packet method is about two times faster than the Chebyshev polynomial expansion method, but the DVR method for direct reaction is less efficient than the second-order split operator [51,52].

4.6. Propagator with explicitly TD Hamiltonian

When the Hamiltonian is time-dependent, such as in a case where there is interaction between laser pulse and molecule or atom, the time ordering in a calculation using short-time propagator for numerical results of high accuracy has to be considered sometimes. For short-time propagators, one can supplement with a Euler-MacLaurin expansion for the time-integrated nonhomogeneous term or with a Magnus expansion for considering the time ordering [53,54]. Tal-Ezer and his coworkers recently introduced a new Chebyshev polynomial expansion method for nonhomogenous time-dependent Schrödinger equation. For more details, one may refer Ref. [55].

5. Boundary conditions

In a solution to TD Schrödinger equation using spatial discretization method, the spatial grid range has to be limited. However, during the propagation time, the wave function usually can extend out of the grid. We have to artificially impose the boundary condition, which is capable of damping the wave function before reaching the end of the grid, but without affecting the interior wave function. A usual choice is by including the absorbing potential in the potential operator or by imposing damping function to the wave function. Smooth exterior complex scaling method was reported to be a more accurate way to impose the boundary conditions in a numerical solution to TD Schrödinger equation, which has been applied in a calculation of electronic dynamics induced by ultrashort laser pulses.

6. Applications

With the pioneer work in the solution to the Schrödinger equation in the field of molecular quantum dynamics, the developments of quantum wave packet now have a general application in the field of atomic physics, molecular physics, molecular reaction dynamics, and the structure simulations of semiconductors or nanomaterials. In the following sections, only those chosen from the author's personal interest are briefly reviewed, which may not be the most representative works.

6.1. Feschbash resonance in the reaction of F + H_2/HD (v_0 = 0, j_0 = 0) and Cl + HD (v_0 = 1, j_0 = 0)

The reactive resonances discover the quasi-bound states of the reaction complex with unique clarity, and they do exist. Recognition of the reactive resonances is helpful for understanding how elementary chemical processes at a single quantum-state level take place. F + H_2 and its isotopic analogs are the most classical prototypes [56].

Interest in the F + H_2 reaction largely results from Lee's benchmark molecular beam studies [57] and early chemical laser work [58]. In 2000, the work on the F + HD reaction of Dong et al. [59] discovered the existence of reactive resonances in crossed beam experiments, where a resonance-enhanced step in the excitation function was observed.

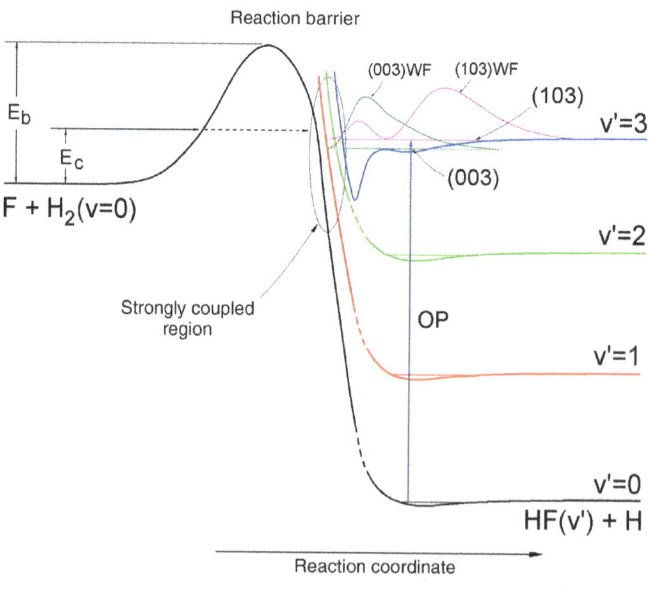

Figure 5. Schematic diagram indicating the resonance-mediated reaction mechanism for the F + H_2 reaction, which has two resonance states trapped in the HF(v' = 3)-H vibrational adiabatic potential (VAP)well. The 1D wave functions of the two resonance states are also given. The (003) state is the lowest ground resonance state; the (103) resonance is the first excited resonance state. Calculated van der Waals states for the lower VAPs are also given. OP, overtone pumping; Eb, barrier height; Ec, collision energy.

In 2006, with both theory and experiment exhibiting consistent behavior on XXZ PES, the work of Qiu et al. [60] reported the evidence for the resonances in the F + H_2 reaction, as shown in

Figure 5. The appearance of this report is enabled by the crossed molecular beam combined with high-resolution Rydberg state tagging technique, the developments on quantum scattering method development, and ab initio method. The sharp forward peak arising at the collision energy of 0.52 kcal/mol is actually due to the interference between the first two Feshbach resonances. With total angular momentum as zero, there are two resonance states at collision energies of 0.26 and 0.46 kcal/mol. Along with increasing J, the resonance energy will shift to higher collision energy. The three-dimensional (3D) scattering wave function at the collision energy of 0.26 kcal/mol suggests the existence of three nodes along the H-F coordinate (which correlates to the HF product) in the HF-H′ complex with zero node along the reaction coordinate. The projection of the $J = 0$ scattering wave function at 0.26 kcal/mol to the HF vibrational states suggests that the dominant character of this wave function is HF ($v' = 3$), with the outgoing waves mostly on HF ($v' = 2$). This implies that the resonance state at 0.26 kcal/mol is the lowest resonance state, (003), which is trapped in the HF ($v' = 3$)-H′ vibrational adiabatic potential (VAP) well.

The 3D scattering wave function for $J = 0$ at the collision energy of 0.46 kcal/mol suggests the existence of three nodes along the HF coordinate (which correlates to the HF product) in the HF-H′ complex, with a single node along the reaction coordinate. The projection of the $J = 0$ scattering wave function at 0.46 kcal/mol to the HF vibrational states indicates that the dominant character in this wave function is HF ($v' = 3$), but with the outgoing waves also mostly on HF ($v' = 2$). This indicates that the resonance state at 0.46 kcal/mol is the excited reaction resonance state trapped in the HF ($v' = 3$)-H′ VAP well. This resonance state can be assigned to the (103) resonance state, which has one-quantum vibration along the reaction coordinate, zero-quantum vibration on the bending motion (or hindered rotation), and three-quanta vibration along the HF stretching. The resonance schemes were shown in **Figure 5**.

Their next work at higher collision energy (0.94 kcal/mol) revealed the tunneling and shape resonance effects [61], other than Feschbash resonance leading to the forward scattering in the reaction of F + H_2, which indicated that the reactive resonances played quite distinguished roles in the same reaction but at different collision energies.

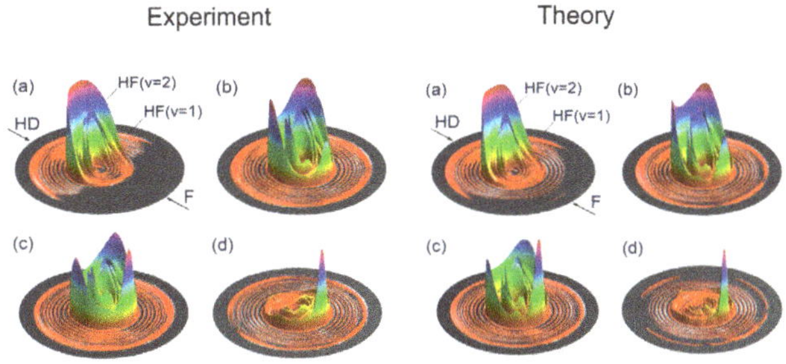

Figure 6. Theoretical and experimental 3D plots for the product of translational energy and angular distributions for the $F(^2P_{3/2})$ + HD($j_0 = 0$) → HF + D reaction at different collision energies: 0.43 kcal/mol (a); 0.48 kcal/mol (b); 0.52 kcal/mol; and (c) 0.71 kcal/mol (d).

In 2008, Ren et al. [62] measured the DCSs at several different collision energies, which suggested strong variation as a function of collision energies, resulting from the existence of strong reactive resonance states. The theoretical DCSs on the new version PES developed by Zhang and his coworkers (FXZ PES) showed good agreement with the experimental observation, as shown in **Figure 6**, which demonstrated that the F + H_2 reaction is the first reaction which can be investigated with spectroscopy accuracy, besides the H + H_2 reaction.

To have a better understanding about the reactive resonance state, the ground resonance state wave function of the F + HD → HF + D, along with the two-dimensional (2D) minimal potential, which was optimized along the angle degree of freedom, is presented in **Figure 7**. It is seen there that the wave function shows features of a semibound state. The outgoing part, corresponding to the HF(v' = 2) product, has two nodes of structure, but the inside peak has three nodes, corresponding to an excited vibrational state of v' = 3.

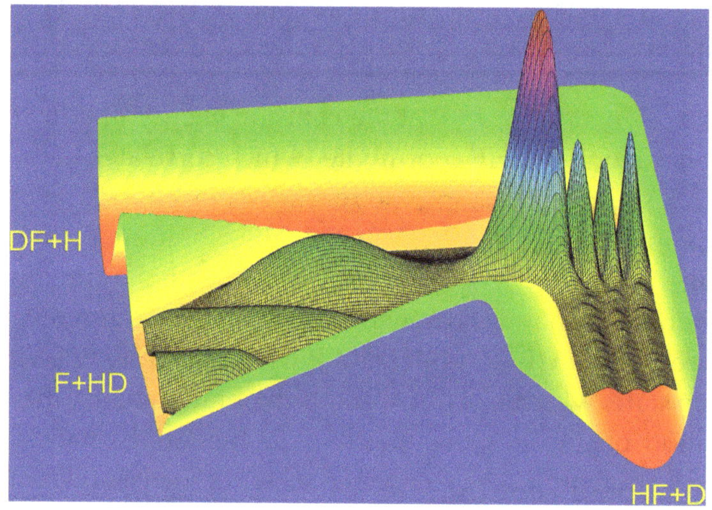

Figure 7. The 3D ground reaction resonance wave function of the F($^2P_{3/2}$) + HD(j_0 = 0) → HF + D reaction, along with the potential, which is optimized along angle degree of freedom.

In 2015, the resonances in the Cl +HD (v_0 = 1, j_0 = 0) reaction were experimentally observed, a first example besides the resonances in F + H_2 and F + HD [63]. The quantum dynamics revealed that the resonances in the Cl + HD reaction come from the bond-softening due to the strong anharmonicity over the transition barrier. Reactive resonances of this kind are expected to arise in many reactions; thus, the existence of reactive resonances is not so rare as we had thought.

6.2. Isotope effects in the O + O_2 reaction

The O + O_2 isotope exchange reactions play an important role in measuring the oxygen isotopic composition of a number of trace gases in the atmosphere. Their temperature dependence and kinetic isotope effects (KIEs) provide important constraints on our understanding of the origin and mechanism of these and other uncommon oxygen KIEs, which are important in the atmosphere [64,65]. Since there is a deep potential well in the process of the reaction and relative large mass of O atom, the rigorous state-to-state quantum dynamics calculation requires much computation. Thanks to the development of the numerical methods and computer techniques, on the most recent Dawes-Lolur-Li-Jiang-Guo (DLLJG) potential energy surface [66], calculation of the state-to-state differential cross sections and thermal reaction rate constant is possible [67–70].

For studying the KIEs in the O + O_2 reaction, using the reactant coordinate-based time-dependent wave packet method, where DVR method was applied for each degree of freedom, the integral cross sections (ICSs) of the $^{18}O + ^{32}O_2$ and $^{16}O + ^{36}O_2$ reactions with the initial states of $(v_0, j_0) = (0, 1)$, $(0, 5)$, $(0, 9)$, and $(0, 21)$ have been calculated explicitly on the newly constructed DLLJG PES. The ICSs for other $j_0 \leq 21$ values were estimated by a j_0-interpolation method and those for $j_0 > 21$ were estimated by extrapolation. These ICSs yield the corresponding initial state-specified reaction rate coefficients. Thermal reaction rate coefficients with Boltzmann, averaging over all relevant initial states were thus calculated approximately, and they exhibit clear negative temperature dependences for both reactions.

The calculated thermal reaction rate coefficients of both the $^{18}O + ^{32}O_2$ and $^{16}O + ^{36}O_2$ reactions on the DLLJG PES show a clear negative temperature dependence, in sharp contrast with the positive temperature dependence obtained on the earlier modified Siebert-Schinke-Bittererova (mSSB) PES [71]. In addition, there is an improved agreement between the calculated KIE and the experiment. These results validated the absence of the "reef" structure in the entrance/exit channels of the DLLJG PES, which is present in the mSSB PES.

The calculated total reaction probabilities of the $^{18}O + ^{32}O_2$ and $^{16}O + ^{36}O_2$ reactions on the DLLJG PES as a function of the total angular momentum (J) suggest that the O + O_2 exchange reactions are dominated by resonances at very low collision energies (<0.2 eV) immediately above the reaction threshold, as shown in **Figure 8**. These resonances depend strongly on the masses of the oxygen atoms involved and/or the zero-point energy difference between the reactant and product diatoms. Though it appears that the isotopic effects in the exchange reactions come from the ZPE difference, the underlying physical mechanism for the isotope effects is shown here to result from strong near-threshold reactive resonances which mediate the reactions. Thus, it was first time pointed out that the accurate characterization of the reactivity for these near-thermoneutral reactions immediately above the reaction threshold is important for correct characterization of the thermal reaction rate coefficients [47].

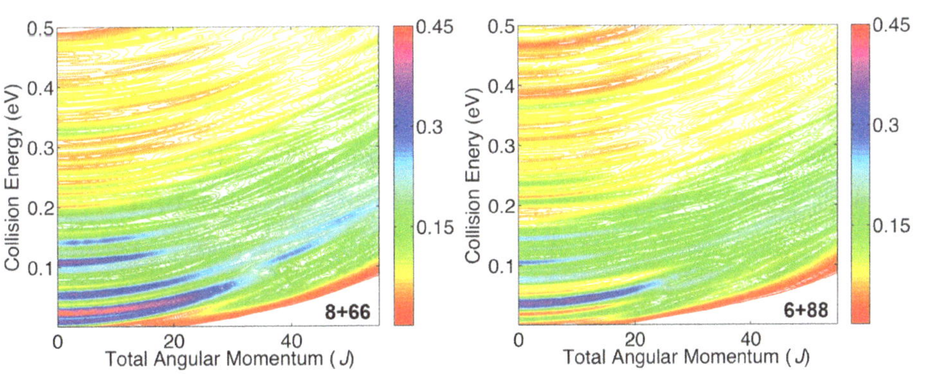

Figure 8. The 2D plot of smoothened total reaction probabilities as a function of collision energy and total angular momentum J of $^{16}O + ^{36}O_2$ (6 + 88) and $^{18}O + ^{32}O_2$ (8 + 66), with the initial state of $j_0 = 1$ on the mSSB PES. The J-shifting rule is clearly observed in both plots, and the reactivity of 8 + 66 is larger at collision energies below 0.15 eV, resulting from the resonance enhancement.

6.3. State-to-state cross sections calculation of a chemical reaction

Theoretical method for a state-to-state calculation is of vital importance in the field of molecular reaction dynamics, which enables a detailed comparison between theoretical results and the experimental differential cross sections by crossed molecular beams [72]. Traditionally, there is only product coordinate-based method to achieve this goal using time-dependent quantum wave packet method in Jacobi coordinates. In 1996, John Zhang and coworkers put forward an efficient reactant-product decoupling method for a direct reaction. Recently, reactant coordinate-based method and transition state wave packet method were also proposed for a state-to-state reaction calculation.

The transition state wave packet (TSWP) method is a relatively new method for computing the reactive S-matrix elements, and has recently been demonstrated to be accurate and efficient for computing $J = 0$ state-to-state reaction probabilities for three, four, and also six atoms [73–79]. The TSWP method is based on the quantum transition state theory of Miller [80,81], who provided a direct way to calculate the cumulative reaction probability (CRP), $N(E)$, and thermal rate coefficient, $k(T)$, from a flux correlation function on the dividing surfaces located near the transition state [82]. Recently, Manthe and coworkers extended this formulism to compute the S-matrix elements from generalized flux correlation functions of TSWPs [49,83]. Specifically, a set of initial TSWPs are calculated as the eigenstates of the thermal flux operator, which is defined in the transition state region as [84]

$$\hat{F}_T = e^{-\hat{H}/2k_BT}\hat{F}e^{-\hat{H}/2k_BT},$$

where T is a reference temperature in Kelvin, and k_B denotes the Boltzmann constant. These TSWPs are then propagated independently into both the reactant and product arrangement channels, and the S-matrix elements are obtained from the resulting cross-correlation functions:

$$S_{\upsilon_p j_p \leftarrow \upsilon_r j_r}(E) = \frac{e^{E/k_BT}}{2\pi\eta^{-*}_{\upsilon_p j_p}(E)\eta^{+}_{\upsilon_r j_r}(E)}\sum_n f_T^n A_{\upsilon_p j_p \leftarrow n}(E) A^*_{n\leftarrow \upsilon_r j_r}(E)$$

where the energy-resolved projection amplitudes in the product (p) and reactant (r) channels, $A_{\upsilon_p j_p \leftarrow n}(E)$ and $A^*_{n\leftarrow \upsilon_r j_r}(E)$, are given as Fourier transforms of the appropriate cross-correlation functions:

$$A_{\upsilon_p j_p \leftarrow n}(E) = \int_{-\infty}^{\infty} dt e^{iEt} C_{\upsilon_p j_p \leftarrow n}(t) = \int_{-\infty}^{\infty} dt e^{iEt} \left\langle \Phi^+_{\upsilon_p j_p} \middle| e^{-i\hat{H}t} \middle| f_T^n \right\rangle,$$

$$A^*_{n\leftarrow \upsilon_r j_r}(E) = \int_{-\infty}^{\infty} dt e^{-iEt} C^*_{n\leftarrow \upsilon_r j_r}(t) = \int_{-\infty}^{\infty} dt e^{-iEt} \left\langle f_T^n \middle| e^{i\hat{H}t} \middle| \Phi^-_{\upsilon_r j_r} \right\rangle.$$

where $\{f_T^n, |f_T^n\rangle\}$ are the thermal flux eigenpairs at the reference temperature of T, and $\eta^+_{v_p j_p}(E)$ and $\eta^-_{v_r j_r}(E)$ are energy-normalizing factors of the asymptotic wave functions ($|\Phi^+_{v_p j_p}\rangle$ and $|\Phi^-_{v_r j_r}\rangle$) for the two channels.

In addition to its conceptual clarity, the TSWP method has several numerical advantages. First, the propagation of each TSWP is essentially an inelastic calculation in the appropriate Jacobi coordinates, in which much smaller bases/grids are needed. One of course needs to transform the initial TSWPs prepared in one coordinate into the other, but this transformation is performed only once in a small region near the transition state. Second, the existence of either a pre-reaction or a post-reaction well has a limited impact on the computational costs, as the analysis plane can be placed deep into the asymptotic region without significantly impeding the numerical efficiency. Third, in contrast to the initial state-specific wave packet (ISSWP) approach, the entire S-matrix can be obtained at all energies in one complete calculation. Finally, the multiple propagations of TSWPs are "embarrassingly" parallel, adding further to its numerical efficiency. However, with the increase of energy, the number of involved TSWPs grows rapidly, especially for those systems with small vibrational frequencies of the activated complex.

The reactant coordinate-based (RCB) method for computing the reactive S-matrix elements has been widely used for the triatomic reaction systems [40,85–89], but it is only recently that the RCB method is extended to tetra-atomic systems [90,91]. The key idea of the RCB method is to propagate the time-dependent initial wave packet deep into the product asymptotic region, where the product states are well defined on an analysis plane. There are generally two schemes to define the product states on the analysis plane: one evaluates product states on the grids of reactant coordinates with an interpolation scheme, while the other projects product states onto a set of intermediate coordinates. The interpolation scheme saves grid points in expressing the product wave functions, thus saving computer memory, and the intermediate coordinate method is numerically more efficient.

In the intermediate coordinate scheme, the projection plane for a particular product channel is designed to include the corresponding scattering coordinate that is defined as the separation of the two products. This enables the expression of the product wave packet in a product form, in which the product wave packet in the scattering coordinate is simply a delta. The remaining degrees of freedom are chosen from reactant coordinate, which renders the projection of time-dependent wave packet on the intermediate coordinate very efficient. The calculation of overlaps between the time-dependent wave packets and the product states is carried out on the intermediate coordinate by transforming them from the respective reactant and product Jacobi coordinates into a set of intermediate coordinates.

Compared to the TSWP method, the RCB method has the advantage of propagating the wave packet in only the reactant Jacobi coordinates, which allows the simultaneous analysis of the two product channels with only a single propagation. However, RCB has a genesis of ISSWP method, and it only calculates a column of the S-matrix.

6.4. Dissociative chemisorption of water on transition-metal surfaces

The interaction of molecular species with metal surfaces is of great importance to many industrial applications. The dissociative chemisorption of water on transition-metal surfaces plays a pivotal role in understanding many industrial heterogeneous processes such as steam reforming, and represents the rate-limiting step in low-temperature WGS reaction on copper catalysts [92]. Tremendous progress has been made for the dissociative chemisorption dynamics.

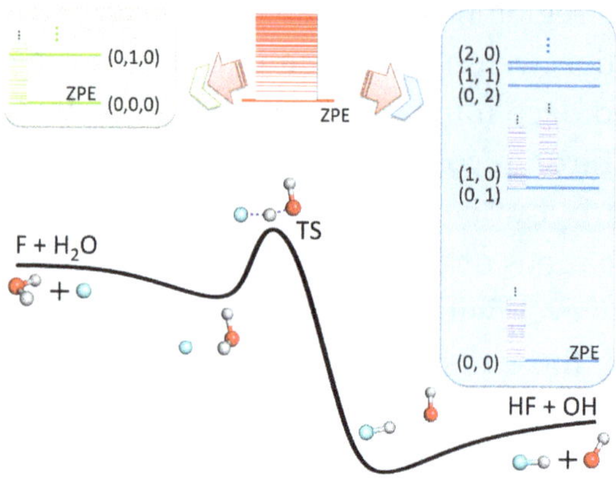

Figure 9. Schematic of the TSWP method. The TSWPs are prepared near the transition state region and then propagated to both the reactant and product sides to resolve the state-to-state information (Adopted from Ref. [4]).

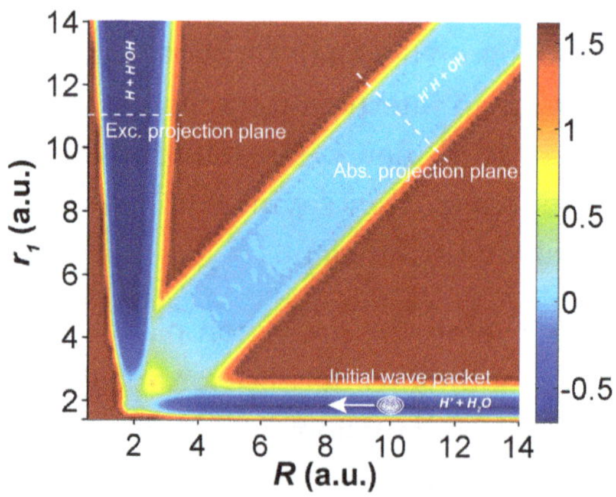

Figure 10. Two-dimensional PES plot in the RCB method for both the H′ + H_2O → H′H + OH abstraction (Abs.) and H′ + H_2O → H + H′OH exchange (Exc.) channels in the reactant Jacobi coordinates R and r_1. Other degrees of freedom are optimized. The initial wave packet is prepared in the reaction-asymptotic region, and two projection planes are located in the asymptotic regions of the corresponding product channels (adopted from Ref. [67]).

A total of 9 degrees of freedom should be considered for the dynamics of H_2O molecule on a rigid surface (as shown in **Figure 11**), rendering it formidable to carry out a fully coupled nine-dimensional (9D) quantum dynamics calculations. Guo and coworkers performed six-dimensional (6D) quantum dynamics calculations to study the mode specificity of H_2O and bond selectivity of HOD on a rigid flat Cu(111) surface [93–95], employing their 6D PESs developed by permutationally invariant polynomials (PIP). The 6D model neglects the effects of impact sites and surface corrugation, because the surface-lateral coordinates (*X* and *Y*) and the azimuthal angle were fixed at the values of transition state. The first quantum-state resolved molecular beam experiment on the dissociative chemisorptions of D_2O on Ni(111) demonstrated a large enhancement in reactivity upon excitation of the asymmetric stretching mode of D_2O [96]. The observed mode specificity was semiquantitatively understood by Guo and coworkers using the 6D quantum dynamical approach on a 6D PES. Recently, Jiang and Guo investigated 6D site-specific dissociation probabilities of this system, but based on a new 9D PES developed by permutation-invariant polynomial-neural network (PIP-NN) method [97].

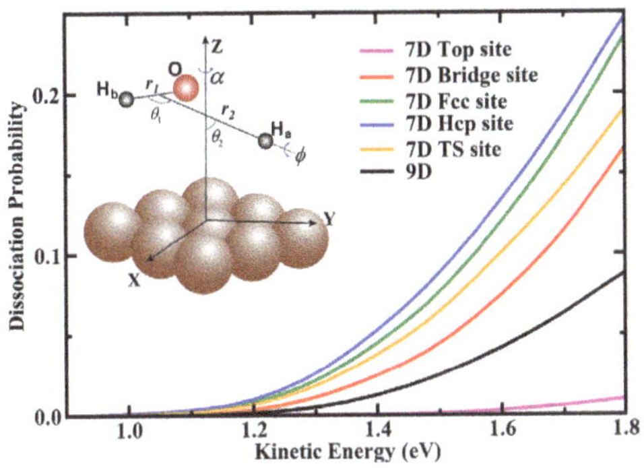

Figure 11. The 9D dissociation probability and 7D site-specific (top, bridge, fcc, hcp, and TS sites) probabilities with H_2O initially in the ground ro-vibrational state (000). The 9D molecular coordinates of the H_2O/Cu(111) system are shown in the inset.

Very recently, we reported the first seven-dimensional (7D) quantum dynamics study for the dissociative chemisorption of H_2O on Cu(111), based on an accurate 9D PES developed by neural network approach [98,99]. The dissociation probabilities exhibit strong azimuthal angle dependence, and large differences were seen between 7D and 6D results, indicating that the 6D quantum model, neglecting the azimuthal angle, can introduce substantial errors. A significant step forward in simulating gas-surface reactions from the first-principles was recently achieved, where we reported the first full-dimensional quantum dynamics study of chemisorption process of H_2O on rigid Cu(111), with all 9 degrees of freedom (9D) fully coupled on a global potential energy surface [100]. We found that the full-dimensional quantum-mechanical reactivity is quite different from the corresponding results obtained by previous reduced-dimensional models, indicating the importance of the challenging full-dimensional quantum-mechanical calculations to achieve a quantitatively accurate understanding of this reaction. The excitations in vibrational modes of H_2O, in particular the stretching modes, are more efficacious than increasing the translational energy in promoting the reaction, much stronger than observed in previous reduced-dimensionality quantum studies. The full-dimensional quantum-mechanical calculations not only offer the dynamic features with the highest accuracy but also allow a conclusive examination of previous dynamical approximations.

6.5. Electronic states of H_2^+ and H_2 by mapped DVR using cylindrical coordinate

For describing the low-lying electronic state of a diatomic molecule, the prolate spheroidal coordinate is a natural choice. However, it may be not an optimal choice for describing high-lying state. The Hamiltonian operator of electrons in a diatomic molecule is very simple and easy for numerical evaluation. It is well known that the Coulomb singularity numerically is very difficult to deal with, and in cylindrical coordinate, it is even more difficult.

To overcome this difficulty, a numerical scheme by a combination of mapped sinc DVR and cosine DVR method was introduced in the work by Lin and Sun [101]. The mapped function for variables is as follows:

$$\rho = f(x_1) = A_0\left(x_1 - \frac{1}{\beta}\arctan \beta x_1\right)$$

and

$$z = g(x_2) = E_0 x_2 - Z_0 log(z_H - x_2 + \sqrt{(z_H - x_2)^2 + \rho_1^2}) - Z_0 log(z_H + x_2 + \sqrt{(z_H + x_2)^2 + \rho_1^2}) - z_0$$

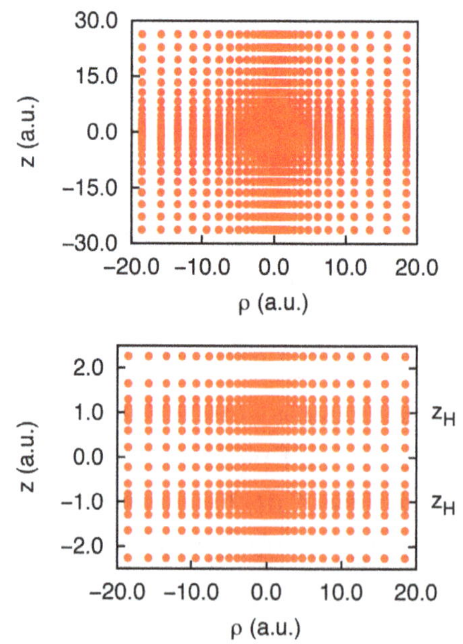

Figure 12. A typical 2D distribution of the grid points of H_2^+ in cylindrical coordinate after variable's mapping. In the bottom, an enlarged part of the distribution is shown.

The parameter A_0 is applied to determine the grid range of ρ, but the parameter β is to adjust the relative distribution of the grid points of ρ degree of freedom. The parameter E_0, which approximately corresponds to the energy of the reference state, following a phase optimization philosophy, is used to adjust the distribution of the grid points of z degree of freedom, along with Z_0 parameter. Z_0 approximately corresponds with the nuclear charges. The parameter z_0 makes function $g(0)=0$, and ρ_1 is the starting grid point of ρ degree of freedom. The evenly distributed grid points in the coordinate x_1 and x_2 are shown in **Figure 12** as a function of ρ and z, where the grid points cluster around the nuclear positions.

With this mapped DVR method, the energy of the lowest electronic state, which is most sensitive to the accuracy of the description of the singularity, can be given with 10 significant numbers only with 65 and 100 grid points for ρ and z. The method also was applied to H_2 molecule, including correlation of electrons. It was found that a straightforward extension of the method for H_2^+ only is capable of giving the energy of ground electronic state with three or four significant digits, with the parameters as follows:

$$N_{\rho 1} = N_{\rho 1} = 20$$

$$N_{z1} = N_{z1} = 60$$

$$N_{\theta} = N_{\theta} = 20$$

The nuclear distance is 1.4 Bohr (a.u.). The grid range for ρ_1 and ρ_2 is [0, 20>] a.u., for z_1 and z_2, it is [-30, 30] a.u., and for θ, it is [0, π].

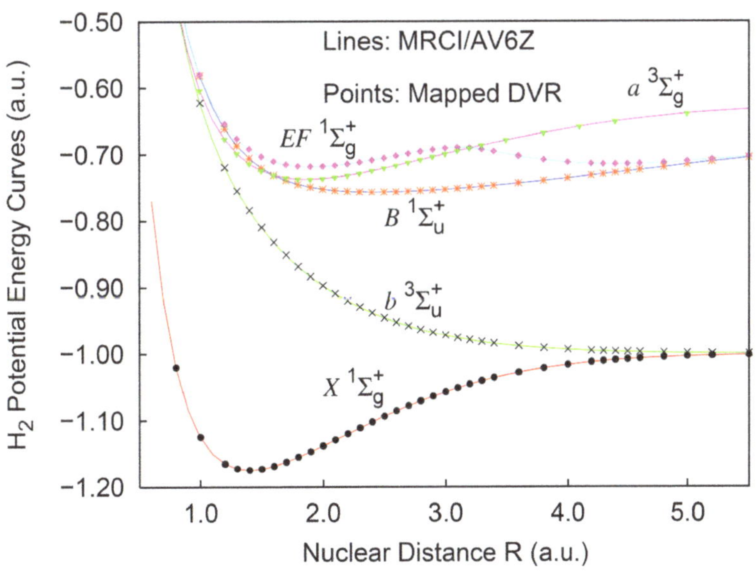

Figure 13. The ground state and low-lying excited states potential energy curves of H_2, calculated by the mapped DVR method and the MRCI/AV6Z method by the Molpro package.

Figure 13 presents the potential energy curves of the ground electronic state and low-lying excited states of H_2 as a function of nuclear distance, calculated by the mapped DVR method (symbols) with the same parameters listed above, along with those (lines) obtained with the MRCI/AV6Z method in the Molpro package [44]. The potential energy curves of the excited electronic states obtained by the mapped DVR method are always lower than those calculated by the MRCI/AV6Z method; however, the curve of the ground electronic state by the mapped DVR method is a little higher than the one obtained by the MRCI/AV6Z method in the dissociation region, because of the strong electronic correlation energy of the ground electronic state. Further numerical studies indicate that by making the grid point distribute more densely around the middle position of the two nuclei, the accuracy of description of the mapped DVR method for the electronic correlation effects can be improved by two orders, which may be enough for a quantum calculation of the electronic dynamics induced by ultrashort laser pulses [102].

6.6. Electronic dynamics in bichromatic circularly polarized attosecond laser fields

Bichromatic circularly polarized attosecond laser pulses are new tools for investigating electron dynamics in strong field ionization of atoms, molecules, and even surfaces, following early experiments on polarization properties of high-intensity high-order harmonic generation [103]. Photoionization in atoms and molecules by various frequency combinations of co-rotating and counter-rotating bichromatic circularly polarized attosecond ultraviolet pulses exhibits signature of spiral interference patterns in photoelectron momentum distributions. The generation of electronic currents such as vortices [104–106] has been proposed as sources of attosecond magnetic field pulses [107,108]. It is found that the helicity of circularly polarized pulses is the essence for the spiral photoelectron distributions. Starace and coworkers [82] have shown that for the cases with two identical frequency circularly polarized attosecond ultra-violet pulses, circularly symmetric patterns independent of the electron ejection angle are produced in photoelectron momentum distributions in atomic He. The corresponding momentum rings are also insensitive to the pulse phases. Electronic vortices can only be obtained by opposite helicity laser pulses.

We numerically solve the three-dimensional time-dependent Schrödinger equation of the aligned molecular ion H_2^+ at equilibrium. A five-point finite difference method combined with fast Fourier transform technique is adopted to describe the molecular Hamiltonian in spatial space [85]. The electron wave function is propagated in time domain with high-order split-operator methods [109,110]. We use an efficient method by calculating a radial flux (electron current density at an asymptotic point) to simulate the ionization spectra [111]. At large asymptotic point, the angular flux distributions can be ignored. As a result, we only need to consider the radial part of the electronic flux along the radial direction. Results show that in bichromatic circular polarization ionization, spiral interference patterns can be observed for both cases with co-rotating and counter-rotating components, which are sensitive to the pulse frequencies, phase, and time delays, as shown in **Figure 14**. We analyze these results by an attosecond perturbation ionization model. Coherent electron wave packets with same kinetic energies created respectively by the two color pulses in the continuum interfere with each other. Photoionization distributions are functions of the photoelectron momentum and the ejection angle, thus giving rise to spiral distributions. Such sensitivity of the spiral electron distributions to the laser parameters offers an ideal means of characterizing these pulses and timing ultrafast photoionization processes by such circular attosecond pulses.

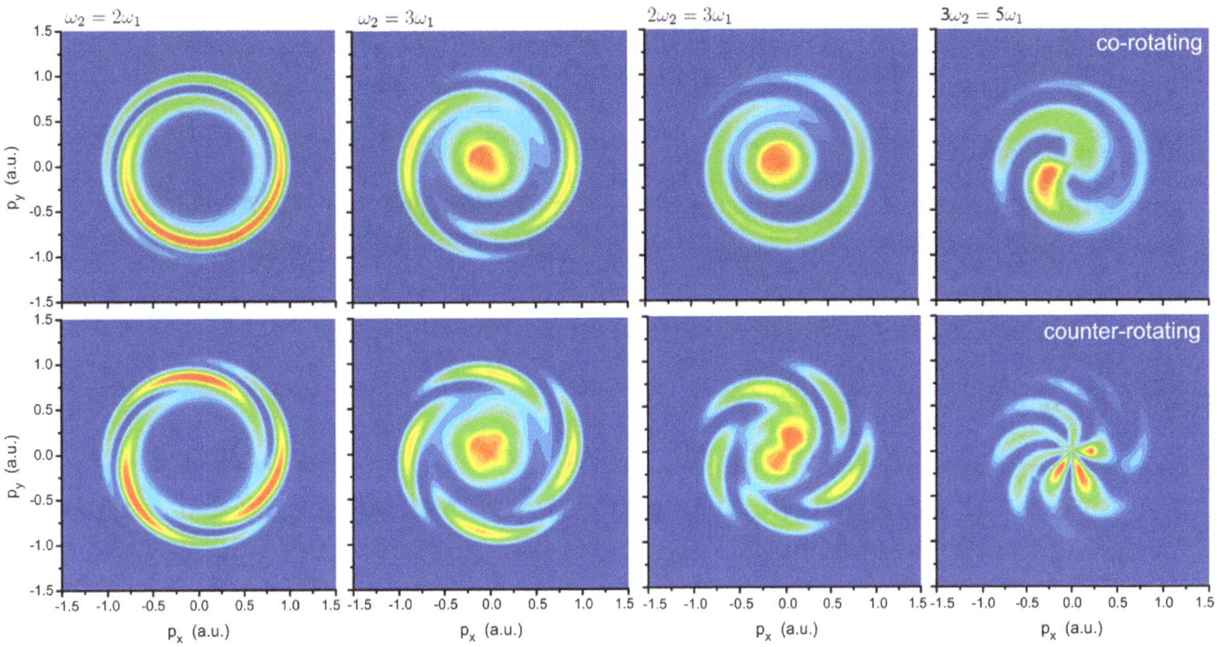

Figure 14. Photoelectron momentum distributions of z-aligned H_2^+ in bichromatic (x, y) circularly polarized attosecond ultraviolet laser pulses with (upper row) co-rotating and (bottom row) counter-rotating components at different frequencies.

7. Conclusion

Recent developments of quantum wave packet in atomic and molecular dynamics and their applications are reviewed. The continuous expansion of computing power and numerical techniques in quantum wave packet method leads to solve the molecular dynamics with increasingly complicated problems. Foreseeably, more general applications of the quantum wave packet method will result in more exciting discoveries in exploration of the quantum dynamics of molecules and atoms.

Acknowledgements

This work was supported by the National Basic Research Program of China (973 program, No. 2013CB922200), the National Natural Science Foundation of China (Grant Nos. 21222308, 21103187, and 21133006), the Chinese Academy of Sciences, and the Key Research Program of the Chinese Academy of Sciences.

Author details

Zhigang Sun

Address all correspondence to: zsun@dicp.ac.cn

State Key Laboratory of Molecular Reaction Dynamics and Center for Theoretical Computational Chemistry, Dalian Institute of Chemical Physics, Chinese Academy of Sciences, Dalian, PR China

References

[1] Kosloff R, Time-dependent quantum-mechanical methods for molecular dynamics. Journal of Physical Chemistry. 1988;92:2087–2100. DOI:10.1021/j100319a003

[2] Kosloff R, Propagation methods for quantum molecular dynamics. Annual Review of Physical Chemistry. 1994;45:145–178. DOI:10.1146/annurev.pc.45.100194.001045

[3] Leforestier C, Bisseling RH, Cerjan C, Feit MD, Friesner R, Guldberg A, Hammerich A, Jolicard G, Karrlein W, Meyer HD, Lipkin N, Roncero O, and Kosloff R, A comparison of different propagation schemes for the time dependent Schrödinger equation. Journal of Computational Physics. 1991;94:59–80. DOI:10.1016/0021-9991(91)90137-A

[4] Guo H, Recursive solutions to large eigenproblems in molecular spectroscopy and reaction dynamics. Reviews in Computational Chemistry. 2007;25:285–347. DOI: 10.1002/9780470189078.ch7

[5] Feit MD, Fleck JA, and Steiger A, Solution of the Schrödinger equation by a spectral method. Journal of Computational Physics. 1982;47:412–433. DOI:10.1016/0021-9991(82)90091-2

[6] Kosloff R and Tal-Ezer H, A direct relaxation method for calculating eigenfunctions and eigenvalues of the schrödinger equation on a grid. Chemical Physics Letters. 1986;127:223–230. DOI:10.1016/0009-2614(86)80262-7

[7] Yao GH and Wyatt RE, Stationary approaches for solving the Schrödinger equation with time-dependent Hamiltonians. Journal of Chemical Physics. 1994;101:1904–1913. DOI:10.1063/1.467700

[8] Peskin U, Kosloff R, and Moiseye N, The solution of the time dependent Schrödinger equation by the (t,t′) method: The use of global polynomial propagators for time dependent Hamiltonians. Journal Chemical Physics. 1994;100:8849–8855. DOI: 10.1063/1.466739

[9] Luo M, Liu QH, and Li ZB, Spectral element method for band structures of two-dimensional anisotropic photonic crystals. Physical Review E. 2009;79:026705. DOI: 10.1103/PhysRevE.79.026705

[10] Rescigno TN and McCurdy CW, Numerical grid methods for quantum-mechanical scattering problems. Physical Review A. 2000;62:032706. DOI:10.1103/PhysRevA.62.032706

[11] Schneider BI and Collins LA, The discrete variable method for the solution of the time-dependent Schrödinger equation. Journal of Non-Crystalline Solids. 2005;351:1551–1558. DOI:10.1016/j.jnoncrysol.2005.03.028

[12] Mazziotti DA, Spectral difference methods for solving differential equations. Chemical Physics Letters. 1999;299:473–480. DOI:10.1016/S0009-2614(98)01324-4

[13] Boyd JP, Sum-accelerated pseudospectral methods: The Euler-accelerated sinc algorithm. Applied Numerical Mathematics. 1991;7:287–296. DOI:10.1016/0168-9274(91)90065-8

[14] Gray SK and Goldfield EM, Dispersion fitted finite difference method with applications to molecular quantum mechanics. Journal of Chemical Physics. 2001;115:8331–8344. DOI:10.1063/1.1408285

[15] Mazziotti DA, Spectral difference methods for solving the differential equations of chemical physics. Journal of Chemical Physics. 2002;117:2455–2468. DOI: 10.1063/1.1490344

[16] Wei GW, Zhang DS, Kouri DJ, and Hoffman DK, Lagrange distributed approximating functionals. Physical Review Letters. 1997;79:775–778. DOI:10.1103/PhysRevLett.79.775

[17] Light JC, Hamilton IP, and Lill JV, Generalized discrete variable approximation in quantum mechanics. Journal of Chemical Physics. 1985;82:1400–1408. DOI: 10.1063/1.448462

[18] Light JC, Discrete-variable representations and their utilization. Advances in Chemical Physics. 2000;114:263–310. DOI:10.1002/9780470141731.ch4

[19] Wang XG and Carrington T, A contracted basis-Lanczos calculation of vibrational levels of methane: Solving the Schrödinger equation in nine dimensions. Journal of Chemical Physics. 2003;119:101–117. DOI:10.1063/1.1574016

[20] Manolopoulos DE and Wyatt RE, Quantum scattering via the log derivative version of the Kohn variational principle. Chemical Physics Letters. 1988;152:23–32. DOI: 10.1016/0009-2614(88)87322-6

[21] Baye D, The Lagrange-mesh method. Physics Report. 2015;565:1–107. DOI:10.1016/j.physrep.2014.11.006

[22] Shizgal BD and Chen HL, The quadrature discretization method (QDM) in the solution of the Schrödinger equation with nonclassical basis functions. Journal of Chemical Physics. 1996;104:4137–4150. DOI:10.1063/1.471225

[23] Chen R and Guo H, Effect of spectral range on the convergence in Lanczos algorithm, a numerical study. Chemical Physical Letters. 2003;369:650–655. DOI:10.1016/S0009-2614(02)02040-7

[24] Lin WB, Kovvali N, and Carin L, Pseudospectral method based on prolate spheroidal wave functions for semiconductor nanodevice simulation. Computer Physics Communications. 2006;175:78–85. DOI:10.1016/j.cpc.2006.02.006

[25] Yu DQ, Cong SL, and Sun ZG, An improved Lobatto discrete variable representation by a phase optimisation and variable mapping method. Chemical Physics. 2015;458:41–51. DOI:10.1016/j.chemphys.2015.07.009

[26] Huang Y, Zhu W, Kouri DJ, and Hoffman DK, Distributed approximating function approach to atom-diatom reactive scattering: Time-dependent and time-independent wavepacket treatments. Journal of Physical Chemistry. 1994;98:1868–1874. DOI: 10.1021/j100058a025

[27] Hoffman DK, Wei GW, Zhang DS, and Kouri DJ, Shannon–Gabor wavelet distributed approximating functional. Chemical Physics Letter. 1998;287:119–124. DOI:10.1016/S0009-2614(98)00130-4

[28] Hoffman DK, Arnold M, and Kouri DJ, Properties of the optimum distributed approximating function class propagator for discretized and continuous wave packet propagations. Journal of Physical Chemistry. 1992;96:6539–6545. DOI:10.1021/j100195a007

[29] Hoffman DK and Kouri DJ, Distributed approximating function theory: A general, fully quantal approach to wave propagation. Journal of Physical Chemistry. 1992;96:1179–1184. DOI:10.1021/j100182a030

[30] Wei GW, Althorpe SC, Zhang DS, Kouri DJ, and Hoffman DK, Lagrange-distributed approximating-functional approach to wave-packet propagation: Application to the time-independent wave-packet reactant-product decoupling method. Physical Review A. 1998;57:3309–3316. DOI:10.1103/PhysRevA.57.3309

[31] Smolyak SA, Quadrature and interpolation formulas for tensor products of certain classes of functions. Soviet Mathematics Doklady. 1963;4:240–243.

[32] Zenger C, Sparse grids. Technical Report 18–90 SFB342, TU Mönchen, 1990.

[33] Griebel M, Schneider M, and Zenger C, A combination technique for the solution of sparse grid problems, in de Groen P, Beauwens R, editors. Iterative Methods in Linear Algebra. North Holland: IMACS, Elsevier; 1992, pp. 263–281.

[34] Thomas PS and Carrington T, Using nested contractions and a hierarchical tensor format to compute vibrational spectra of molecules with seven atoms. Journal of Physical Chemistry A. 2015;119:13074–13091. DOI:10.1021/acs.jpca.5b10015

[35] Shen J, Wang YW, and Yu HJ, Efficient spectral-element methods for the electronic schrödinger equation. Sparse Grids and Applications – Stuttgart. 2014;2014:265–289.

[36] Carrington JRT, Two new methods for computing vibrational energy levels. Canadian Journal of Chemistry. 2015;93:589–593. DOI:10.1139/cjc-2014-0590

[37] Manzhos S, Dawes R, and Carrington JRT, Neural network-based approaches for building high dimensional and quantum dynamics-friendly potential energy surfaces. International Journal of Quantum Chemistry. 2014;115:1012–1020. DOI:10.1002/qua.24795

[38] Avila G and Carrington JRT, Solving the Schroedinger equation using Smolyak interpolants. Journal of Chemical Physics. 2013;139:134114. DOI:10.1063/1.4821348

[39] Avila G and Carrington JRT, Using multi-dimensional Smolyak interpolation to make a sum-of-products potential. Journal of Chemical Physics. 2015;143:044106. DOI:10.1063/1.4926651

[40] Avila G and Carrington JRT, Solving the vibrational Schrödinger equation using bases pruned to include strongly coupled functions and compatible quadratures. Journal of Chemical Physics. 2012;137:174108. DOI:10.1063/1.4764099

[41] Gradinaru V, Fourier transform on sparse grids: Code design and the time dependent Schrödinger equation. Computing. 2007;80:1–22. DOI:10.1007/s00607-007-0225-3

[42] Beck MH, Jäckle A, Worth GA, and Meyer HD, The multiconfiguration time-dependent Hartree (MCTDH) method: A highly efficient algorithm for propagating wavepackets. Physics Reports. 2000;324:1–105. DOI:10.1016/S0370-1573(99)00047-2

[43] Li GY, Rosenthal C, and Rabitz H, High dimensional model representations. Journal of Physical Chemistry A. 2001;105:7765–7777. DOI:10.1021/jp010450t

[44] Sun ZG, Yang WT, and Zhang DH, Higher-order split operator schemes for solving the Schrödinger equation in the time-dependent wave packet method: Applications to triatomic reactive scattering calculations. Physical Chemistry Chemical Physics. 2012;14:1827–1845. DOI:10.1039/C1CP22790D

[45] Liu WT, Zhang DH, and Sun ZG, Efficient fourth-order split operator for solving the triatomic reactive schrödinger equation in the time-dependent wavepacket approach. Journal of Physical Chemistry A. 2014;118:9801–9810. DOI:10.1021/jp5074158

[46] Sun ZG and Yang WT, An exact short-time solver for the time-dependent Schrödinger equation. Journal of Chemical Physics. 2011;134:041101. DOI:10.1063/1.3549570

[47] Park TJ and Light JC, Unitary quantum time evolution by iterative Lanczos reduction. Journal of Chemical Physics. 1986;85:5870–5878. DOI:10.1063/1.451548

[48] Blanes S, Casas F, and Murua A, Symplectic splitting operator methods for the time-dependent Schrödinger equation. Journal of Chemical Physics. 2006;124:234105. DOI:10.1063/1.2203609

[49] Chen RQ and Guo H, Evolution of quantum system in order domain of Chebyshev operator. Journal of Chemical Physics. 1996;105:3569–3581. DOI:10.1063/1.472228

[50] Gray SK and Balint-Kurti GG, Quantum dynamics with real wave packets, including application to three-dimensional (J=0) D+H2→DH+H reactive scattering. Journal of Chemical Physics. 1998;108:950–962. DOI:10.1063/1.475495

[51] Kroes GJ and Neuhauser D, Performance of a time-independent scattering wave packet technique using real operators and wave functions. Journal of Chemical Physics. 1996;105:8690–8701. DOI:10.1063/1.472650

[52] Sun ZG, Guo H, and Zhang DH, Comparison of second-order split operator and Chebyshev propagator in wave packet based state-to-state reactive scattering calculations. Journal of Chemical Physics. 2009;130:174102. DOI:10.1063/1.3126363

[53] van Dijk W, Brown J, and Spyksma K, Efficiency and accuracy of numerical solutions to the time-dependent Schrodinger equation. Physical Review E. 2011;84:056703. DOI: 10.1103/PhysRevE.84.056703

[54] Blanes S, Casas F, Oteo JA, and Ros J, The Magnus expansion and some of its applications. Physics Reports. 2009;470:151–238. DOI:10.1016/j.physrep.2008.11.001

[55] Tal-Ezer H, Kosloff R, and Schaefer I, New, highly accurate propagator for the linear and nonlinear Schrodinger equation. Journal of Scientific Computation. 2012;53:211–221. DOI:10.1007/s10915-012-9583-x

[56] Sun Z, Zhao B, Liu S, and Zhang DH, Reactive scattering and resonance, in Gatti F, editor. Molecular Quantum Dynamics: From Theory to Applications. Berlin: Springer; 2014, pp. 81–116. DOI:10.1007/978-3-642-45290-1_4

[57] Schaefer TP, Siska PE, Parson JM, Tully FP, Wong YC, and Lee Y, Crossed molecular beam study of F + D_2. Journal of Chemical Physics. 1970;53:3385–3387. DOI: 10.1063/1.1674500

[58] Polanyi JC and Schreiber JL, The reaction of F + H_2 →HF + H. A case study in reaction dynamics. Faraday Discussions of the Chemical Society. 1977;62:267–290. DOI:10.1039/DC9776200267

[59] Dong F, Lee SH, and Liu K, Reactive excitation functions for F+p-H_2/n-H_2/D_2 and the vibrational branching for F + HD. Journal of Chemical Physics. 2000;113:3633–3640. DOI:10.1063/1.1287840

[60] Qiu M, Ren Z, Che L, Dai D, Harich SA, Wang X, Yang X, Xu C, Xie D, Gustafsson M, Skoedje RT, Sun Z, and Zhang D-H, Observation of feshbach resonances in the F + H2 → HF + H reaction. Science. 2006;311:1440–1443. DOI:10.1126/science.1123452

[61] Wang XA, Dong WR, Qiu MH, Ren ZF, Che L, Dai DX, Wang XY, Yang XM, Sun ZG, Fu B, Lee SY, Xu X, and Zhang DH, HF(v′ = 3) forward scattering in the F → H2 reaction: Shape resonance and slow-down mechanism. Proceedings of the National Academy of

Sciences of the United States of America. 2008;105:6227–6231. DOI:10.1073/pnas.0710840105

[62] Ren ZF, Che L, Qiu MH, Wang XA, Dong WR, Dai DX, Wang XY, Yang XM, Sun ZG, Fu B, Lee SY, Xu X, and Zhang DH, Probing the resonance potential in the f atom reaction with hydrogen deuteride with spectroscopic accuracy. Proceedings of the National Academy of Sciences of the United States of America. 2008;105:12662–12666. DOI:10.1073/pnas.0709974105

[63] Yang TG, Chen J, Huang L, Wang T, Xiao CL, Sun ZG, Dai DX, Yang XM, and Zhang DH, Extremely short-lived reaction resonances in Cl + HD (v = 1) → DCl + H due to chemical bond softening. Science. 2015;347:60–63. DOI:10.1126/science.1260527

[64] Fleurat-Lessard P, Grebenshchikov SY, Schinke R, Janssen C, and Krankowsky D, Isotope dependence of the $O+O_2$ exchange reaction: Experiment and theory. Journal of Chemical Physics. 2003;119:4700–4712. DOI:10.1063/1.1595091

[65] Gao YQ and Marcus RA, Strange and unconventional isotope effects in ozone formation. Science. 2001;293:259–263. DOI:10.1126/science.1058528

[66] Dawes R, Lolur P, Li A, Jiang B, and Guo H, Communication: An accurate global potential energy surface for the ground electronic state of ozone. Journal of Chemical Physics. 2013;139:201103. DOI:10.1063/1.4837175

[67] Li Y, Sun Z, Jiang B, Xie D, Dawes R, and Guo H, Communication: Rigorous quantum dynamics of $O + O_2$ exchange reactions on an ab initio potential energy surface substantiate the negative temperature dependence of rate coefficients. Journal of Chemical Physics. 2014;141:081102. DOI:10.1063/1.4894069

[68] Sun Z, Liu L, Lin SY, Schinke R, Guo H, and Zhang DH, State-to-state quantum dynamics of $O + O_2$ isotope exchange reactions reveals nonstatistical behavior at atmospheric conditions. Proceedings of the National Academy of Sciences of the United States of America. 2010;107:555–558. DOI:10.1073/pnas.0911356107

[69] Xie W, Liu L, Sun Z, Guo H, and Dawes R, State-to-state reaction dynamics of $^{18}O + ^{32}O_2$ studied by a time-dependent quantum wavepacket method. Journal of Chemical Physics. 2015;142:064308. DOI:10.1063/1.4907229

[70] Sun Z, Yu DQ, Xie WB, Hou JY, Dawes R, and Guo H, Kinetic isotope effect of the $^{16}O + ^{36}O_2$ and $^{18}O + ^{32}O_2$ isotope exchange reactions: Dominant role of reactive resonances revealed by an accurate time-dependent quantum wave packet study. Journal of Chemical Physics. 2015;142:174312. DOI:10.1063/1.4919861

[71] Siebert R, Schinke R, and Bittererova M, Spectroscopy of ozone at the dissociation threshold: Quantum calculations of bound and resonance states on a new global potential energy surface. Physical Chemistry Chemical Physics. 2001;3:1795–1798. DOI: 10.1039/B102830H

[72] Xiao CL, Xu X, Liu S, Wang T, Dong WR, Yang TG, Sun Z, Dai DX, Xu X, Zhang DH, and Yang XM, Experimental and theoretical differential cross sections for a four-atom reaction: HD+OH→H_2O+D. Science. 2011;333:440–442. DOI:10.1126/science.1205770

[73] Welsch R, Huarte-Larrañaga F, and Manthe U, State-to-state reaction probabilities within the quantum transition state framework. Journal of Chemical Physics. 2012;136:064117. DOI:10.1063/1.3684631

[74] Zhao B, Sun Z, and Guo H, Calculation of state-to-state differential and integral cross sections for atom-diatom reactions with transition-state wave packets. Journal of Chemical Physics. 2014;140:234110. DOI:10.1063/1.4883615

[75] Zhao B, Sun Z, and Guo H, Calculation of the state-to-state S-matrix for tetra-atomic reactions with transition-state wave packets: $H_2/D_2 + OH \rightarrow H/D + H_2O/HOD$. Journal of Chemical Physics. 2014;141:154112. DOI:10.1063/1.4898100

[76] Zhao B and Guo H, Modulations of transition-state control of state-to-state dynamics in the $F + H_2O \rightarrow HF + OH$ reaction. Journal of Physical Chemistry Letters. 2015;6:676–680. DOI:10.1021/acs.jpclett.5b00071

[77] Zhao B, Sun Z, and Guo H, Communication: State-to-state dynamics of the $Cl + H_2O \rightarrow HCl + OH$ reaction: Energy flow into reaction coordinate and transition-state control of product energy disposal. Journal of Chemical Physics. 2015;142:241101. DOI: 10.1063/1.4922650

[78] Welsch R and Manthe U, Loss of memory in $H + CH_4 \rightarrow H_2 + CH_3$ state-to-state reactive scattering. Journal of Physical Chemistry Letters. 2015;6:338–342. DOI:10.1021/jz502525p

[79] Zhao B, Sun Z, and Guo H, State-to-state mode specificity: Energy sequestration and flow gated by transition state. Journal of the American Chemical Society. 2015;137:15964–15970. DOI:10.1021/jacs.5b11404

[80] Miller WH, Quantum-mechanical transition-state theory and a new semiclassical model for reaction-rate constants. Journal of Chemical Physics. 1974;61:1823–1834. DOI: 10.1063/1.1682181

[81] Miller WH, Schwartz SD, and Tromp JW, Quantum-mechanical rate constants for bimolecular reactions. Journal of Chemical Physics. 1983;79:4889–4899. DOI: 10.1063/1.445581

[82] Miller WH, "Direct" and "correct" calculation of canonical and microcanonical rate constants for chemical reactions. Journal of Physical Chemistry A. 1998;102:793–806. DOI:10.1021/jp973208o

[83] Manthe U and Welsch R, Correlation functions for fully or partially state-resolved reactive scattering calculations. Journal of Chemical Physics. 2014;140:244113. DOI: 10.1063/1.4884716

[84] Park TJ and Light JC, Quantum flux operators and thermal rate constant: Collinear H $+H_2$. Journal of Chemical Physics. 1988;88:4897–4912. DOI:10.1063/1.454702

[85] Wang T, Chen J, Yang T, Xiao C, Sun Z, Huang L, Dai D, Yang X, and Zhang DH, Dynamical resonances accessible only by reagent vibrational excitation in the F + HD → HF + D reaction. Science. 2013;342:1499–1502. DOI:10.1126/science.1246546

[86] Gómez-Carrasco S and Roncero O, Coordinate transformation methods to calculate state-to-state reaction probabilities with wave packet treatments. Journal of Chemical Physics. 2006;125:054102. DOI:10.1063/1.2218337

[87] Sun Z, Guo H, and Zhang DH, Extraction of state-to-state reactive scattering attributes from wave packet in reactant Jacobi coordinates. Journal of Chemical Physics. 2010;132:084112. DOI:10.1063/1.3328109

[88] Dai JQ and Zhang JZH, Time-dependent wave packet approach to state-to-state reactive scattering and application to $H+O_2$ reaction. Journal of Physical Chemistry. 1996;100:6898. DOI:10.1021/jp9536662

[89] Sun Z, Lin X, Lee SY, and Zhang DH, A reactant-coordinate-based time-dependent wave packet method for triatomic state-to-state reaction dynamics: Application to the H + O_2 reaction. Journal of Physical Chemistry A. 2009;113:4145–4154. DOI:10.1021/jp810512j

[90] Zhao B, Sun Z, and Guo H, A reactant-coordinate-based wave packet method for full-dimensional state-to-state quantum dynamics of tetra-atomic reactions: Application to both the abstraction and exchange channels in the H + H_2O reaction. Journal of Chemical Physics. 2016;144:064104. DOI:10.1063/1.4941671

[91] Zhao B, Sun Z, and Guo H, Reactant-coordinate-based wave packet method for full-dimensional state-to-state quantum dynamics of tetra-atomic reactions: Application to both the abstraction and exchange channels in the H + H_2O reaction, Journal Chemical Physics. 2016;144:064104. DOI:10.1063/1.4941671

[92] Chorkendorff I and Niemantsverdriet JW, Concepts of Modern Catalysis and Kinetics. Weinheim: Wiley-VCH; 2003.

[93] Jiang B, Xie D, and Guo H, Vibrationally mediated bond selective dissociative chemisorption of HOD on Cu(111). Chemical Science. 2013;4:503–508. DOI:10.1039/C2SC21393A

[94] Jiang B, Li J, Xie D, and Guo H, Effects of reactant internal excitation and orientation on dissociative chemisorption of H_2O on Cu(111): Quasi-seven-dimensional quantum dynamics on a refined potential energy surface. Journal of Chemical Physics. 2013;138:044704. DOI:10.1063/1.4776770

[95] Jiang B, Ren X, Xie D, and Guo H, Enhancing dissociative chemisorption of H_2O on Cu(111) via vibrational excitation. Proceedings of the National Academy of Sciences of the United States of America. 2012;109:10224–10227. doi:10.1073/pnas.1203895109

[96] Hundt PM, Jiang B, van Reijzen, Maarten E, Guo H, and Beck RD, Vibrationally promoted dissociation of water on Ni(111). Science. 2014;344:504–507. DOI:10.1126/science.1251277

[97] Jiang B and Guo H, Quantum and classical dynamics of water dissociation on Ni(111): A test of the site-averaging model in dissociative chemisorption of polyatomic molecules. Journal of Chemical Physics. 2015;143:164705. DOI:10.1063/1.4934357

[98] Liu T, Zhang ZJ, Fu B, Yang X, and Zhang DH, A seven-dimensional quantum dynamicsstudy of the dissociative chemisorption of H_2O on Cu(111): Effects of azimuthal angles and azimuthal angle-averaging. Chemical Science. 2016;7:1840–1845. DOI:10.1039/C5SC03689E

[99] Liu T, Zhang ZJ, Fu B, Yang X, and Zhang DH, Mode specificity for the dissociative chemisorption of H_2O on Cu(111): A quantum dynamics study on an accurately fitted potential energy surface. Physical Chemistry Chemical Physics. 2016;18:8537–8544. DOI:10.1039/C6CP00034G

[100] Zhang ZJ, Liu T, Fu B, Yang X, and Zhang DH, First-principles quantum dynamical theory for the dissociative chemisorption of H_2O on rigid Cu(111). Nature Communation. Under revision, 2016.

[101] Lin XS and Sun ZG, Accurate mapped trigonometric discrete variable representations for Coulomb singularities in molecules: Applications with H2+ and H2 in cylindrical coordinates. Chemical Physics Letters. 2015;621:35–40. DOI:10.1016/j.cplett.2014.12.043

[102] Lin XS and Sun ZG, To be published.

[103] Weihe FA, Dutta SK, Korn G, Du D, Bucksbaum PH, and Shkolnikov PL, Polarization of high-intensity high-harmonic generation. Physical Review A. 1995;51:R3433. DOI:10.1103/PhysRevA.51.R3433

[104] Barth I, Manz J, Shigeta Y, and Yagi, K, Unidirectional electronic ring current driven by a few cycle circularly polarized laser pulse: Quantum model simulations for Mg-porphyrin. Journal of the American Chemical Society. 2006;128:7043. DOI:10.1021/ja057197l

[105] Djiokap JMN, Hu SX, Madsen LB, Manakov NL, Meremianin AV, and Starace AF, Electron vortices in photoionization by circularly polarized attosecond pulses. Physical Review Letters. 2015;115:113004. DOI:10.1103/PhysRevLett.115.113004

[106] Yuan KJ, Lu H, and Bandrauk AD, High-order-harmonic generation in molecular sequential double ionization by intense circularly polarized laser pulses. Physical Review A. 2015;92:023415. DOI:10.1103/PhysRevA.92.023415

[107] Yuan KJ and Bandrauk AD, Attosecond-magnetic-field-pulse generation by electronic currents in bichromatic circularly polarized UV laser fields. Physical Review A. 2015;92:023415. DOI:10.1103/PhysRevA.92.063401

[108] Bandrauk AD and Lu HZ, High-Dimensional Partial Differential Equations in Science and Engineering, CRM Lecture Series, Vol. 41. Philadelphia: American Math. Soc.; 2007, pp. 1–15.

[109] Bandrauk AD and Shen H, Exponential split operator methods for solving coupled time-dependent Schrödinger equations. Journal of Chemical Physics. 993;99:1185. DOI: 10.1063/1.465362

[110] Bandrauk AD and Lu HZ, Exponential propagators (integrators) for the time-dependent Schrödinger equation. Journal of Theoretical and Computational Chemistry. 2013;12:1340001. DOI:10.1142/S0219633613400014

[111] Zhang DH and Zhang JZH, Quantum reactive scattering with a deep well: Time-dependent calculation for $H+O_2$ reaction and bound state characterization for HO_2. Journal of Chemical Physics. 1994;101:3671. DOI:10.1063/1.467551

7

Quantum Walks

Takuya Machida

Abstract

Quantum walks are quantum counterparts of random walks. While probability distributions of random walks diffusively spread out as the walkers are updating, quantum walks have ballistic behavior. Some of the ballistic behaviors have been revealed in long-time limit theorems and their probability distributions are all far away from the Gaussian distributions, which are known as limit distributions of random walks. In this chapter, we are going to be seeing limit distributions for a standard quantum walk on the line and two kinds of time-dependent quantum walk on the line.

Keywords: Quantum walk, limit theorem, Fourier analysis, time-dependent walk, line

1. Introduction

Quantum walks (QWs) are mathematical models on graphs whose systems repeatedly update according to time-evolution rules. They have been in an emerging field which describes the quantum world. Experts in mathematics, physics, and information theory have been interested in them and to study QWs, and a lot of fascinating properties of QWs have been discovered. Historically, QWs were independently introduced in science from several view points; mathematics in 1988 [1], physics in 1993 [2], and computer science in 1996 [3]. After a while, they began to get attention around 2000. Since QWs can be considered as quantum counterparts of random walks in mathematics, they are also called quantum random walks. The dynamics of QWs are similar to those of random walks in mathematical terms. But, whereas a random walker moves on a graph at random, a quantum walker spreads out as a wave on a graph. Although random walks are stochastic processes, QWs are different. They are unitary processes because the systems of quantum walkers get updated with unitary operators. In quantum physics, the update rules of QWs are interpreted as discretized models of Dirac equations. High dimensional Dirac equations are hard to solve even in numerics due to their complexities, and then we expect QWs to become alternative systems to solve the equations on computer. QWs also play an important role in quantum computers because they are quantum algorithms themselves. Indeed, some quantum algorithms based on QWs show quadratic speed-up, compared to the corresponding classical algorithms [4]. Such algorithms imply that quantum computers could give rise to excellent performance.

In this chapter, we are going to be seeing mathematical aspects of the QWs, which will be described as limit theorems. We first observe a standard QW on the line in Sec.2 Then we shift our focus to time-dependent QWs on the line in Secs. 3 and 4.

2. A quantum walk on the line

We start off with the description of a standard QW on the line. The system of the QW is defined in a tensor space of two Hilbert spaces. One is a Hilbert space $\mathcal{H}_p$ spanned by an orthogonal normalized basis $\{|x\rangle : x \in \mathbb{Z}\}$, the other is a Hilbert space $\mathcal{H}_c$ spanned by an orthogonal normalized basis $\{|0\rangle, \{|1\rangle\}$. We can consider $|0\rangle$ and $|1\rangle$ in the Hilbert space $\mathcal{H}_c$ as the down-spin and the up-spin states of a quantum particle, respectively. The Hilbert space $\mathcal{H}_p$ is called a position space and the Hilbert space $\mathcal{H}_c$ is called a coin space. A QW on the line at time t (=0, 1, 2 ...) is expressed in the tensor Hilbert space,

$$|\Psi_t\rangle = \sum_{x\in\mathbb{Z}} |x\rangle \otimes |\psi_t(x)\rangle \in \mathcal{H}_p \otimes \mathcal{H}_c. \tag{1}$$

Customarily, $|\psi_t(x)\rangle \in \mathcal{H}_c$ is called a coin state or an amplitude at position x at time t. Given an initial state $|\Psi_0\rangle$, the walker is repeatedly updating,

$$|\Psi_{t+1}\rangle = SC|\Psi_t\rangle, \tag{2}$$

with

$$C = \sum_{x\in\mathbb{Z}} |x\rangle\langle x| \otimes U, \tag{3}$$

$$S = \sum_{x\in\mathbb{Z}} |x-1\rangle\langle x| \otimes |0\rangle\langle 0| + |x+1\rangle\langle x| \otimes |1\rangle\langle 1|, \tag{4}$$

where U is a unitary operator. In this chapter, we employ a form of the operator

$$U = \cos\theta |0\rangle\langle 0| + \sin\theta |0\rangle\langle 1| + \sin\theta |1\rangle\langle 0| - \cos\theta |1\rangle\langle 1| \quad (\theta \in [0, 2\pi)), \tag{5}$$

and an initial state

$$|\Psi_0\rangle = |0\rangle \otimes (\alpha|0\rangle + \beta|1\rangle), \tag{6}$$

assuming the complex numbers α and β satisfy the condition $|\alpha|^2 + |\beta|^2 = 1$. The reason that we assigned unitarity to the operator U and assumed the constraint $|\alpha|^2 + |\beta|^2 = 1$ is that we define the probability that the quantum walker is observed at position x at time t,

$$\mathbb{P}(X_t = x) = \langle\Psi_t| \left(|x\rangle\langle x| \otimes \sum_{j=0}^{1} |j\rangle\langle j| \right) |\Psi_t\rangle, \tag{7}$$

where the random variable X_t is regarded as the position of the walker at time t. Thanks to the assignment and the constraint, the right side of Eq. (7) certainly becomes a probability distribution. **Figure 1** shows the probability distribution $\mathbb{P}(X_t = x)$ at time 500 when the walker starts updating with $\alpha = 1/\sqrt{2}$ and $\beta = i/\sqrt{2}$. In these pictures, only positive values of the probability are plotted. While random walkers normally show diffusive behavior, it is known that the quantum walker acts ballistic as time goes up. We see for certain the ballistic behavior of the probability distribution $\mathbb{P}(X_t = x)$ in **Figure 2** when we take $\alpha = 1/\sqrt{2}$ and $\beta = i/\sqrt{2}$. As shown in **Figure 1**, the probability distribution $\mathbb{P}(X_t = x)$ holds two sharp peaks and where they occur strongly depends on the value of the parameter θ. We get more detailed information about that from **Figure 3**.

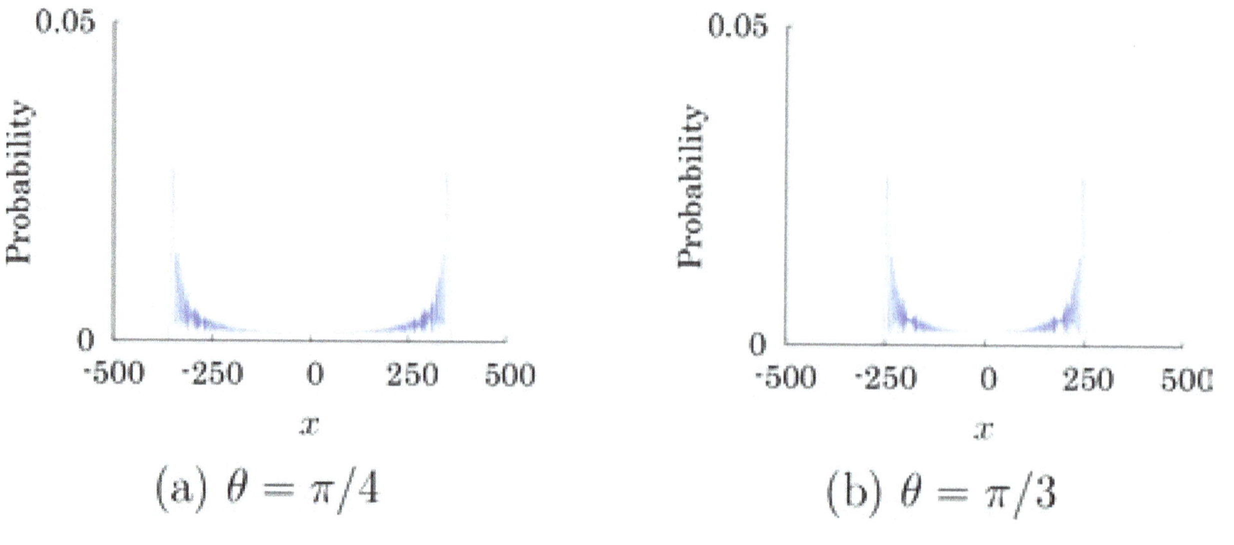

Figure 1. Probability distribution at time 500.

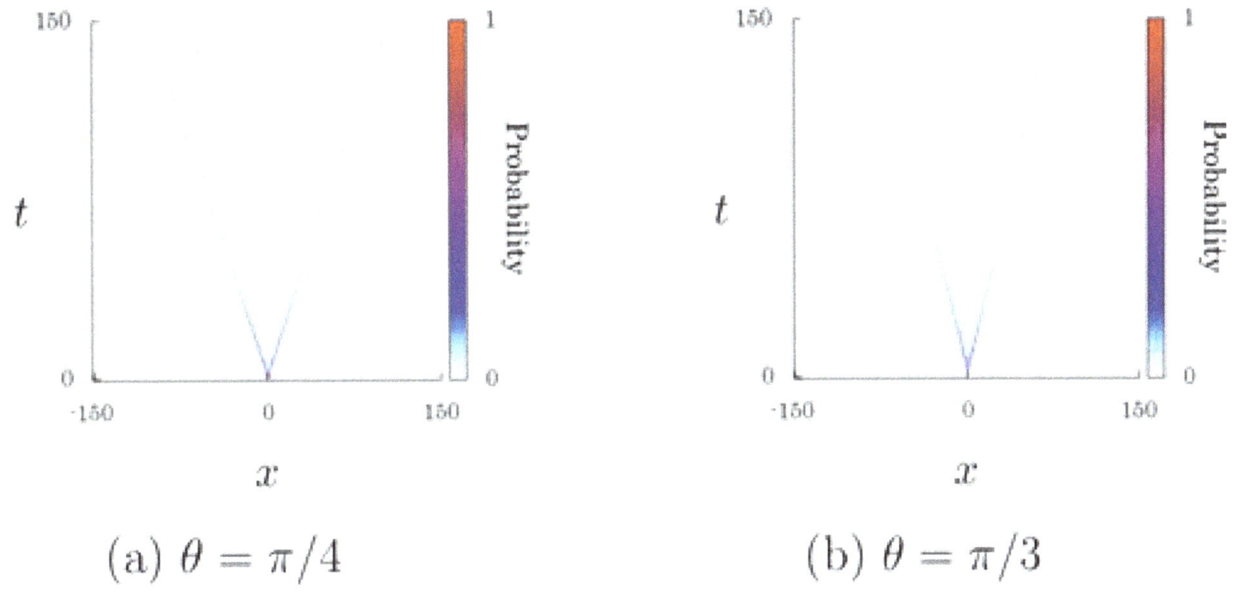

Figure 2. Time evolution of the probability distribution $\mathbb{P}(X_t = x)$.

The definition of the standard QW on the line has been done. So, what are we curious about? One of the major studies on QWs is to know how their probability distributions behave after they have updated a lot of times. For the probability distribution of the QW defined in this section, one can assert a limit theorem which tells us an approximate behavior of the probability distribution $\mathbb{P}(X_t = x)$ after time t goes enough up. First, we see a limit theorem (Theorem 1) when the value of the parameter θ, which is embedded in the operator U, is picked in the interval $[0, \pi)$. Then we will extend it for $\theta \in [0, 2\pi)$ (Theorem 2) which is easily proved by making the most of Theorem 1.

Theorem 1*Assume that $\theta \in [0, \pi)$ and $\theta \neq 0, \pi/2$. For a real number x, we have*

$$\begin{aligned}\lim_{t\to\infty} \mathbb{P}\left(\frac{X_t}{t} \leq x\right) = \int_{-\infty}^{x} &\frac{\sin\theta}{\pi(1-y^2)\sqrt{\cos^2\theta - y^2}} \\ &\times\left[1 - \left\{|\alpha|^2 - |\beta|^2 + \frac{\sin\theta(\alpha\overline{\beta} + \overline{\alpha}\beta)}{\cos\theta}\right\} y\right] I_{(-|\cos\theta|, |\cos\theta|)}(y)\,dy,\end{aligned} \tag{8}$$

where $I_A(x)$ is an indicator function such that

$$I_A(x) = \begin{cases} 1 & (x \in A) \\ 0 & (otherwise) \end{cases}. \tag{9}$$

This limit theorem, to be exact which was a theorem for a general unitary operator $U \in U(2)$, was proven for the first time by a combinatorial method in 2002 [5]. It was also possible to obtain by Fourier analysis introduced by Grimmett et al. [6]. Here we use the second method to derive the limit theorem. Let $|\hat{\psi}_t(k)\rangle$ be the Fourier transform of the quantum walker in the form

$$|\hat{\psi}_t(k)\rangle = \sum_{x \in \mathbb{Z}} e^{-ikx} |\psi_t(x)\rangle \quad (k \in [-\pi, \pi)). \tag{10}$$

Oppositely, we can obtain the coin states by inverse Fourier transformation,

$$|\psi_t(x)\rangle = \frac{1}{2\pi} \int_{-\pi}^{\pi} e^{ikx} |\hat{\psi}_t(k)\rangle dk. \tag{11}$$

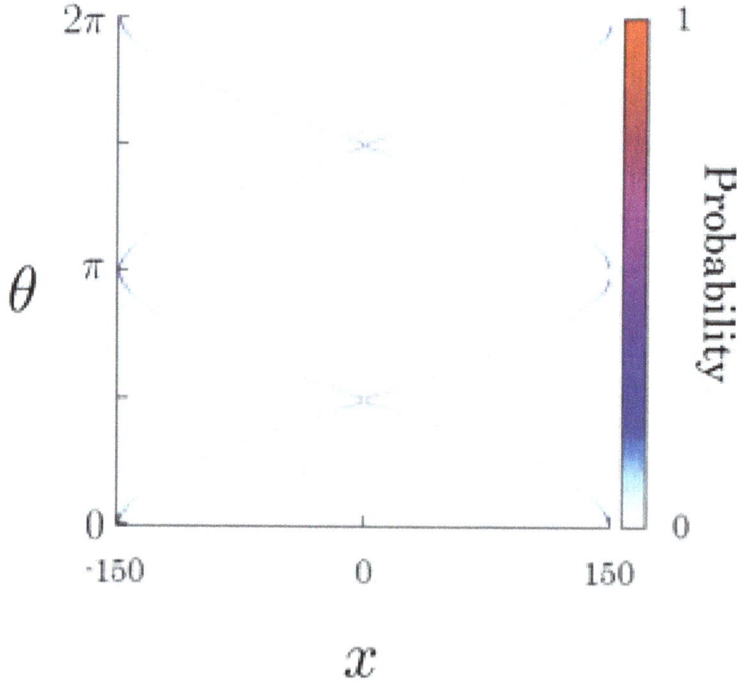

Figure 3. This picture shows how the probability distribution at time 150 depends on the value of the parameter θ. ($\alpha = 1/\sqrt{2},\ \beta = i/\sqrt{2}$).

Equation (2) gives the evolution of the Fourier transform,

$$|\hat{\psi}_{t+1}(k)\rangle = R(k)U|\hat{\psi}_t(k)\rangle, \tag{12}$$

with

$$R(k) = e^{ik} \mid 0\rangle \langle 0 \mid + e^{-ik} \mid 1\rangle \langle 1 \mid. \tag{13}$$

The iteration by Eq. (12) connects the system at time t to the initial state,

$$\mid \hat{\psi}_t(k)\rangle = (R(k)U)^t \mid \hat{\psi}_0(k)\rangle. \tag{14}$$

To prove the theorem, we concentrate on a convergence

$$\lim_{t\to\infty} \mathbb{E}[(X_t / t)^r] = \int_{-\infty}^{\infty} x^r f(x) dx \quad (r = 0,1,2,\ldots), \tag{15}$$

with

$$\begin{aligned} f(x) = & \frac{\sin\theta}{\pi(1-x^2)\sqrt{\cos^2\theta - x^2}} \\ & \times \left[1 - \left\{ \mid \alpha \mid^2 - \mid \beta \mid^2 + \frac{\sin\theta(\alpha\overline{\beta} + \overline{\alpha}\beta)}{\cos\theta} \right\} x \right] I_{(-|\cos\theta|, |\cos\theta|)}(x). \end{aligned} \tag{16}$$

It is known from probability theory that the convergence guarantees Eq. (8). Before we compute the limit, let us depict the r-th moment $\mathsf{E}(X_t^r) = \sum_{x=-\infty}^{\infty} x^r \mathbb{P}(X_t = x)$ in Fourier picture. Since the initial state is given by the form of Eq. (6), the Fourier transform at time t is rewritten in a finite sum

$$\mid \hat{\psi}_t(k)\rangle = \sum_{x=-\infty}^{\infty} e^{-ikx} \mid \psi_t(x)\rangle = \sum_{x=-t}^{t} e^{-ikx} \mid \psi_t(x)\rangle. \tag{17}$$

Noting that

$$\frac{d^r}{dk^r} \mid \hat{\psi}_t(k)\rangle = \sum_{x=-t}^{t} (-ix)^r e^{-ikx} \mid \psi_t(x)\rangle, \tag{18}$$

we have

$$\begin{aligned}\langle\hat{\psi}_t(k)|\,i^r\frac{d^r}{dk^r}|\hat{\psi}_t(k)\rangle &= \Big(\sum_{x=-t}^{t}e^{ikx}\langle\psi_t(x)|\Big)\Big(\sum_{y=-t}^{t}y^r e^{-iky}|\psi_t(y)\rangle\Big)\\ &= \sum_{x=-t}^{t}\sum_{y=-t}^{t}x^r e^{ik(x-y)}\langle\psi_t(x)|\psi_t(y)\rangle.\end{aligned} \tag{19}$$

Integrating Eq. (19) over the interval $[-\pi,\pi)$, one can generate the r-th moment of the random variable X_t,

$$\begin{aligned}&\int_{-\pi}^{\pi}\langle\hat{\psi}_t(k)|\,i^r\frac{d^r}{dk^r}|\hat{\psi}_t(k)\rangle dk = \int_{-\pi}^{\pi}\sum_{x=-t}^{t}\sum_{y=-t}^{t}y^r e^{ik(x-y)}\langle\psi_t(x)|\psi_t(y)\rangle dk\\ &= 2\pi\sum_{x=-t}^{t}x^r\langle\psi_t(x)|\psi_t(x)\rangle = 2\pi\sum_{x=-\infty}^{\infty}x^r\langle\psi_t(x)|\psi_t(x)\rangle\\ &= 2\pi\sum_{x=-\infty}^{\infty}x^r\mathbb{P}(X_t=x) = 2\pi\mathbb{E}(X_t^r),\end{aligned} \tag{20}$$

from which the r-th moment results in a representation in Fourier picture,

$$\mathbb{E}(X_t^r) = \frac{1}{2\pi}\int_{-\pi}^{\pi}\langle\hat{\psi}_t(k)|\,i^r\frac{d^r}{dk^r}|\hat{\psi}_t(k)\rangle dk. \tag{21}$$

Here, let $\lambda_j(k)\,(j=1,2)$ be the eigenvalues of the matrix $R(k)U$, and $|v_j(k)\rangle$ be the normalized eigenvector associated to the eigenvalue $\lambda_j(k)$. Then the initial state of the Fourier transform is decomposed by the normalized eigenvectors,

$$|\hat{\psi}_0(k)\rangle = \langle v_1(k)|\hat{\psi}_0(k)\rangle|v_1(k)\rangle + \langle v_2(k)|\hat{\psi}_0(k)\rangle|v_2(k)\rangle. \tag{22}$$

The Fourier transform at time t is, therefore, expressed with the eigenvalues and the eigenvectors,

$$|\hat{\psi}_t(k)\rangle = (R(k)U)^t|\hat{\psi}_0(k)\rangle = \sum_{j=1}^{2}\lambda_j(k)^t\langle v_j(k)|\hat{\psi}_0(k)\rangle|v_j(k)\rangle, \tag{23}$$

which also gives a description of the derivative

$$\frac{d^r}{dk^r}|\hat{\psi}_t(k)\rangle = (t)_r\cdot\sum_{j=1}^{2}\lambda_j(k)^{t-r}\lambda_j'(k)^r\langle v_j(k)|\hat{\psi}_0(k)\rangle|v_j(k)\rangle + O(t^{r-1}), \tag{24}$$

where $(t)_r = t(t-1)(t-2) \times \cdots \times (t-r+1) = \prod_{j=t-r+1}^{t} j$ and $\lambda_j'(k) = (d/dk)\lambda_j(k)$. Recalling the initial state in Eq. (6), we now have the Fourier transform at time 0 in the form

$$|\hat{\psi}_0(k)\rangle = \alpha|0\rangle + \beta|1\rangle. \tag{25}$$

Equations (23) and (24) change the integral picture in Eq. (21),

$$\begin{aligned}\mathbb{E}(X_t^r) &= \frac{1}{2\pi}\int_{-\pi}^{\pi}\langle\hat{\psi}_t(k)|\, i^r \frac{d^r}{dk^r}|\hat{\psi}_t(k)\rangle dk \\ &= (t)_r \cdot \sum_{j=1}^{2}\frac{1}{2\pi}\int_{-\pi}^{\pi}\left(\frac{i\lambda_j'(k)}{\lambda_j(k)}\right)^r |\langle v_j(k)|\hat{\psi}_0(k)\rangle|^2 dk + O(t^{r-1}).\end{aligned} \tag{26}$$

Dividing Eq. (26) by t^r, we reach a representation

$$\frac{\mathbb{E}(X_t^r)}{t^r} = \frac{(t)_r}{t^r} \cdot \sum_{j=1}^{2}\frac{1}{2\pi}\int_{-\pi}^{\pi}\left(\frac{i\lambda_j'(k)}{\lambda_j(k)}\right)^r |\langle v_j(k)|\hat{\psi}_0(k)\rangle|^2 dk + \frac{O(t^{r-1})}{t^r}, \tag{27}$$

and obtain a convergence as $t \to \infty$,

$$\lim_{t\to\infty}\mathbb{E}\left[\left(\frac{X_t}{t}\right)^r\right] = \lim_{t\to\infty}\frac{\mathbb{E}(X_t^r)}{t^r} = \sum_{j=1}^{2}\frac{1}{2\pi}\int_{-\pi}^{\pi}\left(\frac{i\lambda_j'(k)}{\lambda_j(k)}\right)^r |\langle v_j(k)|\hat{\psi}_0(k)\rangle|^2 dk. \tag{28}$$

Now what we need more has made sense. It is the eigensystem of the operator $R(k)U$. To make a computation about it, we give a standard basis to the Hilbert space $\mathcal{H}_c$,

$$|0\rangle = \begin{bmatrix}1\\0\end{bmatrix}, \quad |1\rangle = \begin{bmatrix}0\\1\end{bmatrix}, \tag{29}$$

from which a matrix representation follows,

$$R(k)U = \begin{bmatrix} e^{ik}\cos\theta & e^{ik}\sin\theta \\ e^{-ik}\sin\theta & -e^{-ik}\cos\theta \end{bmatrix}. \tag{30}$$

The matrix contains two different eigenvalues

$$\lambda_j(k) = -(-1)^j \sqrt{1-c^2\sin^2 k} + ic\sin k \quad (j=1,2), \tag{31}$$

in which c and s are short for $cos\theta$ and $sin\theta$, respectively. Differentiating Eq. (31) with respect to the variable k, we get

$$\lambda_{j'}(k) = -(-1)^j \frac{ic\cos k}{\sqrt{1-c^2\sin^2 k}} \left\{-(-1)^j \sqrt{1-c^2\sin^2 k} + ic\sin k\right\}, \tag{32}$$

from which the function $i\lambda_j'(k)/\lambda_j(k)$ is computed,

$$\frac{i\lambda_j'(k)}{\lambda_j(k)} = (-1)^j \frac{c\cos k}{\sqrt{1-c^2\sin^2 k}}. \tag{33}$$

The normalized eigenvector associated to the eigenvalue $\lambda_j(k)$ has a form

$$|v_j(k)\rangle = \frac{1}{\sqrt{N_j(k)}} \begin{bmatrix} se^{ik} \\ -c\cos k - (-1)^j \sqrt{1-c^2\sin^2 k} \end{bmatrix}, \tag{34}$$

with its normalized factor

$$N_j(k) = 2\sqrt{1-c^2\sin^2 k}\left\{\sqrt{1-c^2\sin^2 k} + (-1)^j c\cos k\right\}. \tag{35}$$

Back in Eq. (28), we put $i\lambda_j'(k)/\lambda_j(k) = x\,(j = 1, 2)$ and then obtain another integral form

$$
\begin{aligned}
\lim_{t\to\infty} \mathbb{E}\left[\left(\frac{X_t}{t}\right)^r\right] &= \sum_{j=1}^{2} \frac{1}{2\pi}\int_{-\pi}^{\pi}\left(\frac{i\lambda_j'(k)}{\lambda_j(k)}\right)^r \left|\langle v_j(k)|\hat{\psi}_0(k)\rangle\right|^2 dk \\
&= \int_{-c}^{c} x^r \frac{|c|s}{\pi c(1-x^2)\sqrt{c^2-x^2}}\left[1-\left\{|\alpha|^2-|\beta|^2+\frac{s(\alpha\overline{\beta}+\overline{\alpha}\beta)}{c}\right\}x\right]dx \\
&= \int_{-|c|}^{|c|} x^r \frac{s}{\pi(1-x^2)\sqrt{c^2-x^2}}\left[1-\left\{|\alpha|^2-|\beta|^2+\frac{s(\alpha\overline{\beta}+\overline{\alpha}\beta)}{c}\right\}x\right]dx \\
&= \int_{-\infty}^{\infty} x^r \frac{s}{\pi(1-x^2)\sqrt{c^2-x^2}} \\
&\quad\times\left[1-\left\{|\alpha|^2-|\beta|^2+\frac{s(\alpha\overline{\beta}+\overline{\alpha}\beta)}{c}\right\}x\right] I_{(-|c|,|c|)}(x)dx.
\end{aligned} \tag{36}
$$

Equation (36) allows us to hold Eq. (8).

Now that we have obtained a limit theorem for the QW whose operator was defined by

$$
U = \cos\theta|0\rangle\langle 0| + \sin\theta|0\rangle\langle 1| + \sin\theta|1\rangle\langle 0| - \cos\theta|1\rangle\langle 1| \quad (\theta\in[0,\pi)), \tag{37}
$$

Theorem 1 can be extended for the parameter $\theta \in [0, 2\pi)$.

Theorem 2*Assume that $\theta \in [0, 2\pi)$ and $\theta \neq 0, \pi/2, \pi, 3\pi/2$. For a real number x, we have*

$$
\begin{aligned}
\lim_{t\to\infty} \mathbb{P}\left(\frac{X_t}{t}\le x\right) &= \int_{-\infty}^{x} \frac{|\sin\theta|}{\pi(1-y^2)\sqrt{\cos^2\theta-y^2}} \\
&\times\left[1-\left\{|\alpha|^2-|\beta|^2+\frac{\sin\theta(\alpha\overline{\beta}+\overline{\alpha}\beta)}{\cos\theta}\right\}y\right] I_{(-|\cos\theta|,|\cos\theta|)}(y)dy.
\end{aligned} \tag{38}
$$

Since we already had the limit theorem for the parameter $\theta \in [0, \pi)$ as Theorem 1, it is enough to prove Theorem 2 for $\theta \in [\pi, 2\pi)$ ($\theta \neq \pi, 3\pi/2$). Do you think we have to carry out the same calculation for such a parameter again? We do not actually have to do that and can avoid the same math by applying a small skill to the operator U. Let us slightly change the form of the operator U, which is described by $U(\theta)$ below,

$$
\begin{aligned}
U(\theta) := U &= \cos\theta|0\rangle\langle 0| + \sin\theta|0\rangle\langle 1| + \sin\theta|1\rangle\langle 0| - \cos\theta|1\rangle\langle 1| \\
&= -\{\cos(\theta-\pi)|0\rangle\langle 0| + \sin(\theta-\pi)|0\rangle\langle 1| \\
&\quad + \sin(\theta-\pi)|1\rangle\langle 0| - \cos(\theta-\pi)|1\rangle\langle 1|\} = -U(\theta-\pi).
\end{aligned} \tag{39}
$$

The negative sign in front of the operator $U(\theta - \pi)$ in Eq. (39) does not affect the probability distribution $\mathbb{P}(X_t = x)$, which means the probability distribution given by the operator $U(\theta)$ is completely same as that given by the operator $U(\theta - \pi)$. Moreover, as long as the parameter θ picks a value in the interval $[\pi, 2\pi)$, the variable $\theta - \pi$ stays in the interval $[0, \pi)$. Since Theorem 1 works on the QW operated by $U(\theta - \pi)(\theta \in [\pi, 2\pi))$, one can assert a limit theorem for the parameter $\theta \in [\pi, 2\pi)$,

$$\begin{aligned}
&\lim_{t\to\infty} \mathbb{P}\left(\frac{X_t}{t} \le x\right) \\
&= \int_{-\infty}^{x} \frac{\sin(\theta-\pi)}{\pi(1-y^2)\sqrt{\cos^2(\theta-\pi)-y^2}} \left[1 - \left\{|\alpha|^2 - |\beta|^2 \right.\right. \\
&\qquad\qquad \left.\left. + \frac{\sin(\theta-\pi)(\alpha\overline{\beta}+\overline{\alpha}\beta)}{\cos(\theta-\pi)}\right\} y\right] I_{(-|\cos(\theta-\pi)|, |\cos(\theta-\pi)|)}(y)\,dy \qquad (40) \\
&= \int_{-\infty}^{x} \frac{-\sin\theta}{\pi(1-y^2)\sqrt{\cos^2\theta - y^2}} \\
&\times \left[1 - \left\{|\alpha|^2 - |\beta|^2 + \frac{\sin\theta(\alpha\overline{\beta}+\overline{\alpha}\beta)}{\cos\theta}\right\} y\right] I_{(-|\cos\theta|, |\cos\theta|)}(y)\,dy
\end{aligned}$$

We should remark that Eq. (40) holds under the condition $\theta - \pi \neq 0, \pi/2$, that is, $\theta \neq \pi, 3\pi/2$. As a consequence of Theorem 1 and Eq. (40), Theorem 2 comes up.

Figure 4 shows an example of the limit density function.

$$\begin{aligned}
&\frac{d}{dx} \lim_{t\to\infty} \mathbb{P}\left(\frac{X_t}{t} \le x\right) \\
&= \frac{|\sin\theta|}{\pi(1-x^2)\sqrt{\cos^2\theta - x^2}} \qquad (41) \\
&\times \left[1 - \left\{|\alpha|^2 - |\beta|^2 + \frac{\sin\theta(\alpha\overline{\beta}+\overline{\alpha}\beta)}{\cos\theta}\right\} x\right] I_{(-|\cos\theta|, |\cos\theta|)}(x)\,dy.
\end{aligned}$$

We see that the limit density function reproduces the features of the probability distribution shown in **Figure 1**.

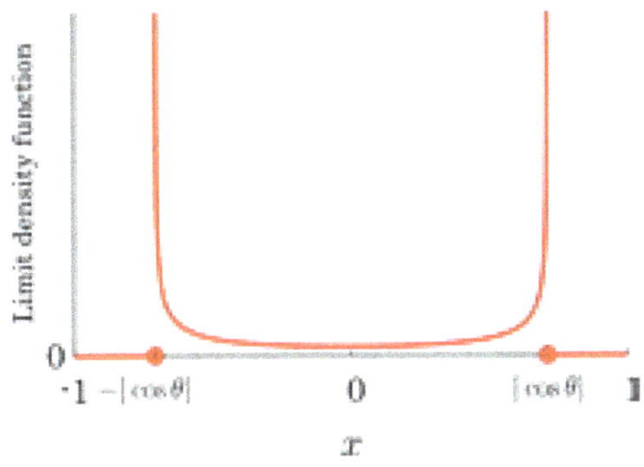

Figure 4. The limit density function ($\alpha = 1/\sqrt{2},\ \beta = i/\sqrt{2}$).

3. Two-period time-dependent QW

In this section, we see a time-dependent QW whose coin-flip operator depends on time. The evolution of the QW is given by two unitary operators,

$$\begin{aligned} U_1 &= \cos\theta_1 |0\rangle\langle 0| + \sin\theta_1 |0\rangle\langle 1| + \sin\theta_1 |1\rangle\langle 0| - \cos\theta_1 |1\rangle\langle 1| \\ &= c_1 |0\rangle\langle 0| + s_1 |0\rangle\langle 1| + s_1 |1\rangle\langle 0| - c_1 |1\rangle\langle 1|, \end{aligned} \tag{42}$$

$$\begin{aligned} U_2 &= \cos\theta_2 |0\rangle\langle 0| + \sin\theta_2 |0\rangle\langle 1| + \sin\theta_2 |1\rangle\langle 0| - \cos\theta_2 |1\rangle\langle 1| \\ &= c_2 |0\rangle\langle 0| + s_2 |0\rangle\langle 1| + s_2 |1\rangle\langle 0| - c_2 |1\rangle\langle 1|, \end{aligned} \tag{43}$$

with $\theta_j \in [0, 2\pi)\,(j = 1, 2)$, and $cos\theta_j$ (resp. $sin\theta_j$) has been briefly written as c_j (resp. s_j). The total system at time t evolves to the next state at time $t + 1$ according to the time evolution rule

$$|\Psi_{t+1}\rangle = \begin{cases} SC_1 |\Psi_t\rangle & (t = 0,2,4,\ldots) \\ SC_2 |\Psi_t\rangle & (t = 1,3,5,\ldots) \end{cases}, \tag{44}$$

where

$$C_j = \sum_{x\in\mathbb{Z}} |x\rangle\langle x| \otimes U_j \quad (j = 1,2). \tag{45}$$

It is plane that the operators C_1 and C_2 are alternately casted on the QW, which means that the unitary operator 2-periodically changes in time-line. If the parameters θ_1 and θ_2 take the same value, then the QW becomes the standard walk defined in Sec.

Now, we are looking at examples of the probability distribution when the walker starts off with $|\Psi_0\rangle = |0\rangle \otimes (1/\sqrt{2}|0\rangle + i/\sqrt{2}|1\rangle)$. **Figure 5** draws the probability distribution at time 500 and two sharp peaks are observed in each picture. In the pictures, only positive values of probability are plotted. As shown in **Figure 6**, the probability distribution is spreading in proportion to time t. We also see how it depends on the parameters θ_1 and θ_2 in **Figure 7**.

The features, which we have seen in **Figures 5** to **7** are caught by a limit theorem.

Theorem 3*Assume that $\theta_1, \theta_2 \neq 0, \pi/2, \pi, 3\pi/2$. For a real number x, we have*

$$\lim_{t\to\infty} \mathbb{P}\left(\frac{X_t}{t} \le x\right) = \int_{-\infty}^{x} \frac{\sqrt{1-\xi(\theta_1,\theta_2)^2}}{\pi(1-y^2)\sqrt{\xi(\theta_1,\theta_2)^2 - y^2}} \times \left[1 - \left\{|\alpha|^2 - |\beta|^2 + \frac{\sin\theta_1(\alpha\overline{\beta} + \overline{\alpha}\beta)}{\cos\theta_1}\right\} y\right] I_{(-\xi(\theta_1,\theta_2),\xi(\theta_1,\theta_2))}(y)dy, \tag{46}$$

with $\xi(\theta_1, \theta_2) = min\{|cos\theta_1|, |cos\theta_2|\}$.

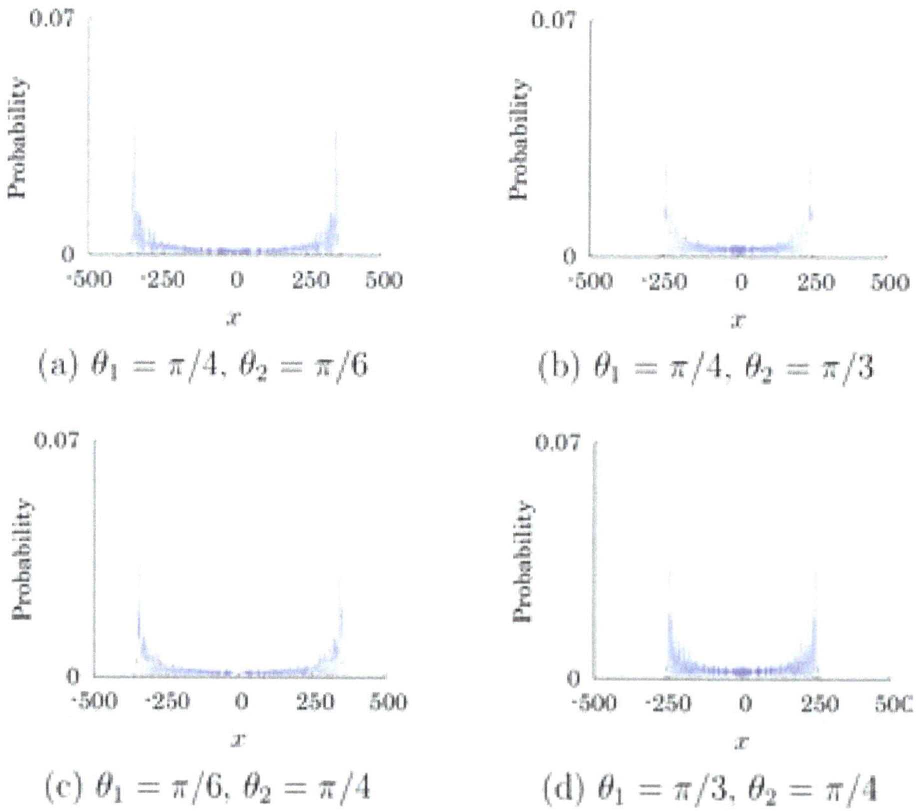

Figure 5. Probability distribution at time 500 ($\alpha = 1/\sqrt{2},\ \beta = i/\sqrt{2}$).

This limit theorem was proved by Fourier analysis in 2010 [7]. First, we find the time evolution of the Fourier transform $|\hat{\psi}_t(k)\rangle = \sum_{x\in\mathbb{Z}} e^{-ikx}|\psi_t(x)\rangle (k \in [-\pi, \pi))$,

$$|\hat{\psi}_{t+1}(k)\rangle = \begin{cases} R(k)U_1 |\hat{\psi}_t(k)\rangle & (t = 0,2,4,\ldots) \\ R(k)U_2 |\hat{\psi}_t(k)\rangle & (t = 1,3,5,\ldots) \end{cases}, \tag{47}$$

which comes from Eq. (44). We should recall $R(k) = e^{ik}|0\rangle\langle 0| + e^{-ik}|1\rangle\langle 1|$. From the recurrence, the transform at each time gets a connection to its initial state,

$$|\hat{\psi}_{2t}(k)\rangle = (R(k)U_2R(k)U_1)^t |\hat{\psi}_0(k)\rangle, \tag{48}$$

$$|\hat{\psi}_{2t+1}(k)\rangle = R(k)U_1(R(k)U_2R(k)U_1)^t |\hat{\psi}_0(k)\rangle. \tag{49}$$

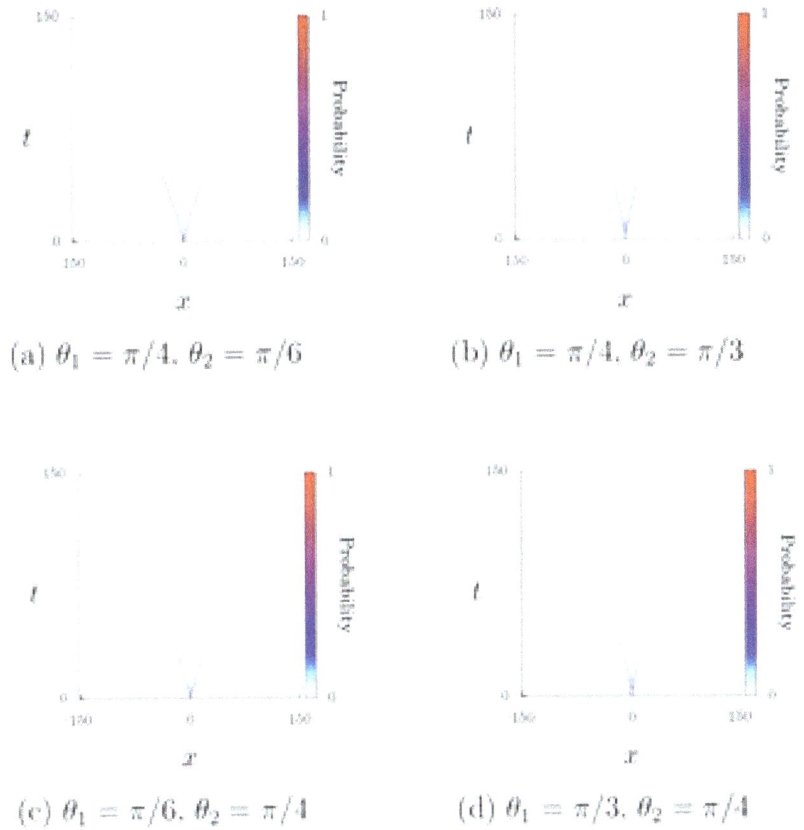

Figure 6. Time evolution of the probability distribution ($\alpha = 1/\sqrt{2},\ \beta = i/\sqrt{2}$).

The eigensystem of the matrix $R(k)U_2R(k)U_1$ reforms Eqs. (48) and (49),

$$|\hat{\psi}_{2t}(k)\rangle = \sum_{j=1}^{2} \lambda_j(k)^t \langle v_j(k)|\hat{\psi}_0(k)\rangle |v_j(k)\rangle, \tag{50}$$

$$|\hat{\psi}_{2t+1}(k)\rangle = R(k)U_1\sum_{j=1}^{2}\lambda_j(k)^t\langle v_j(k)|\hat{\psi}_0(k)\rangle|v_j(k)\rangle. \tag{51}$$

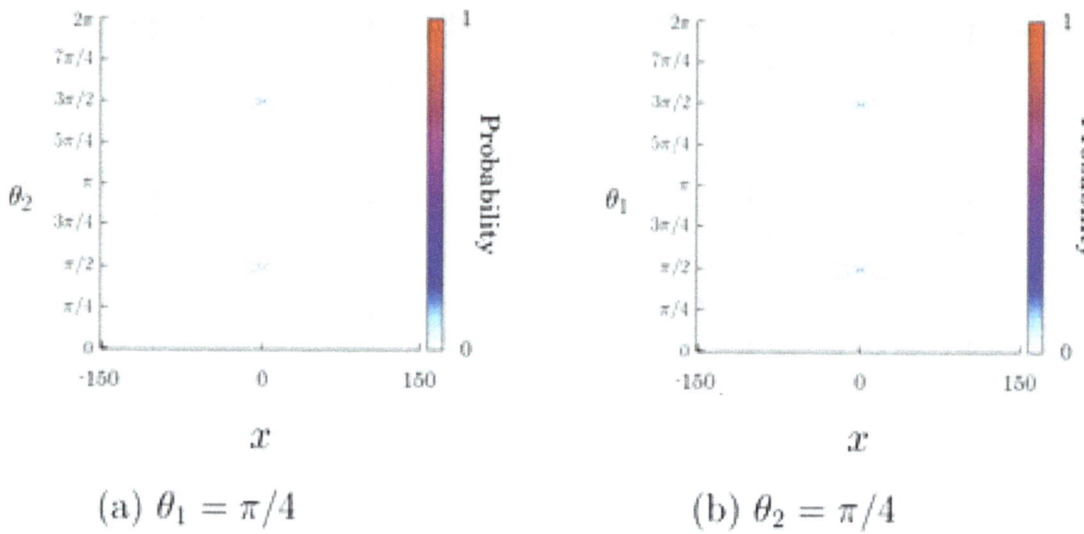

Figure 7. These pictures show how the probability distribution at time 150 depends on the value of the parameters θ_1 and θ_2 ($\alpha = 1/\sqrt{2},\ \beta = i/\sqrt{2}$).

Arranging the r-th derivatives ($r = 0, 1, 2, \ldots$) of the Fourier transform on the scale of time t,

$$\frac{d^r}{dk^r}|\hat{\psi}_{2t}(k)\rangle = (t)_r\cdot\sum_{j=1}^{2}\lambda_j(k)^{t-r}\lambda_{j'}(k)^r\langle v_j(k)|\hat{\psi}_0(k)\rangle|v_j(k)\rangle + O(t^{r-1}), \tag{52}$$

$$\frac{d^r}{dk^r}|\hat{\psi}_{2t+1}(k)\rangle = (t)_r\cdot R(k)U_1\sum_{j=1}^{2}\lambda_j(k)^{t-r}\lambda_{j'}(k)^r\langle v_j(k)|\hat{\psi}_0(k)\rangle|v_j(k)\rangle + O(t^{r-1}), \tag{53}$$

we obtain the representations of the r-th moment of the random variable X_t,

$$\mathbb{E}(X_{2t}^r) = (t)_r\cdot\sum_{j=1}^{2}\frac{1}{2\pi}\int_{-\pi}^{\pi}\left(\frac{i\lambda_{j'}(k)}{\lambda_j(k)}\right)^r\left|\langle v_j(k)|\hat{\psi}_0(k)\rangle\right|^2 dk + O(t^{r-1}), \tag{54}$$

$$\mathbb{E}(X_{2t+1}^r) = (t)_r\cdot\sum_{j=1}^{2}\frac{1}{2\pi}\int_{-\pi}^{\pi}\left(\frac{i\lambda_{j'}(k)}{\lambda_j(k)}\right)^r\left|\langle v_j(k)|\hat{\psi}_0(k)\rangle\right|^2 dk + O(t^{r-1}). \tag{55}$$

Dividing these equations by time $2t$ or $2t + 1$ followed by taking a limit makes the same expression,

$$\lim_{t\to\infty}\frac{\mathbb{E}(X_{2t}^r)}{(2t)^r}=\sum_{j=1}^{2}\frac{1}{2\pi}\int_{-\pi}^{\pi}\left(\frac{i\lambda_j'(k)}{2\lambda_j(k)}\right)^r\left|\langle v_j(k)\,|\,\hat{\psi}_0(k)\rangle\right|^2dk, \tag{56}$$

$$\begin{aligned}\lim_{t\to\infty}\frac{\mathbb{E}(X_{2t+1}^r)}{(2t+1)^r}&=\lim_{t\to\infty}\frac{\mathbb{E}(X_{2t+1}^r)}{(2t)^r}\cdot\left(\frac{2t}{2t+1}\right)^r\\&=\sum_{j=1}^{2}\frac{1}{2\pi}\int_{-\pi}^{\pi}\left(\frac{i\lambda_{j'}(k)}{2\lambda_j(k)}\right)^r|\langle v_j(k)\,|\,\hat{\psi}_0(k)\rangle|^2dk,\end{aligned} \tag{57}$$

which are combined as

$$\begin{aligned}\lim_{t\to\infty}\mathbb{E}\left[\left(\frac{X_{2t}}{2t}\right)^r\right]&=\lim_{t\to\infty}\mathbb{E}\left[\left(\frac{X_{2t+1}}{2t+1}\right)^r\right]\\&=\sum_{j=1}^{2}\frac{1}{2\pi}\int_{-\pi}^{\pi}\left(\frac{i\lambda_j'(k)}{2\lambda_j(k)}\right)^r\left|\langle v_j(k)\,|\,\hat{\psi}_0(k)\rangle\right|^2dk.\end{aligned} \tag{58}$$

As a result, we have

$$\lim_{t\to\infty}\mathbb{E}\left[\left(\frac{X_t}{t}\right)^r\right]=\sum_{j=1}^{2}\frac{1}{2\pi}\int_{-\pi}^{\pi}\left(\frac{i\lambda_{j'}(k)}{2\lambda_j(k)}\right)^r\left|\langle v_j(k)\,|\,\hat{\psi}_0(k)\rangle\right|^2dk. \tag{59}$$

The Hilbert space $\mathcal{H}_c$ spanned by Eq. (29) gives a matrix representation to the operator $R(k)U_2R(k)U_1$,

$$R(k)U_2R(k)U_1=\begin{bmatrix}c_1c_2e^{2ik}+s_1s_2 & s_1c_2e^{2ik}-c_1s_2\\-s_1c_2e^{-2ik}+c_1s_2 & c_1c_2e^{-2ik}+s_1s_2\end{bmatrix}, \tag{60}$$

and one can find its eigenvalues

$$\lambda_j(k)=c_1c_2\cos 2k+s_1s_2-(-1)^j i\sqrt{1-(c_1c_2\cos 2k+s_1s_2)^2}\quad (j=1,2). \tag{61}$$

The normalized eigenvector associated to the eigenvalue $\lambda_j(k)$ takes a form

$$|v_j(k)\rangle = \frac{1}{\sqrt{N_j(k)}}\begin{bmatrix} s_1c_2e^{2ik} - c_1s_2 \\ i\left\{-c_1c_2\sin 2k - (-1)^j\sqrt{1-(c_1c_2\cos 2k + s_1s_2)^2}\right\}\end{bmatrix}, \tag{62}$$

with

$$\begin{aligned} N_j(k) = 1 - (c_1^2 - s_1^2)(c_2^2 - s_2^2) - 4c_1c_2s_1s_2\cos 2k + 2c_1^2c_2^2\sin^2 2k \\ + (-1)^j 2c_1c_2\sin 2k\sqrt{1-(c_1c_2\cos 2k + s_1s_2)^2}. \end{aligned} \tag{63}$$

Here we compute

$$\frac{i\lambda_j'(k)}{2\lambda_j(k)} = (-1)^j\frac{c_1c_2\sin 2k}{\sqrt{1-(c_1c_2\cos 2k + s_1s_2)^2}}, \tag{64}$$

from Eq. (61). Putting $i\lambda_j'(k)/2\lambda_j(k) = x\,(j = 1, 2)$ gives rise to another expression of Eq. (59),

$$\begin{aligned} \lim_{t\to\infty}\mathbb{E}\left[\left(\frac{X_t}{t}\right)^r\right] &= \int_{-\infty}^{\infty} x^r \frac{\sqrt{1-\xi(\theta_1,\theta_2)^2}}{\pi(1-x^2)\sqrt{\xi(\theta_1,\theta_2)^2 - x^2}} \\ &\times\left[1 - \left\{|\alpha|^2 - |\beta|^2 + \frac{\sin\theta_1(\alpha\overline{\beta} + \overline{\alpha}\beta)}{\cos\theta_1}\right\}x\right] I_{(-\xi(\theta_1,\theta_2),\xi(\theta_1,\theta_2))}(x)dx. \end{aligned} \tag{65}$$

For the same reason as the proof for Theorem 1, this convergence promises Theorem 3. As mentioned earlier, if the parameters θ_1 and θ_2 take the same value θ, then the 2-period time-dependent QW is the standard walk shown in Sec.2. In that case, Theorem 3 is in agreement with Theorem 2. Indeed, inserting a value, which is supposed to be θ now, to both θ_1 and θ_2 in Theorem 2 produces the limit probability distribution below,

$$\begin{aligned} \lim_{t\to\infty}\mathbb{P}\left(\frac{X_t}{t} \le x\right) &= \int_{-\infty}^{x} \frac{|\sin\theta|}{\pi(1-y^2)\sqrt{\cos^2\theta - y^2}} \\ &\times\left[1 - \left\{|\alpha|^2 - |\beta|^2 + \frac{\sin\theta(\alpha\overline{\beta} + \overline{\alpha}\beta)}{\cos\theta}\right\}y\right] I_{(-|\cos\theta|,|\cos\theta|)}(y)dy. \end{aligned} \tag{66}$$

Given an initial state with $\alpha = 1/\sqrt{2}$, $\beta = i/\sqrt{2}$, **Figure 8** draws the limit density function

$$\frac{d}{dx}\lim_{t\to\infty}\mathbb{P}\left(\frac{X_t}{t}\le x\right)=\frac{\sqrt{1-\xi(\theta_1,\theta_2)^2}}{\pi(1-x^2)\sqrt{\xi(\theta_1,\theta_2)^2-x^2}}I_{(-\xi(\theta_1,\theta_2),\xi(\theta_1,\theta_2))}(x). \tag{67}$$

We confirm that the function has two singularities at the points $\pm\,\xi(\theta_1, \theta_2)$ which correspond to two sharp peaks in **Figure 5**.

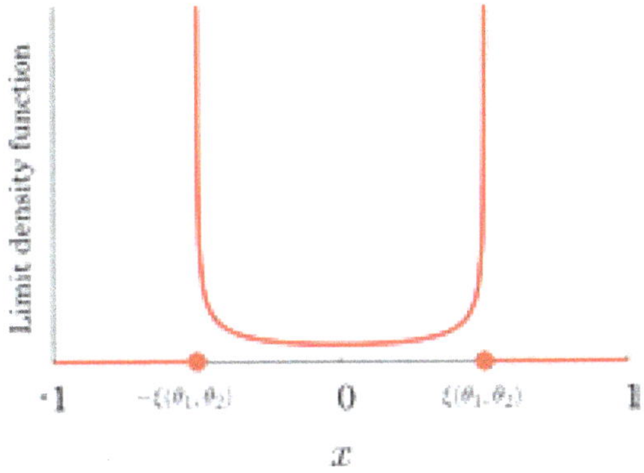

Figure 8. Limit density function ($\alpha=1/\sqrt{2},\ \beta=i/\sqrt{2}$).

4. Three-period time-dependent QW

The standard QW in Sec.2 and the two-period time-dependent QW in Sec.3 have the same type of limit density function. In the final section, we see a three-period time-dependent QW and its limit density function. As a result, a different type of limit density function will be discovered. With a unitary operator $U \in U(2)$, the system of three-period time-dependent QW is periodically updating,

$$|\Psi_{t+1}\rangle=\begin{cases}SC|\Psi_t\rangle & (t=0,3,5,\ldots)\\ SC|\Psi_t\rangle & (t=1,4,6,\ldots)\\ S|\Psi_t\rangle & (t=2,5,7,\ldots)\end{cases}, \tag{68}$$

where

$$C=\sum_{x\in\mathbb{Z}}|x\rangle\langle x|\otimes U, \tag{69}$$

$$S=\sum_{x\in\mathbb{Z}}|x-1\rangle\langle x|\otimes|0\rangle\langle 0|+|x+1\rangle\langle x|\otimes|1\rangle\langle 1|. \tag{70}$$

The 3-period time-dependent QW was studied by Grünbaum and Machida [8] when the unitary operator U was of the form

$$\begin{aligned} U &= \cos\theta|0\rangle\langle 0| + \sin\theta|0\rangle\langle 1| + \sin\theta|1\rangle\langle 0| - \cos\theta|1\rangle\langle 1| \\ &= c|0\rangle\langle 0| + s|0\rangle\langle 1| + s|1\rangle\langle 0| - c|1\rangle\langle 1|, \end{aligned} \tag{71}$$

with $\theta \in [0, 2\pi)$. Note that we have abbreviated $\cos\theta$ and $\sin\theta$ to c and s in Eq. (71), respectively. Let us view the probability distribution $\mathbb{P}(X_t = x)$ when the initial state is given by $|\Psi_0\rangle = |0\rangle \otimes (1/\sqrt{2}|0\rangle + i/\sqrt{2}|1\rangle)$. The operator in Eq. (71) can give rise to probability distributions which have four sharp peaks, as shown in **Figure 9**. These four peaks can also be observed at relatively small time in **Figure 10**. Seeing **Figure 11**, we guess some values of the parameter θ when the number of sharp peaks is three. They are $\pi/3$, $2\pi/3$, $4\pi/3$, and $5\pi/3$, and these values can be exactly estimated by a limit theorem which will be introduced later.

We find a long-time limit theorem in the paper [8] and it asserts the convergence of a random variable rescaled by time t.

Theorem 4*Assume that $\theta \neq 0, \pi/2, \pi, 3\pi/2$. For a real number x, we have*

$$\begin{aligned} \lim_{t\to\infty} \mathbb{P}\left(\frac{X_t}{t} \leq x\right) = \int_{-\infty}^{x} [&\{1 - \nu(\alpha,\beta;y)\} f(y) I_{\left(\frac{1-4c^2}{3}, \frac{\sqrt{1+8c^2}}{3}\right)}(y) \\ &+ \{1 + \nu(\alpha,\beta;-y)\} f(-y) I_{\left(-\frac{\sqrt{1+8c^2}}{3}, -\frac{1-4c^2}{3}\right)}(y)] dy, \end{aligned} \tag{72}$$

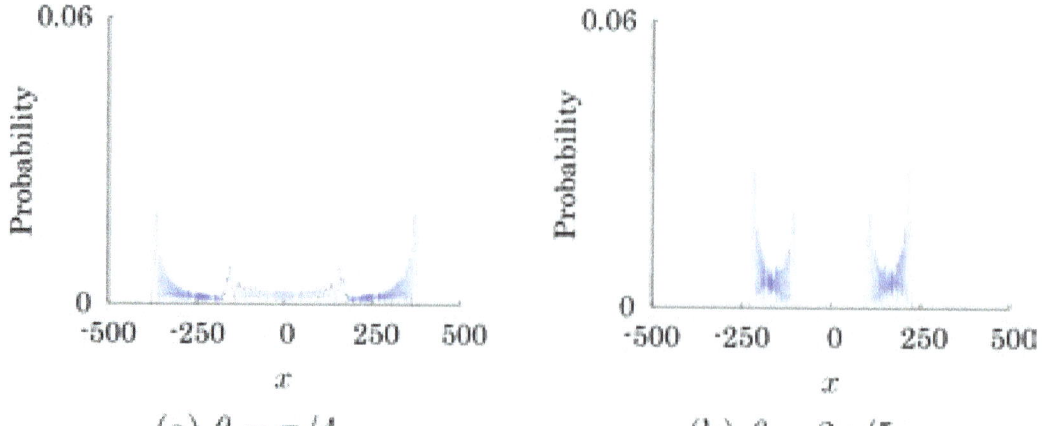

(a) $\theta = \pi/4$ (b) $\theta = 2\pi/5$

Figure 9. Probability distribution at time 500 $(\alpha = 1/\sqrt{2},\ \beta = i/\sqrt{2})$.

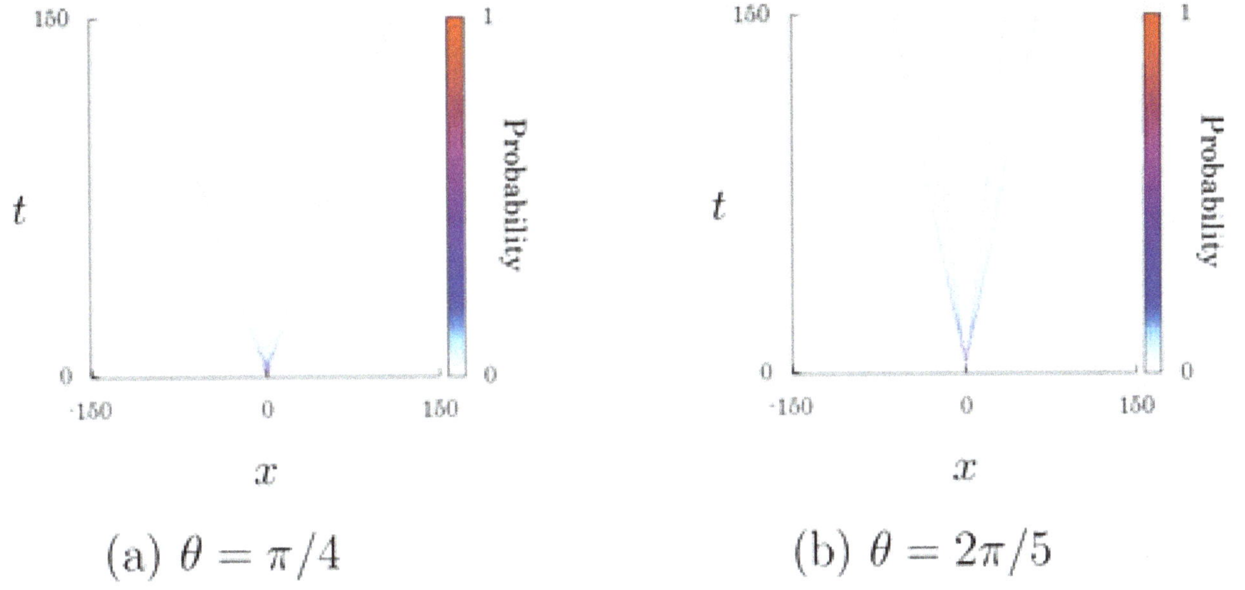

Figure 10. Time evolution of the probability distribution $(\alpha = 1/\sqrt{2},\ \beta = i/\sqrt{2})$.

where

$$f(x) = \frac{|s|\left(|s|x + \sqrt{D(x)}\right)^2}{\pi(1-x^2)\sqrt{W_+(x)}\sqrt{W_-(x)}\sqrt{D(x)}}, \tag{73}$$

$$\begin{aligned}\nu(\alpha,\beta;x) =& \frac{1}{c(1+8c^2)}\left\{9c^3(|\alpha|^2-|\beta|^2)+3s(1+6c^2)\Re(\alpha\overline{\beta})\right\}x \\ &+\frac{s}{c|s|(1+8c^2)}\left\{cs(|\alpha|^2-|\beta|^2)-(1+2c^2)\Re(\alpha\overline{\beta})\right\}\sqrt{D(x)},\end{aligned} \tag{74}$$

$$D(x) = 1 + 8c^2 - 9c^2x^2, \tag{75}$$

$$W_+(x) = -(1-4c^2) + 3(1-2c^2)x^2 + 2|s|x\sqrt{D(x)}, \tag{76}$$

$$W_-(x) = 1 + 8c^2 - 3(1+2c^2)x^2 - 2|s|x\sqrt{D(x)}, \tag{77}$$

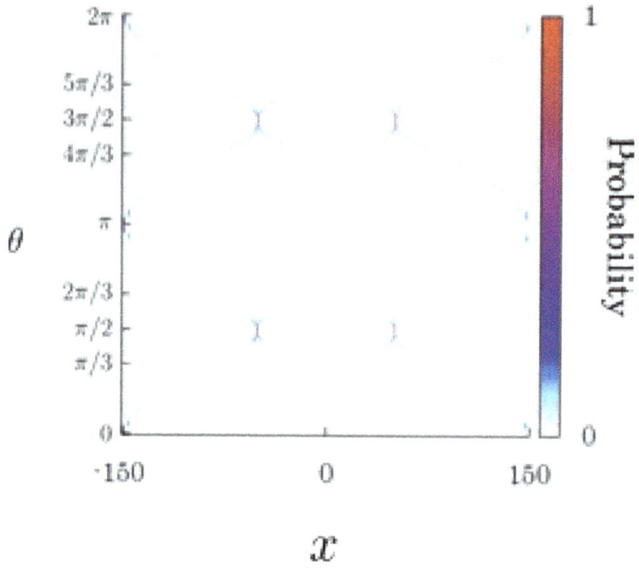

Figure 11. This picture shows how the probability distribution at time 150 depends on the value of the parameter θ. $(\alpha=1/\sqrt{2},\ \beta=i/\sqrt{2})$.

and $\Re(z)$ denotes the real part of the complex number z.

This limit theorem can be derived by Fourier analysis as well. For the Fourier transform $|\hat{\Psi}_t(k)\rangle=\sum_{x\in\mathbb{Z}} e^{-ikx}|\psi_t(x)\rangle (k\in[-\pi,\pi))$, Eq. (68) produces a time evolution of the Fourier transform,

$$\begin{aligned}|\hat{\psi}_{3t}(k)\rangle &= \left\{R(k)(R(k)U)^2\right\}^t |\hat{\Psi}_0(k)\rangle,\\ |\hat{\psi}_{3t+1}(k)\rangle &= R(k)U\left\{R(k)(R(k)U)^2\right\}^t |\hat{\Psi}_0(k)\rangle,\\ |\hat{\psi}_{3t+2}(k)\rangle &= (R(k)U)^2\left\{R(k)(R(k)U)^2\right\}^t |\hat{\Psi}_0(k)\rangle.\end{aligned} \tag{78}$$

Given an orthogonal normalized basis such as Eq. (29), the matrix

$$R(k)(R(k)U)^2 = \begin{bmatrix} c^2e^{3ik}+s^2e^{ik} & cse^{3ik}-cse^{ik} \\ -cse^{-3ik}+cse^{-ik} & c^2e^{-3ik}+s^2e^{-ik}\end{bmatrix} \tag{79}$$

has two eigenvalues

$$\lambda_j(k) = c^2\cos 3k + s^2\cos k - (-1)^j i\sqrt{1-(c^2\cos 3k+s^2\cos k)^2} \quad (j=1,2). \tag{80}$$

A possible expression of the normalized eigenvector associated to the eigenvalue $\lambda_j(k)$ is

$$| v_j(k)\rangle = \frac{1}{\sqrt{N_j(k)}}\begin{bmatrix} -2cse^{2ik}\sin k \\ c^2\sin 3k + s^2\sin k + (-1)^j\sqrt{1-(c^2\cos 3k + s^2\cos k)^2} \end{bmatrix}, \tag{81}$$

where $N_j(k)$ is the normalization factor

$$\begin{aligned} N_j(k) = 2\{&1-(c^2\cos 3k + s^2\cos k)^2 \\ &+(-1)^j(c^2\sin 3k + s^2\sin k)\sqrt{1-(c^2\cos 3k + s^2\cos k)^2}\}. \end{aligned} \tag{82}$$

With a decomposition $|\hat{\Psi}_{3t}(k)\rangle = \sum_{j=1}^{2}\lambda_j^t(k)\langle v_j(k)|\hat{\Psi}_0(k)\rangle|v_j(k)\rangle$, we get representations in the eigenspace,

$$|\hat{\psi}_{3t}(k)\rangle = \sum_{j=1}^{2}\lambda_j(k)^t\langle v_j(k)|\hat{\psi}_0(k)\rangle|v_j(k)\rangle, \tag{83}$$

$$|\hat{\psi}_{3t+1}(k)\rangle = R(k)U\sum_{j=1}^{2}\lambda_j(k)^t\langle v_j(k)|\hat{\psi}_0(k)\rangle|v_j(k)\rangle, \tag{84}$$

$$|\hat{\psi}_{3t+2}(k)\rangle = (R(k)U)^2\sum_{j=1}^{2}\lambda_j(k)^t\langle v_j(k)|\hat{\psi}_0(k)\rangle|v_j(k)\rangle, \tag{85}$$

and compute their derivatives

$$\frac{d^r}{dk^r}|\hat{\psi}_{3t}(k)\rangle = (t)_r\sum_{j=1}^{2}\lambda_j(k)^{t-r}\lambda_{j'}(k)^r\langle v_j(k)|\hat{\psi}_0(k)\rangle|v_j(k)\rangle + O(t^{r-1}), \tag{86}$$

$$\begin{aligned} \frac{d^r}{dk^r}|\hat{\psi}_{3t+1}(k)\rangle = (t)_r R(k)U\sum_{j=1}^{2}\lambda_j(k)^{t-r}\lambda_j'(k)^r\langle v_j(k)|\hat{\psi}_0(k)\rangle|v_j(k)\rangle \\ + O(t^{r-1}), \end{aligned} \tag{87}$$

$$\begin{aligned} \frac{d^r}{dk^r}|\hat{\psi}_{3t+2}(k)\rangle = (t)_r (R(k)U)^2\sum_{j=1}^{2}\lambda_j(k)^{t-r}\lambda_j'(k)^r\langle v_j(k)|\hat{\psi}_0(k)\rangle|v_j(k)\rangle \\ + O(t^{r-1}). \end{aligned} \tag{88}$$

The moments $\mathsf{E}(X_{3t}^r)$, $\mathsf{E}(X_{3t+1}^r)$, and $\mathsf{E}(X_{3t+2}^r)$ turn out to be of the form

$$(t)_r \cdot \sum_{j=1}^{2} \frac{1}{2\pi} \int_{-\pi}^{\pi} \left(\frac{i\lambda_j'(k)}{\lambda_j(k)} \right)^r \left| \langle v_j(k) \mid \hat{\psi}_0(k) \rangle \right|^2 dk + O(t^{r-1}), \tag{89}$$

and we see

$$\begin{aligned} &\lim_{t\to\infty} \mathbb{E}\left[\left(\frac{X_{3t}}{3t} \right)^r \right] = \lim_{t\to\infty} \mathbb{E}\left[\left(\frac{X_{3t+1}}{3t+1} \right)^r \right] = \lim_{t\to\infty} \mathbb{E}\left[\left(\frac{X_{3t+2}}{3t+2} \right)^r \right] \\ &= \sum_{j=1}^{2} \frac{1}{2\pi} \int_{-\pi}^{\pi} \left(\frac{i\lambda_{j'}(k)}{3\lambda_j(k)} \right)^r \left| \langle v_j(k) \mid \hat{\psi}_0(k) \rangle \right|^2 dk, \end{aligned} \tag{90}$$

which is put together as

$$\lim_{t\to\infty} \mathbb{E}\left[\left(\frac{X_t}{t} \right)^r \right] = \sum_{j=1}^{2} \frac{1}{2\pi} \int_{-\pi}^{\pi} \left(\frac{i\lambda_j'(k)}{3\lambda_j(k)} \right)^r \left| \langle v_j(k) \mid \hat{\psi}_0(k) \rangle \right|^2 dk, \tag{91}$$

where

$$\frac{i\lambda_j'(k)}{3\lambda_j(k)} = (-1)^j \frac{3c^2 \sin 3k + s^2 \sin k}{3\sqrt{1-(c^2\cos 3k + s^2 \cos k)^2}}. \tag{92}$$

Setting $i\lambda_j'(k)/3\lambda_j(k) = x\ (j=1,\ 2)$ in Eq. (91) leads us to an integral expression of the limit,

$$\begin{aligned} \lim_{t\to\infty} \mathbb{E}\left[\left(\frac{X_t}{t} \right)^r \right] = &\int_{-\infty}^{\infty} x^r [\{1-\nu(\alpha,\beta;x)\} f(x) I_{\left(\frac{1-4c^2}{3},\frac{\sqrt{1+8c^2}}{3}\right)}(x) \\ &+ \{1+\nu(\alpha,\beta;-x)\} f(-x) I_{\left(-\frac{\sqrt{1+8c^2}}{3},-\frac{1-4c^2}{3}\right)}(x)] dx, \end{aligned} \tag{93}$$

which guarantees Theorem 4. **Figure 12** shows the limit density function $(d/dx)\mathbb{P}(X_t/t \le x)$ when $\alpha = 1/\sqrt{2}$, $\beta = i/\sqrt{2}$, and we view the features of **Figure 9** in the limit density function. The density function contains singular points at $\pm(1-4c^2)/3$, $\pm\sqrt{1+8c^2}/3$. When $\theta = \pi/4$, they are found at

$\pm 1/3(=\pm 0.333\cdots)$, $\pm\sqrt{5}/3(=\pm 0.745\cdots)$ in **Figure 12-(a)**, and when $\theta = 2\pi/5$, at $\pm(\sqrt{5}-1)/6(=\pm 0.206\cdots)$, $\pm\sqrt{4-\sqrt{5}}/3(=\pm 0.442\cdots)$ in **Figure 12-(b)**.

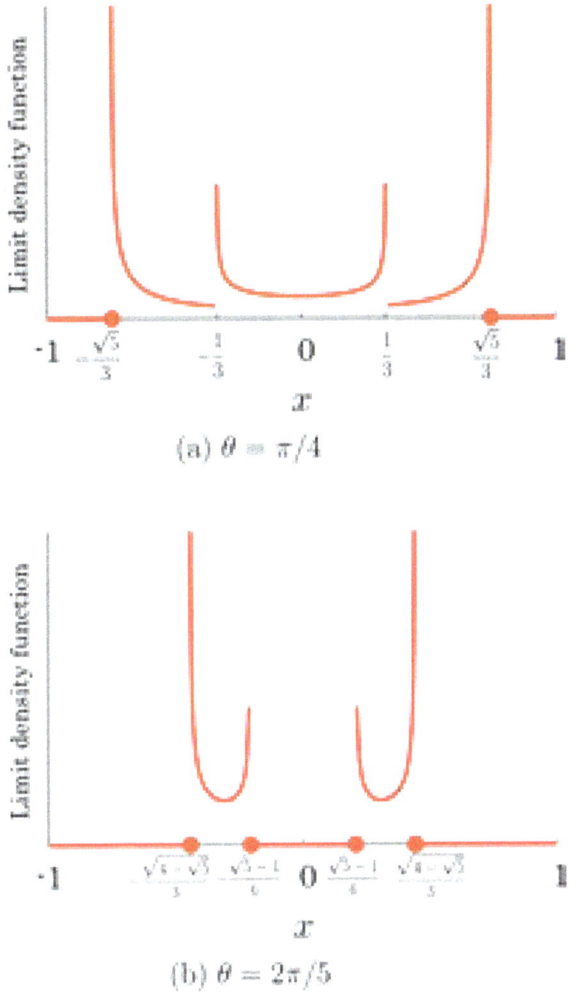

Figure 12. Limit density function $(\alpha = 1/\sqrt{2},\ \beta = i/\sqrt{2})$.

Acknowledgements

The author is supported by JSPS Grant-in-Aid for Young Scientists (B) (No. 16K17648).

Author details

Takuya Machida

Address all correspondence to: tz14040@meiji.ac.jp

College of Industrial Technology, Nihon University, Narashino, Chiba, Japan

References

[1] S. P. Gudder (1988), *Quantum probability*, Academic Press. Probability and Mathematical Statistics.

[2] Y. Aharonov, L. Davidovich and N. Zagury (1993), *Quantum random walks*, Phys. Rev. A, 48(2), pp. 1687–1690.

[3] D. A. Meyer (1996), *From quantum cellular automata to quantum lattice gases*, Journal of Statistical Physics, 85(5–6), pp. 551–574.

[4] S. E. Venegas-Andraca (2012), *Quantum walks: a comprehensive review*, Quantum Information Processing, 11(5), pp. 1015–1106.

[5] N. Konno (2002), *Quantum random walks in one dimension*, Quantum Information Processing, 1(5), pp. 345–354.

[6] G. Grimmett, S. Janson and P.F. Scudo (2004), *Weak limits for quantum random walks*, Phys. Rev. E, 69(2), p. 026119.

[7] T. Machida and N. Konno (2010), *Limit theorem for a time-dependent coined quantum walk on the line*, F. Peper et al. (Eds.): IWNC 2009, Proceedings in Information and Communications Technology, 2, pp. 226–235.

[8] F. A. Grünbaum and T. Machida (2015), *A limit theorem for a 3-period time-dependent quantum walk*, Quantum Information and Computation, 15(1& 2), pp. 50–60.

8

A Highly Ordered Radiative State in a 2D Electron System

Parakhonsky A, Lebedev M, Dremin A and Kukushkin I

Abstract

A two-dimensional (2D) electron system in a perpendicular magnetic field acts like a macroscopic strongly correlated source of radiation with inherent temporal and spatial correlations. The formation of the correlated macroscopic state of 2D electrons passes through uniform and nonuniform states of the electron subsystem. These are reflected in two types of noise of photoresponse in the 2D system. The strange attractor characterized by small-sized dynamics in the phase space of a photoluminescence of 2D electrons is revealed.

Keywords: giant optical fluctuations, highly ordered correlated state of electrons, nonlinear dynamics, quantum Hall effect, strange attractor

1. Introduction

The formation of a highly ordered correlated state in low-dimensional electron systems is well known to produce fluctuations of their physical parameters (a potential fluctuation, spin and charge fluctuations, phase fluctuations of wave functions, etc.) [1]. The fluctuation character of various kinds of phenomena can be implicitly indicative of their fundamental nature. In particular, the studies of quantum shot noise made it possible to reveal the fractional electron effective charge [2,3]. The strong correlation effects may occur in a quasi-two-dimensional (2D) electron system in a transverse magnetic field, such as the fractional quantum Hall effect (FQHE) [4] and the pinned Wigner solid [5]. However, to date, quantum-sized electron systems have not been investigated in terms of the fluctuations in their optical response. In that context, the optical methods could provide information that is beyond the reach of the magnetoresistance studies. These are the single-particle density of states immediately below the Fermi level, single-particle energy gaps, some information on a random potential (the amplitude and characteristic length scales), interference-enhanced effects, polarization correlations, coherence time, and so on. It is also important to emphasize here that quantum localization effects are not the strong restriction for the optical technique in the QHE regime [6].

In a series of articles [7–17], we reported that the giant optical fluctuations (GOFs) occur in a photoexcited 2D electron system (GaAs/AlGaAs quantum wells) in the QHE regime. Initially, it was revealed that the rated value of a variance of an intensity of radiative recombination of 2D electrons with photoexcited holes increases by several orders of magnitude for certain values of the Landau-level filling factor ν [the ratio of a 2D electron concentration n_s to the multiplicity of the Landau-level degeneracy $(2\pi L^2_B)^{-1}$; $L_B=(\hbar/eB)^{1/2}$ is the magnetic length, $\hbar$ is the Plank constant, B is a magnetic field, and e is the elementary charge] [7]. The fluctuations are of a low-frequency character with characteristic times on the order of tens, hundreds, and even thousands of seconds. The follow-up study revealed that these fluctuations occur due to the strong interaction between elements of the photoexcited 2D system and have large-scale spatial and temporal correlations [8,9]. Moreover, we revealed that the signal intensity of intersubband inelastic light scattering in the QHE regime [the spin-density excitation (SDE) mode] exhibits giant fluctuations similar to the luminescence intensity fluctuations [10].

Fluctuations of optical signals are considered traditionally in the theory of optical coherence [18]. It is thus considered that the source of radiation is in a stationary state close to thermodynamic equilibrium and only field fluctuations are taken into account. However, besides optical coherence, intrinsic times of quantum dynamics of an electron subsystem can be perceptible in an optical signal. It seems that the consideration of source fluctuations under thermodynamically nonequilibrium conditions is a reasonable generalization of the existing theory. In particular, the source can be in a deterministic chaos regime, as it must occur at critical points and phase transitions. QHE is the macroscopic quantum effect that is manifested in the quantization of the Hall resistance ρ_{xy} and disappearance of the diagonal resistance ρ_{xx}. In that context, ρ_{xx} valleys can be considered as the system states in the vicinity of a critical point. It is reasonable to expect in the circumstances that the radiation of a source will be not chaotic. Indeed, we found that the radiation of the GOF source has correlations [8,12–14]. This radiation is not chaotic *a priori* and has nonlinear dynamics of photon counts as revealed from the time series analysis [11,15]. At that, there are conditions in a vicinity of the filling factor $\nu=2$, where the system behaves as the macroscopic correlated ("coherent") source of radiation [16,17].

2. GOF detection under QHE conditions: sample structure and initial experimental conditions

High-quality samples containing a 250-Å-wide GaAs/$Al_{0.3}Ga_{0.7}As$ quantum well were used as a standard object where the effect of GOF is observed [7]. **Figure 1** shows the sample structure in which the radiative recombination of 2D electrons with photoexcited holes was studied.

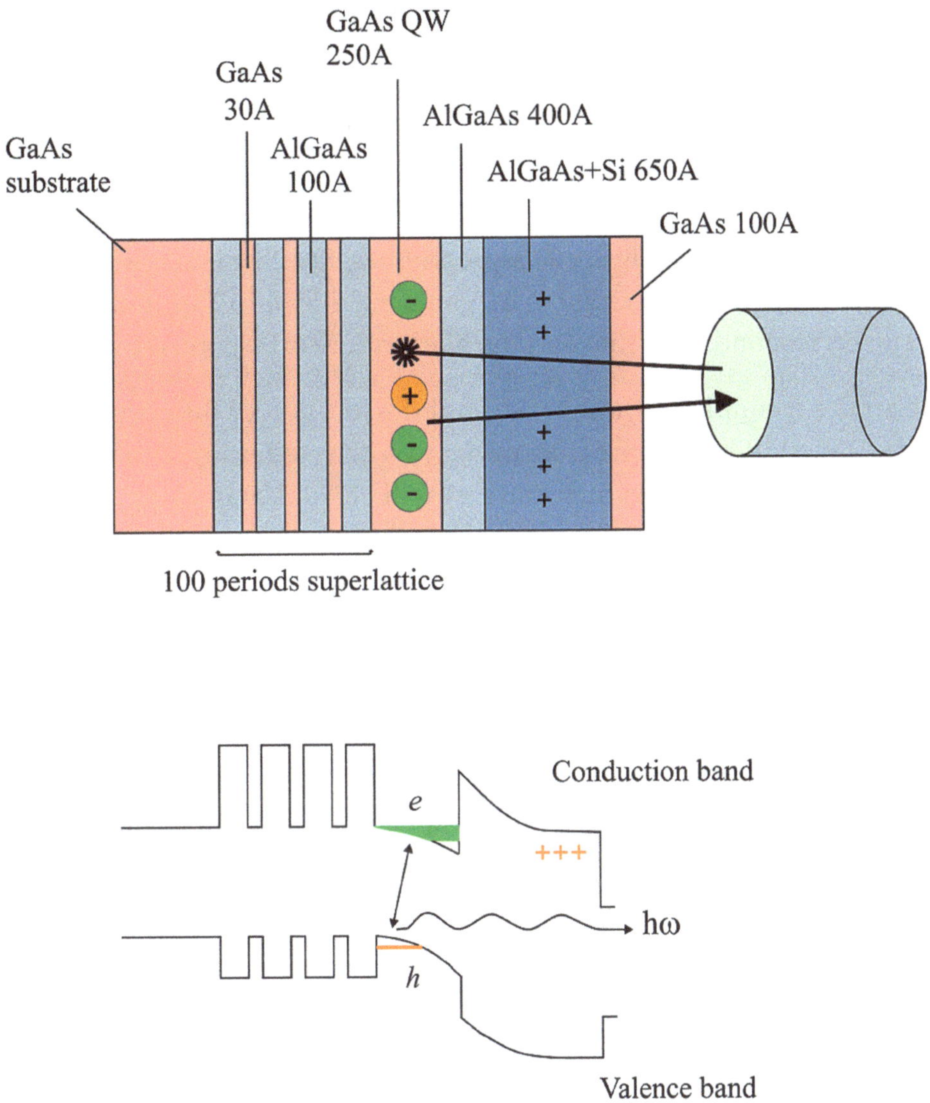

Figure 1. Layer-by-layer structure of the sample studied and the corresponding schematic energy band structure.

The samples were grown by molecular beam epitaxy on a GaAs substrate by the following scheme: GaAs buffer layer 3000 Å thick, undoped GaAs/AlGaAs (30/100 Å) superlattice 13,000 Å in total thickness, GaAs quantum well 250 Å thick, AlGaAs spacer 400 Å thick, doped AlGaAs:Si layer (doping level, 10^{18} cm^{-3}) 650 Å thick, and GaAs cap layer 100 Å thick. The characteristic mobility of 2D electrons and the concentration in these structures were μ=1.3×10^6 $cm^2/(V\ s)$ and n_s=3.8×10^{11} cm^{-2}, respectively. The sample under study was immersed into a liquid helium cryostat with the superconducting solenoid provided magnetic field B from 0 to 12 T. The temperature of the sample T was varied from 1.5 to 4.2 K. In early experiments, the system was optically excited by a laser diode with photon energy E_L=1.653 eV and output power time instability of less than 10^{-4}. The photoexcitation was delivered and the photoluminescence (PL) signals were collected using optical fibers. Here, it is appropriate to mention another advantage of the magneto-optics over magnetotransport measurements. The fact is that, under the conditions of nonequilibrium electron-hole pair injection, along with the reduction of the concentration of 2D electrons, their mobility considerably increases. This is because the concentration of charged centers decreases near a heteroboundary under illumination [19]. In early experiments in GOF, we used an optical scheme with a small photoexcitation spot (d_{PE}≤3 mm). Herein, the excitation of PL signals and their collection was accomplished using one of the registering fibers. The GOF effect is characterized as follows under these conditions. These are telegraph noise of the spectral position of a 2D PL line and a giant noise of the PL intensity. **Figure 2a** represents a set of 2D PL intensity spectra in a color scale measured at T=1.65 K in the range of magnetic field B=0–10 T at small steps of 0.1 T: the higher intensity corresponds to white and the black color corresponds to a weak intensity.

Figure 2a illustrates a simple method to find out the location of the required filling factor ν and to get the GOF conditions. Each PL spectrum was detected using a semiconductor charge-coupled detector (CCD; Princeton Instruments) with 1340×100 imaging array and the high quantum efficiency (QE=70%) available in the near-infrared regions of the spectrum. A "Monospec" spectrometer was used as a spectral instrument providing a spectral resolution of 0.03 meV. This allowed the entire luminescence spectrum to be measured simultaneously in the wavelength region of our interest. The CCD matrix response speed (1 spectrum per second) was quite sufficient for studying 2D PL fluctuations. The positions of filling factors

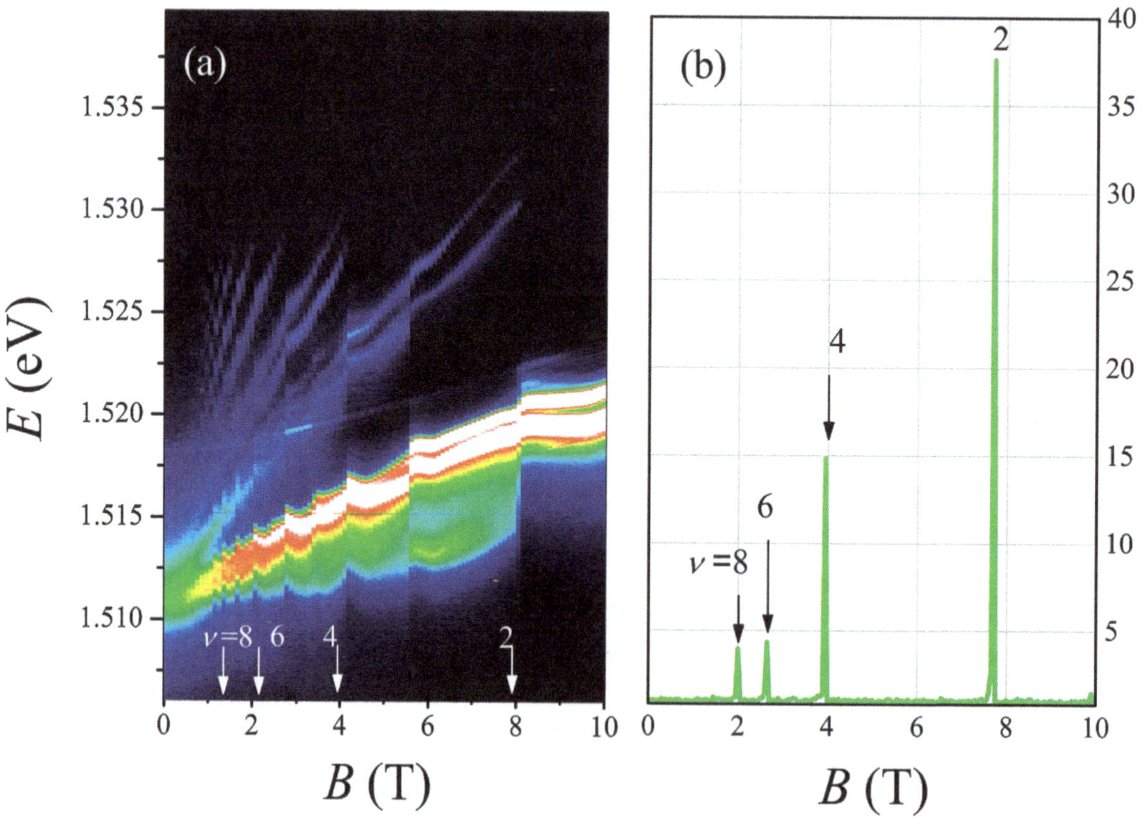

Figure 2. (a) 2D PL spectra measured in the range of magnetic field B=0–10 T (the color scale) and (b) ratio of the variance to the mean intensity $D/\langle I\rangle$. One can see that the PL statistic acts deviate significantly from the Poisson distribution and $D/\langle I\rangle >> 1$ at integer even filling factors ν=8, 6, 4, and 2 (T=1.65 K).

ν=8, 6, 4, and 2 correspond to very sharp jumps in the spectral positions of 2D PL lines. The transport characteristics of the samples were measured in parallel with the optical studies (under continuous laser illumination) to control the magnetoresistance at the Shubnikov-de Haas oscillation minima corresponding to the QHE (see [7]). **Figure 2b** illustrates the magnetic field dependence of the ratio of the variance $D=\langle I^2\rangle - \langle I\rangle^2$ to the mean intensity $\langle I\rangle$ (the mean number of photon counts) of a PL signal from the lowest Landau level in the sample studying. The spectra were processed mathematically simultaneously with their recording. It is seen that this parameter increases by an order of magnitude at these values of ν:

$$\frac{D}{\langle I\rangle} >> 1. \tag{1}$$

Thus, this distribution of photon counts differs substantially from the Poisson distribution and the magnitude of fluctuations has a maximum in the vicinity of the filling factor ν=2.

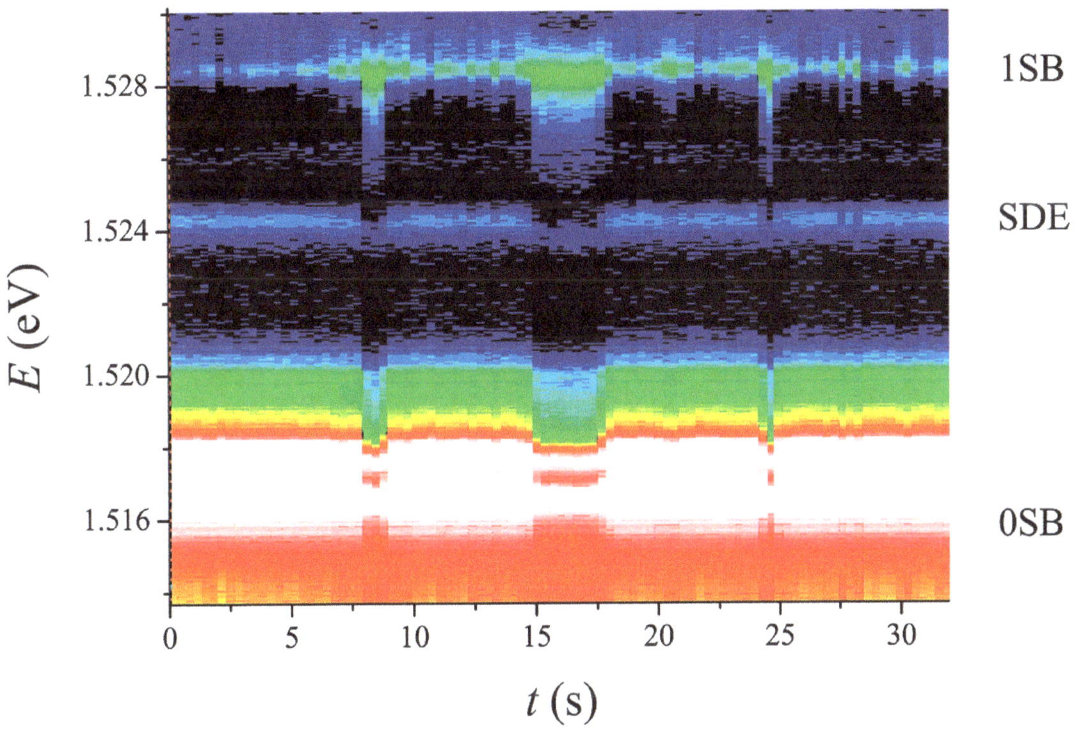

Figure 3. Time dependences of the spectral positions of the SDE line and the PL lines of the ground (0SB) and the first excited (1SB) size-quantized subbands.

It should be noted here that such a regime occurs also for ν=1 and for fractional factors ν as well at lower temperatures [11]. By studying the inelastic light scattering spectra under the QHE conditions, we revealed that the intensity of intersubband Raman scattering (SDE mode) undergoes giant fluctuations analogous to the PL intensity fluctuations [10]. In addition to the fluctuations in the inelastic light scattering intensity, we also revealed the telegraph noise of the SDE spectral line analogue to the PL spectral lines of the ground (0SB) and the first excited (1SB) size-quantized subbands in the vicinity of ν=2 (**Figure 3**). Fluctuations of the SDE mode intensity are observed simultaneously and in phase with fluctuations of the 2D PL intensity, and the recombination and inelastic light scattering processes proceed consistently over a long period of time. This fact indicates that, under the QHE conditions, the electronic system is in a macroscopic correlated state.

3. Macroscopic character of giant fluctuations of 2D electrons: a multifiber scheme technique

We applied the correlation analysis to the study of formation and characteristic aspects of the revealed macroscopic state in a 2D electron system. Correlation functions are quite often used in the analysis of noise due to the ease of interpretation. Correlation spectroscopy was for the first time applied to study of intensity fluctuations by Hanbury Brown and Tviss [20]. Subsequently, it was gradually resulting in a situation where the correlation spectroscopy became the traditional tool of quantum optics. However, as mentioned above, this method had rarely been used for studying fluctuations near critical points or phase transitions. Meanwhile, the correlation spectroscopy could be used successfully for spatial and temporal analysis of optical fluctuations of complex systems. In a critical point, a characteristic range of correlations (correlation radius) significantly exceeds the interparticle distance and a system is becoming susceptible to changes of local concentrations [21]. If the photoexcited 2D electron system is in the quasi-equilibrium under IQHE conditions, then the intensity of recombination radiation consists of a convolution of 2D electron and photoexcited hole distribution functions [6]:

$$I(\hbar\omega) \propto \int_0^\infty f_e(E) D_e(E) f_h(\hbar\omega - E) D_h(\hbar\omega - E) dE. \tag{2}$$

The width of the experimentally measured hole distribution function f_h turns out to be small; therefore, the 2D line spectrum reflects the energy distribution of the single-particle 2D electron density of states $D_e(E) \propto dn_s/dE$ [6]. Thus, if we consider fluctuations of 2D PL intensity under the existing conditions, they will depend on the fluctuations of a local concentration of 2D electrons n_s. The connection between $D_e(E)$иn_s will remain in the case of interparticle interaction as well, but the energy spectrum will be significantly complicated and it is necessary to apply for the Green function approach. This means that a random potential screening will play a significant role in nonequilibrium processes and the 2D electron redistribution. In this context, high electron concentrations in doped quantum wells (as in the case of our samples) are sensitive to optical excitation [22] and deviations from the single-particle scenario of the integer QHE can occur resulting in that fluctuations and correlation effects are enhanced. The exciton effects are of little importance in this case because of the screening of exciton states by a system of 2D electrons in the doped systems. It was observed in both luminescence and reflectance spectra [23].

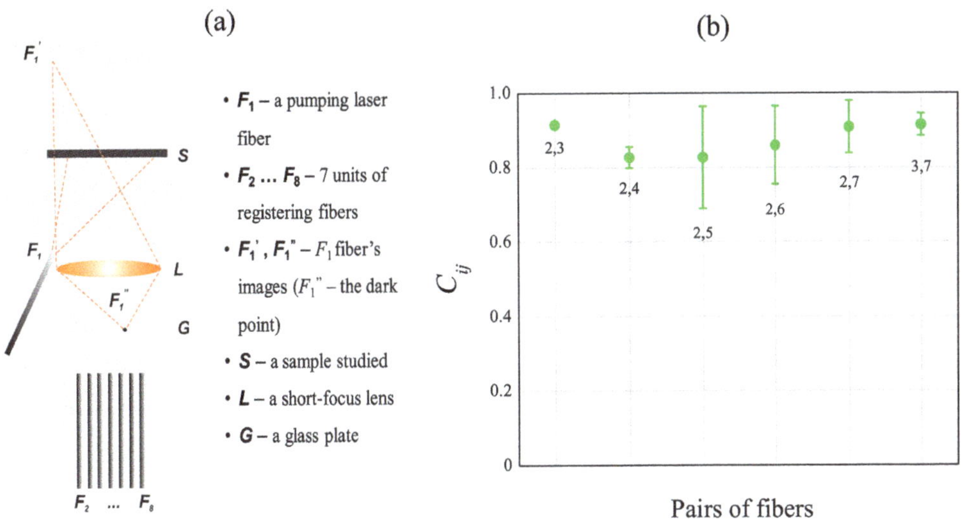

Figure 4. (a) An optical multifiber scheme used to record PL signals from seven sections of the surface; the transparent glass plate G was placed between the multifiber set and the lens to eliminate the laser light reflectance (a dark spot on the glass plate at the point-focused image). (b) C_{ij} for different pairs of fibers at the magnetic field B=7.78 (T=1.65 K).

In steady-state conditions, time dependence of the radiation intensity represents a realization of a stationary random process. The intensity correlation functions depend only on the time difference. The information on the concentration redistribution is contained in correlation functions of intensity fluctuations. In analyzing the autocorrelation function, it is possible to study periodic processes to find the characteristic time of a relaxation and dynamics of attractors. Cross-correlation function can provide information on spatial correlations in the system of interacting carriers.

We used special multifiber optical scheme to record and analyze the spatial correlation between the radiation intensities of PL signals from different points of a large sample ($S \propto 1$ cm^2) in the lateral direction (**Figure 4**) [14]. Pumping a fiber F_1 with a core of 0.4 mm in diameter gave a light spot of approximately 1 cm in diameter on the sample surface (the spot appeared larger than the sample size). Short-focus (f=15 mm) lens L formed an image of the sample with a magnification close to unity in the plane where the edges of the registering fibers (F_2-F_8) were arranged. These fibers with the same core in diameter were arranged right up to each other with distances between their centers of approximately 1.2 mm, which did not overlap with each other. Sufficiently long sequences of the photo count time dependences were recorded at a step of 1 s. These sequences were simultaneously measured from two different fibers F_i and F_j. The subscript indexes i and j mean any pair of a number of F_2-F_8. The 2D system was optically excited by a continuous wave (CW) semiconductor laser Uptronic with photon energy of 1.536 eV and an output power of 75 mW. The correlation coefficients C_{ij} were calculated between the radiation intensities (I_i and I_j):

$$C_{ij} = \frac{\langle \Delta I_i \cdot \Delta I_j \rangle}{\sqrt{D_i \cdot D_j}}, \tag{3}$$

where $\Delta I_{i(j)}=I_{i(j)} - \langle I_{i(j)} \rangle$, D_i and D_j are variances of I_i and I_j respectively. The values of C_{ij} ranged from 0.80 to 0.97 for different sets of indexes *i,j* at ν=2 (**Figure 4b**). This result is quite surprising, indicating that local electron concentrations at different points in the entire area of the 2D system are essentially equal. That is, all areas of the 2D electron system spaced at approximately 1 cm can emit light in a strongly correlated way. This is the point of uniformity of the electronic density (*B*=7.78 T, ν=2). Both types of noise (intensity fluctuations and noise of the spectral position) are simultaneously observed in this point. The existence of a common optical excitation spot is essential for the correlations observed [8]. The resonant photoexcitation (it was carried out by means of a tunable diode laser DL pro 780 with an output power of 15 mW) in the first excited subband (E_{1SB}) of a quantum well has shown that this state formation occurs due to the exchange of electrons between the neighboring regions of a 2D layer [16]. **Figure 5** shows spectra measured in two different points of a sample separated by a distance of 6 mm (*i*=2, *j*=7). It is easily seen that PL lines, originally not coinciding by spectral positions, converge to the same value of energy E_j=1.517 eV (a point of uniformity). In the work [16], a 2D electron system near ν=2 is considered as being in one of the radiant states: uniform electron density in 2D layer and a nonuniform one. The switching between these two states represents a telegraph noise where the autocorrelation function is given by:

$$A(\tau)=\langle I(t)\cdot I(t+\tau)\rangle = I_0^2 e^{-\frac{2|\tau|}{\tau_s}}, \tag{4}$$

where $|I_0|$ is a fixed value of intensity, and τ_s is the average switching time. This value can reach hundreds of seconds at $T < 1.7$ K.

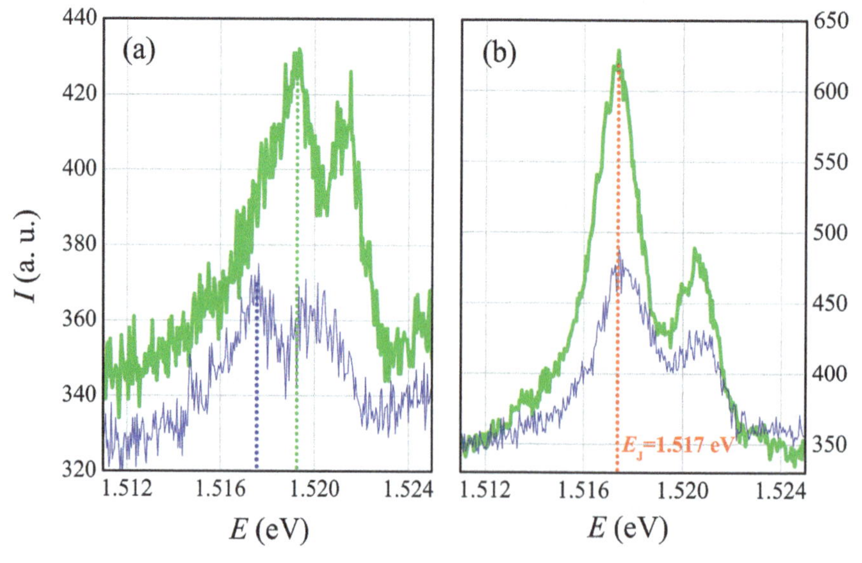

Figure 5. 2D electron PL spectra measured in two different points of a sample separated by a distance of 6 mm [fiber 2 (green curves) and fiber 7 (blue curves)] at *B*=7.55 T (a) and *B*=7.78 T (b) (ν=2; *T*=1.65 K). PL spectra (in color scale; bottom) have identical spectral positions at ν=2 for all points of a sample.

With further increase in magnetic field ($B > 7.78$ T, $\nu \leq 2$), the telegraph noise disappears and the intensity noise amplitude decreases by more than one order of magnitude.

The range of an intensity noise covers a much wider range of a magnetic field (~2 T; see **Figure 6**) than it occurs in the case of a small spot of photoexcitation (0.005–0.01 T). We decided to analyze the noise region in terms of possible regularities in its dynamics and turned to the analysis of autocorrelation and cross-correlation functions of 2D PL signals in this region.

It was revealed that time dependences of the PL autocorrelation $A_S(\tau)$ and cross-correlation $C_S(\tau)$ functions have a periodic component and they practically coincide at B=7.78 T [16]:

$$A_S(\tau) \approx C_S(\tau) \propto \left(1 + e^{-2|\tau|/\tau_s}\right)\cos\left(\frac{2\pi\tau}{P}\right)e^{-|\tau|/\tau_d}. \tag{5}$$

Here, P is a period of oscillations and τ_d is a damping time. A term referred to as the damping time τ_d was introduced in formula (5), where the experimental dependences need to be adequately described [16].

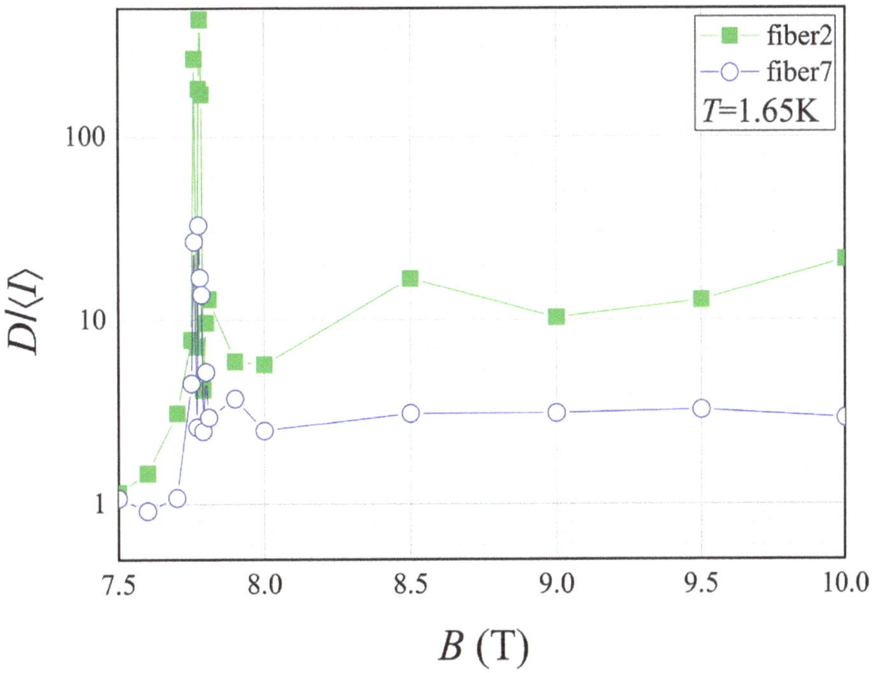

Figure 6. Ratio $D\ /\langle I \rangle$ calculated for a 2D PL intensity measured in two different points of the 2D plane (fibers 2 and 7) simultaneously in the range of magnetic fields of B=7.5–10 T at T=1.65 K.

Here, it must be taken into consideration that damped autocorrelation function is one of the criteria for the strange attractor of a dynamical system (see the next section). **Figure 7a** shows these correlations. Practically complete coincidence of the functions $A_S(\tau)$ and $C_S(\tau)$ means that the processes in two spatially spaced points of a sample proceed consistently. This confirms the fact that we have some kind of a macroscopic correlated radiative source. With the increase in magnetic field ($B > 7.78$ T), the first exponential term in (5) disappears, and cross-correlation functions have only oscillating terms and a damped component:

$$C_S(\tau) \propto \cos\left(\frac{2\pi\tau}{P} + \Delta\Phi\right) e^{-|\tau|/\tau_d}, \tag{6}$$

where $\Delta\Phi=\Phi_j-\Phi_i$ is a difference in phase of stationary waves in corresponding points of the sample.

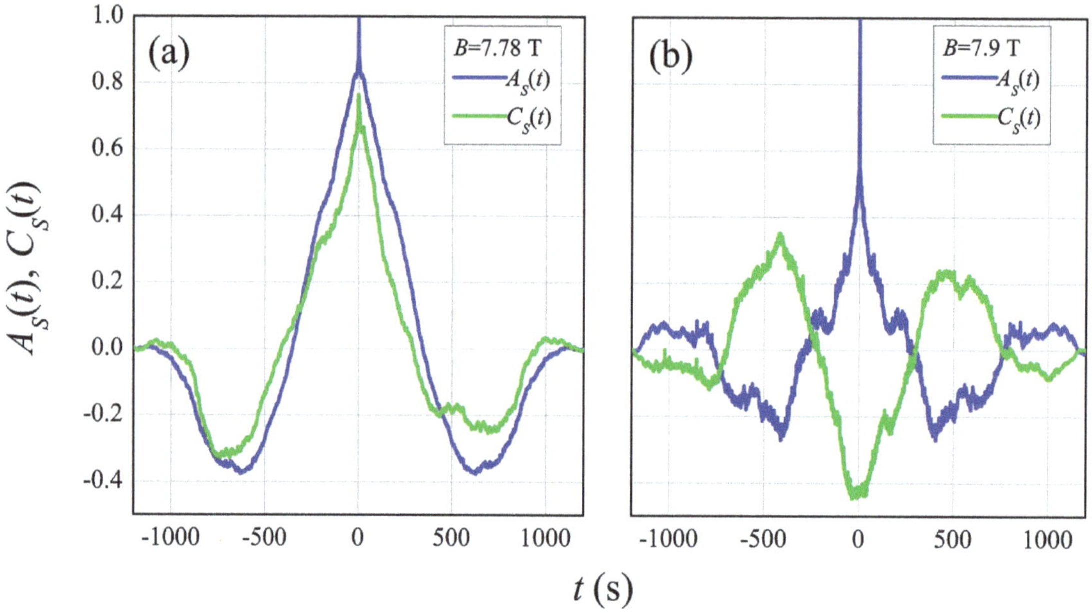

Figure 7. Autocorrelation functions of 2D PL signals from the fifth fiber and cross-correlation functions for the pair of fibers (2 and 5) at the magnetic fields of B=7.78 T (a) and B=7.9 T (b).

Thus, cross-correlation functions are sensitive to this difference in phase of stationary waves in corresponding points of the sample (**Figure 7b**). A phase difference indicates that the phase is not a random variable, but it has a fixed value, unalterable by the switching. It appears that a 2D PL intensity is modulated by "a standing wave", whose phase is defined by experimental conditions. An example of 2D electron systems in which a standing wave can be generated is plasmon excitation in the microwave (MW) field [[24]

4. Phase space portrait of the GOFs: beginning of the instability in the 2D system in a vicinity of $\nu = 2$; overview of the GOF effect and its possible mechanisms

It is well known that the comprehensive analysis of the time-series data of fluctuating signals may give useful information on the evolution of many kinds of dynamical systems. The study of GOF time dependences, using this idea, showed that there are complex regimes of a motion in our 2D electron system in the vicinity of ν=2 [14].

In particular, it was natural to expect a specific dynamics of PL intensity fluctuations with the presence of a modulating standing wave in a point of uniformity of the electron density. The phase portrait of a possible strange attractor of dissipative dynamical systems can be received using the methods described in the works [25–27].

It was shown in [26] that a time series of measurements of a single observable $x(t)$ of the dynamic system trajectory can be used to reconstruct qualitative features of the strange attractor in phase space. In our case, such a component is the time dependence of 2D electron PL intensity $I(t)$.

Sufficiently long (1–3 h) time sequences $I(t)$ with the steps of Δt=1 s at different values of magnetic field in the vicinity of $\nu = 2$ were recorded and analyzed. The sequence of the m-phase space vectors of one component $x(t)$ is given by:

$$x(t) = \left\{I(t), I(t+\tau), I(t+2\tau), \ldots, I(t+(m-1)\tau)\right\}. \quad (7)$$

Here, m is the embedding dimension and τ is the time shift. The parameter τ is taken to be the temporal correlation radius, whereby the autocorrelation function $A(t)$ of a time sequence goes to zero or has the first minimum. We used system (7) for the reconstruction of a pseudo-attractor of the fluctuating 2D system at various values of magnetic field. We revealed that, at the value m=3, there is such mode of noise in the point of uniformity of the electronic density (B=7.78 T) where the value of vectors (7) are eventually grouped in the vicinity of each of three axes of phase space (**Figure 8**, green curve). Moreover, it was found that the pronounced multiplying effect occurs in the phase space volume in this regime, that is, this phase portrait has distinctive features of a strange attractor [for the time sequences measured at the other magnetic, a strange attractor was not observed (**Figure 8**, red curve)]. The central 3D graph (the blue curve) demonstrates the result of a similar reconstruction when re-recording the noise under the close conditions; however, it has no such a characteristic phase portrait as shown in the top figure. A set of trajectories in phase space forms a structureless sphere in this case. A noise character differs for these two cases (**Figure 8**, left). In the first instance, fluctuations have discrete regions where the noise amplitude is insignificant. In the latter, the GOFs occur in a continuous mode, practically throughout all time series $I(t)$. It is necessary to notice that a noise character is essential to the understanding of this regime of a motion: fluctuations have discrete regions where the noise amplitude is insignificant. A strange attractor is absent in the system when the fluctuations occur in a continuous mode. An evaluation of correlation dimension requires some adjustment in the presence of data, taking into account transient states in the dynamics of attractors. We have provided a qualitative picture of the attractors here. An extensive statistical analysis including a consideration of these transient processes and the correlation entropy will be the subject of subsequent works. Nevertheless, we note that there is the process stage with nonlinear small-sized dynamics in a point of uniformity. The process of GOFs is not random in this regime but is governed by the limited number of control parameters. Such nonlinear small-sized dynamics is defined by the coherence of elements of 2D electron system. The described analysis points to a certain determinism in the behavior of a 2D electron system under the GOF conditions. Such determinism takes place at the beginning of instability in the system [25].

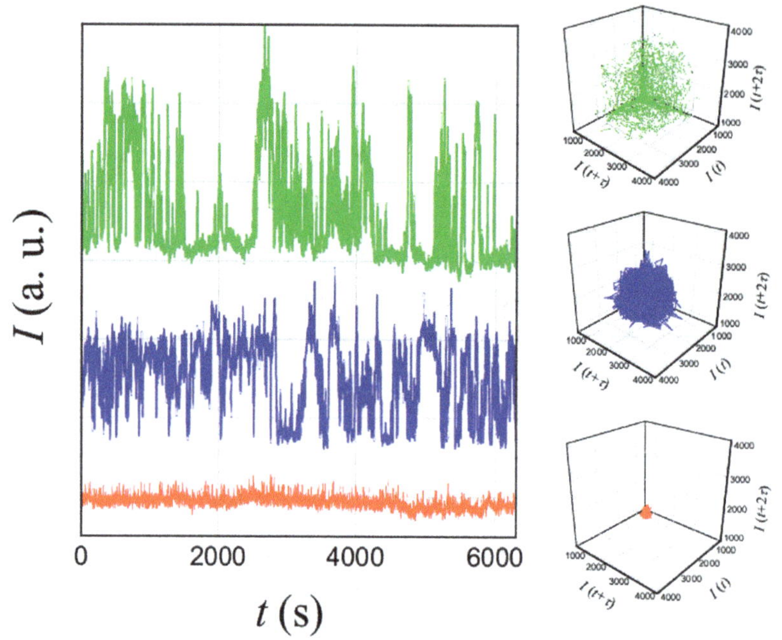

Figure 8. Time dependences of a 2D PL intensity measured at $\nu = 2$ (green and blue curves) and $\nu < 2$ (red curve; left). The measurements are carried out for the maximum of a PL line of the lowest Landau level; corresponding phase space portraits (right).

As an autocorrelation function of a periodic function is a function with the same period, the characteristic time t_{in} of the beginning of system instability can be derived. $A_S(t)$ data set at ν=2 showed that t_{in} greatly varies from several seconds to dozens of minutes. The study of the 2D PL spectral power density (SPD) obtained through the fast Fourier transform of $A_S(t)$ functions in the vicinity of ν=2 has revealed that this time can be as long as 20 min (**Figure 9**).

This time may be called an electronic "coherence" time. It should be noted that optical coherence time in the system studied is very short and it is defined by a spectral line width ($t \infty 1$ ps). **Figure 9** demonstrates the initiating step of the GOF process since before a point of uniformity. An SPD of noise is at a maximum value in the GOF regime. Following the factor ν=2, SPD decreased by several times within a narrow magnetic field interval (ΔB=0.02 T).

Thus, there is a mechanism that makes 2D electron concentration uniformly distributed across the all illuminated surface ($S \infty 1$ cm^2) under photoexcitation in the QHE regime. This attenuation of the electron density along a sample surface occurs due to the exchange of electrons between the neighboring regions of a 2D layer.

One possible scenario of this specific ordered 2D electron state formation under QHE conditions is the phase transition in the 2D electron system. A phenomenological model, describing the correlation functions of 2D PL signals, has been developed based on experimental data [17].

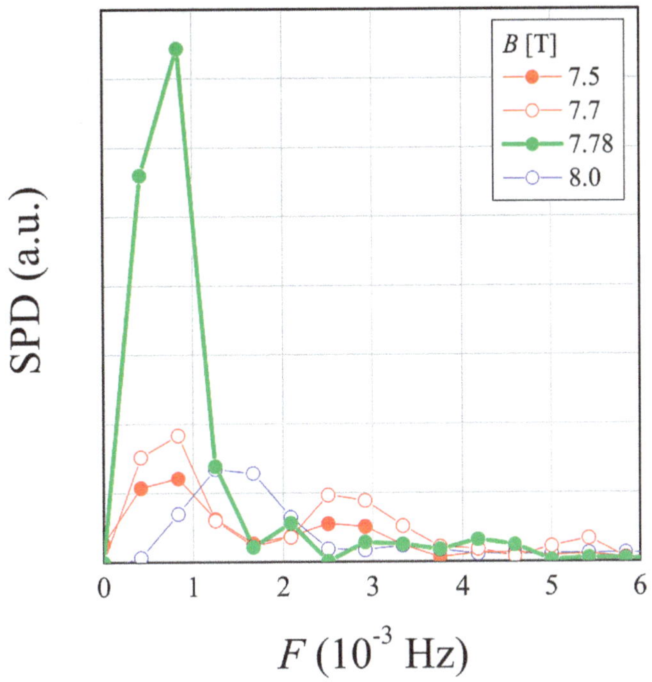

Figure 9. SPD for the second fiber at different magnetic fields ($\nu \geq 2$, $\nu=2$, and $\nu \leq 2$). The maximum time value of the beginning of system instability t_{in}=1370 s.

It gives the explanation for various phase conditions of a 2D electron system in a vicinity of the filling factor ν=2 in terms of the formation of incompressible quantum liquid. In that case, once the electron concentration is attenuated along the entire surface, a random impurity potential is no longer to be screened. Therefore, a large electrostatic energy is attributing to that state. This uniform state is unstable and there are spontaneous transitions between the Hall insulator states and a conducting state in the system that results in the two types of a noise in 2D system. When a magnetic field is over the uniformity point (ν=2), the system is divided into regions where there are drops of Hall liquid and an electron gas. Optical fluctuations go on, but the correlation between the regions decreases. With a further increase in a magnetic field, the drops gradually disappear. Vacuum fluctuations of the electromagnetic field can become an additional contribution to the system noise [10].

Under QHE conditions, small electron-density fluctuations can give rise to giant fluctuations of the conductivity of the 2D electron system and lead to fluctuation metal-insulator transitions. Another possibility involves the occurrence of a new coherent macroscopic state of the electron system described by a common wave function with a unified phase similar to superfluidity. Such coherent ordered electron states could occur due to the quantum leakage, such as the steady-state Josephson effect.

5. Summary

Thus, a system of 2D electrons in a perpendicular magnetic field in a vicinity of the filling factor ν=2 exhibits GOFs. The photon count statistics deviates appreciably from the Poisson description and a 2D electron layer in this regime cannot be considered as a system consisting of a high number of independent systems. A technique for the study of spatial correlations of optical fluctuations in 2D layer by means of the multifiber set under conditions of a big photoexcitation spot ($S \infty 1\ cm^2$) was developed. The study of the optical fluctuations, using this technique, showed that there are strong correlations between the radiation intensities from different points of the large sample. All areas of the 2D electron system emit light in a strongly correlated way under these circumstances, that is, the correlations are restricted by the dimensions of a sample. Hence, a 2D electron system acts like a macroscopic light source. Correlation analysis of the fluctuations showed that, in a 2D electron system, the uniform state with low PL intensity (ν=2) or the nonuniform state with high intensity ($\nu < 2$) could exist. The autocorrelation and cross-correlation functions of the fluctuations are periodic and practically coincide at ν=2 (a point of uniformity of the electron density). Cross-correlation functions are sensitive to the difference in the phase of stationary waves in corresponding points of the sample and the phase is strongly defined by the experimental conditions. When the filling factor ν=2, the electron density equalizing along the sample surface, as a consequence, we observe correlations of the PL signals at macroscopic distances. The noise correlation time (electronic "coherence" time) varies from several seconds to dozens of minutes. A photoexcited 2D electron system can be considered as an open dissipative system being far from equilibrium because it continuously gains energy due to laser excitation and consumes energy through the recombination of 2D electrons with photoexcited holes. The use of a mathematical analysis tool of the theory of nonlinear dynamic systems showed that the GOFs are a manifestation of complex dynamics in the system of interacting 2D electrons. The behavior of a 2D electron system at ν=2 corresponds to the regime of deterministic chaos: the strange attractor characterized by small-sized dynamics was revealed in the phase space of a 2D PL intensity in this regime. Thus, in this study, we deal with the radiation source in the 2D electron system in the vicinity of a critical point. This source is not chaotic and has a specific shape and intrinsic time of correlations. In the following studies, the question to be answered is whether these correlations resulted from the phase transition in a 2D electron system or the formation of a strongly correlated superfluidity-like quantum state.

Acknowledgements

The work was supported by the Russian Foundation for Basic Research.

Author details

Parakhonsky A*, Lebedev M, Dremin A and Kukushkin I

*Address all correspondence to: alpar@issp.ac.ru

Institute of Solid State Physics, Russian Academy of Sciences, Chernogolovka, Moscow Region, Russia

References

[1] Lerner V, Althsuler B, Fal'ko V, Giamarchi T, editors. Strongly Correlated Fermions and Bosons in Low-Dimensional Disordered Systems. 1st ed. London (UK): Springer; 2002. 398 p. DOI: 10.1007/978-94-010-0530-2

[2] Saminadayar L, Glattli D, Jin Y, Etienne B. Observation of the *e*/3 fractionally charged Laughlin quasiparticle. Phys. Rev. Lett. 1997; 79(13): 2526–2529http://dx.org/10.1103/PhysRevLett.79.2526

[3] De-Picciotto R, Reznikov M, Heiblum M, Umansky V, Bunin G, Mahalu D. Direct observation of a fractional charge. Nature. 1997; 389: 162–164. DOI: 10.1038/38241

[4] Tsui D, Stormer H, Gossard A. Two-dimensional magnetotransport in the extreme quantum limit. Phys. Rev. Lett. 1982; 48(22): 1559–1562. DOI: http://dx.doi.org/10.1103/PhysRevLett.48.1559

[5] Pudalov V, D'Iorio M, Kravchenko S, Campbell J. Zero-magnetic-field collective insulator phase in a dilute 2D electron system. Phys. Rev. Lett. 1993; 70(12): 1866–1869. DOI: http://dx.doi.org/10.1103/PhysRevLett.70.1866

[6] Kukushkin I, Timofeev V. Magneto-optics of strongly correlated two-dimensional electrons in single heterojunctions. Adv. Phys. 1996; 45(3): 147–242. DOI: 10.1080/00018739600101487

[7] Volkov O, Kukushkin I, Lebedev M, Lesovik G, von Klitzing K, Eberl K. Anomalous fluctuations of 2D-electron recombination radiation intensity in the quantum Hall regime. JETP Lett. 2000; 71(9): 383–386. DOI: 10.1134/1.568360

[8] Lebedev M, Kukushkin I , Volkov O, Parakhonskii A, Yu Smet, von Klitzing K. Spatial correlations of recombination radiation intensities of two-dimensional electrons under the conditions of the quantum Hall effect. JETP Lett. 2003; 77(6): 295–299. DOI: 10.1134/1.1577760

[9] Lebedev M, Volkov O, Parakhonskii A, Koval'skii V, Kukushkin I. Giant fluctuations of radiation intensity of two-dimensional electrons in double quantum wells. JETP Lett. 2004; 80(5): 317–320. DOI: 10.1134/1.1825113

[10] Lebedev M, Kukushkin I, Parakhonskii A, Kirpichev V, Volkov O, von Klitzing K. Fluctuations of the intersubband splitting energy and the potential well shape of two-dimensional electrons in the quantum Hall effect regime. JETP Lett. 2005; 82(3): 124–128. DOI: 10.1134/1.2086128

[11] Parakhonskiĭ A, Lebedev M, Kukushkin I, Yu Smet, von Klitzing K. Giant fluctuations in the radiation intensity of two-dimensional electrons under quantum Hall effect conditions. Phys. Solid State. 2007; 49(5): 976–983. DOI: 10.1134/S1063783407050289

[12] Parakhonsky A, Lebedev M, Kirpichev V, Kukushkin I. Hysteresis of giant intensity fluctuations of emission for two-dimensional electrons in the conditions for the integer

quantum Hall effect. Semiconductors. 2009; 43(1):76–81. DOI: 10.1134/S106378260901014X

[13] Parakhonsky A, Lebedev M, Dremin A. The effect of He-Ne illumination on giant optical fluctuations in double quantum wells under QHE conditions. Physica E. 2010; 42(5): 1610–1612. DOI: 10.1016/j.physe.2010.01.006

[14] Parakhonsky A, Lebedev M, Dremin A. Macroscopic lateral correlations of the giant optical fluctuations under the quantum Hall effect conditions. Physica E. 2011; 43(8): 1449–1452. DOI: 10.1016/j.physe.2011.04.004

[15] Parakhonsky A, Lebedev M, Dremin A, Kukushkin I. The strange attractor of giant optical fluctuations of 2D electrons in the quantum Hall effect regime. Physica E. 2012; 44(7–8): 1653–1656. DOI: 10.1016/j.physe.2012.04.014

[16] Parakhonsky A, Lebedev M, Dremin A, Kukushkin I. The study of spatial-temporal correlations of giant optical fluctuations of 2D electrons. Physica E. 2014; 56: 319–325. DOI: 10.1016/j.physe.2013.09.011

[17] Parakhonsky A, Lebedev M, Dremin A, Kukushkin I. Correlation and spectral aspects of giant optical fluctuations of 2D electrons. Physica E. 2015; 65: 1–3. DOI: 10.1016/j.physe.2014.07.011

[18] Mandel L, Wolf E. Optical Coherence and Quantum Optics. 1st ed. Epping (UK): Cambridge University Press; 1995. 1194 pp. DOI: 10.1017/CBO9781139644105

[19] Plaut A, Kukushkin I, von Klitzing K, Ploog K. Magneto-optics in GaAs-AlxGa1-xAs single heterojunctions. Phys. Rev. B. 1990; 42(9): 5744–5750. DOI: 10.1103/PhysRevB.42.5744

[20] Hanbury Brown R, Twiss R. Interferometry of the intensity fluctuations in light. I. Basic theory: the correlation between photons in coherent beams of radiation. Proc. R. Soc. A. 1957; 242(1230): 300–324. DOI: 10.1098/rspa.1957.0177

[21] Kadanoff L, Swift J. Transport coefficients near the liquid-gas critical point. Phys. Rev. 1968; 166(1): 89–101. DOI: 10.1103/PhysRev.166.89

[22] Van der Meulen H, Sarkar D, Calleja J, Hey R, Friedland K, Ploog K. Free versus localized hole magneto-photoluminescence in semiconductor heterojunctions near integer filling factors. Phys. Rev. B. 2004; 70(15): 155314 (1–7). DOI: 10.1103/PhysRevB.70.155314

[23] Gubarev S, Kukushkin I, Tovstonog S, Akimov M, Smet J, Von Klitzing K, Wegscheider W . JETP Lett. 2000; 72(6): 324–328. DOI: 10.1134/1.1328448

[24] Takhtamirov E, Melnik R. Excitation of plasmons in a two-dimensional electron gas with defects by microwaves: wake-field method. Phys. Rev. B. 2011; 84: 045313 (1–11). DOI: 10.1103/PhysRevB.84.045313

[25] Grassberger P, Procaccia I. Characterization of strange attractors. Phys. Rev. Lett. A. 1983; 50(5): 346–349. DOI: 10.1103/PhysRevLett.50.346

[26] Takens F. Detecting strange attractors in turbulence. In: Rand D, Young L-S, editors. Dynamical Systems and Turbulence. 1st ed. Berlin: Springer-Verlag; 1981. pp. 366–381. DOI: 10.1007/BFb0091924

[27] Schuster H, Just W. Deterministic Chaos: An Introduction. 2nd ed. Weinheim: Physik-Verlag; 1984. 248 pp.

9

Dynamic Resonant Tunneling

Er'el Granot and Gilad Zangwill

Abstract

The physics of dynamic resonant tunneling is investigated.

First, the resonant tunneling effect through an opaque barrier via a delta-function well is illustrated. Then, it is shown that, even in the adiabatic regime, where the dynamics can be governed by an analytic solution, the particle can be activated to higher energies. If the well varies quickly enough that the particle cannot escape from the well during the energetic elevation, the activation can be enhanced, as was anticipated by Azbel. However, and this is the main result of this work, the quasi-bound state of the well can even "reduce" the activation. In fact, because the resonant energy of the well matches twice the incoming particle's energy, and if the contribution to the wave function from both parts destructively interferes, then the particle cannot dwell in the well and activation is suppressed.

This effect can be utilized in frequency-controlled transistors, and it is even speculated that it may explain the reason that humans can distinguish between tens of thousands of different odors with merely few hundreds of odor receptors.

Lastly, the short time dynamics of a very fast perturbative well is also discussed.

Keywords: Resonant Tunneling, Dynamic Tunneling, Vibrational Tunneling, Odor detection, Olfactory, Forbidden activation, Selected Activation

1. Introduction

Resonant tunneling is a fascinating quantum phenomenon. It manifests the ability of quantum particles to pass with high probability through an opaque barrier by traveling via a semibound state [1–3].

The high sensitivity of the current on the bound-state parameters suggested harnessing this effect to heterostructure devices in general and transistors in particular [4–10].

Resonant tunneling is usually described as a one-dimensional (1D) phenomenon; however, resonant tunneling in higher number of dimensions was also investigated (see, for example, [11, 12]). Nevertheless, because the main features of resonant tunneling appear in 1D, most of the research was concentrated on the simplest 1D systems.

Tunneling and resonant tunneling are rarely stationary processes. They are affected by thermal noises, and clearly, the accumulation of particles in the bound state varies the potential. It is well known that tunneling in the presence of an oscillating barrier can cause activation (higher energy) and therefore can increase substantially the tunneling current.

This phenomenon was investigated in electronics [13], nanotechnology [14–16], the foundations of quantum mechanics [17–33], and even biology and biochemistry [34–39].

The resonant tunneling effect occurs when the incoming particle's energy coincides with the eigenenergy of the quasi-bound state. In case the barrier is very opaque, the particle remains inside the well at the quasi-bound state for exponentially long time. Therefore, when the particle is quasi-trapped inside the well, its state has to vary with the changes in the well, and its energy varies with the eigenenergy of the quasi-bound state because it does not have the time to escape from the well. Therefore, it was conjectured (see refs. [17, 20]) that a decrease in the perturbation time-scale will enhance the activation. However, not in any energy the particle can remain within the well. Destructive interference can prevent particle trapping and therefore suppress particle activation [23, 33]. In the next several sections, we will elaborate on the delicate structure of these effects.

2. Stationary tunneling

Let us begin with the propagation of a quantum particle through an opaque but stationary barrier. The Schrödinger equation is then

$$-\frac{\partial^2}{\partial x^2}\psi(x,t)+U(x)\psi(x,t)=i\frac{\partial\psi(x,t)}{\partial t} \tag{1}$$

Hereinafter for simplicity, we adopt the units, where the electron mass is half and the reduced Planck constant is unity (i.e., $m=1/2$ and $\hbar=1$).

In the stationary case (i.e., when the potential is time independent), there is no change in the incoming particle's energy. For any incoming energy ω, the generic stationary solution looks like

$$\psi(x,t)=\varphi_\omega(x)\exp(-i\omega t) \tag{2}$$

where $\varphi_\omega(x)$ are the solutions of the stationary Schrödinger equation:

$$-\frac{\partial^2}{\partial x^2}\varphi_\omega(x)+[U(x)-\omega]\varphi_\omega(x)=0. \tag{3}$$

Therefore, every solution can be written as a superposition of incoming $[\varphi_\omega^+(x)]$ and outgoing $[\varphi_\omega^-(x)]$ solutions, where

$$\varphi_\omega^+(x)=\begin{cases}\exp\left(i\sqrt{\omega}x\right)+r_\omega\exp\left(-i\sqrt{\omega}x\right) & x\to-\infty\\ t_\omega\exp\left(i\sqrt{\omega}x\right) & x\to\infty\end{cases}$$
$$\varphi_\omega^-(x)=\begin{cases}t_\omega\exp\left(-i\sqrt{\omega}x\right) & x\to-\infty\\ \exp\left(-i\sqrt{\omega}x\right)+r_\omega\exp\left(i\sqrt{\omega}x\right) & x\to\infty\end{cases} \tag{4}$$

Physically, $\varphi_\omega^+(x)$ and $\varphi_\omega^-(x)$ stand for beams of particles coming from the left and right respectively. t_ω and r_ω are the transmission and reflection coefficients of the barrier for energy ω, respectively. In the WKB approximation (see, for example, ref. [19]), the transmission coefficient can be evaluated as

$$t_\omega\sim\exp\left(-\int_{x_L}^{x_R}[U(x)-\omega]dx\right) \tag{5}$$

where x_L and x_R are the left and right boundaries of the barrier. For a rectangular barrier, an exact expression can be derived [19]:

$$t_\omega=\frac{\exp(-2ikL)}{\cosh(2KL)+\frac{1}{2}i(K/k-k/K)\sinh(2KL)} \tag{6}$$

where $k\equiv\sqrt{\omega}$ and $K\equiv\sqrt{U-\omega}$. Therefore, when the barrier is opaque [i.e., $\sqrt{U-\omega}L \gg 1$ (high and/or wide)], the transmission is exponentially small.

3. Resonant tunneling via a delta-function well

Let us introduce a delta-function well in the barrier (at $x = x_0$; in **Figure 1**, x_0=0). Then, the Schrödinger equation is

$$-\frac{\partial^2}{\partial x^2}\varphi_\omega(x)+[U(x)-\omega-f_0\delta(x-x_0)]\varphi_\omega(x)=0 \tag{7}$$

It is convenient to use the outgoing Green function $G_\omega^+(x,x_0)$, which is a solution of the equation:

$$-\frac{\partial^2}{\partial x^2}G_\omega^+(x,x_0)+[U(x)-\omega]G_\omega^+(x,x_0)=\delta(x-x_0) \tag{8}$$

with the boundary conditions:

$$G_\omega^+(x,x_0)\sim\exp\left(i\sqrt{\omega}|x-x_0|\right)\text{ for }|x|\to\infty. \tag{9}$$

Therefore, the outgoing Green function reads

$$G_\omega^+(x,x_0)=\begin{cases}\dfrac{\varphi_\omega^-(x_0)\varphi_\omega^+(x)}{\varphi_\omega^+(x_0)\varphi_\omega^{-\prime}(x_0)-\varphi_\omega^-(x_0)\varphi_\omega^{+\prime}(x_0)} & x>x_0\\ \dfrac{\varphi_\omega^-(x)\varphi_\omega^+(x_0)}{\varphi_\omega^+(x_0)\varphi_\omega^{-\prime}(x_0)-\varphi_\omega^-(x_0)\varphi_\omega^{+\prime}(x_0)} & x<x_0\end{cases}. \tag{10}$$

Using the Green function, we can easily construct a solution for the wave equation with the combined potential (the barrier with the delta-function well). In which case, the solution reads

$$\varphi_\omega(x)=\varphi_\omega^+(x)+\frac{f_0\varphi_\omega^+(x_0)}{1-f_0G_\omega^+(x_0,x_0)}G_\omega^+(x,x_0), \tag{11}$$

which for $x>x_0$ is

$$\varphi_\omega(x)=\frac{\varphi_\omega^+(x>x_0)}{1-f_0G_\omega^+(x_0,x_0)}=\frac{\varphi_\omega^+(x>x_0)}{1-f_0/\left(\varphi_\omega^{-\prime}(x_0)/\varphi_\omega^-(x_0)-\varphi_\omega^{+\prime}(x_0)/\varphi_\omega^+(x_0)\right)}.$$

In case of a rectangular barrier, i.e.,

$$U(x)=\begin{cases}U & |x|\le L\\ 0 & |x|>L\end{cases} \tag{12}$$

then

$$\varphi_\omega^+(x>L)=t_\omega\exp(ikx) \tag{13}$$

where $2L$ is the width of the barrier, and t_ω is taken from (6), i.e.,

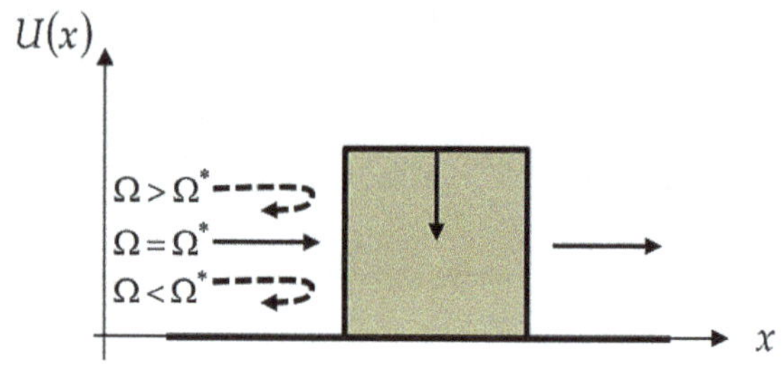

Figure 1. In a stationary resonant tunneling process, only when the particle's energy is equal to the quasi-eigenstate energy can the particle penetrate the barrier with high probability.

$$\varphi_\omega(x)=i\frac{\exp\left(ik(x-2L)-i\arctan\left[\left(K^2-k^2\right)/2kK\right]\right)}{1+i\left(1-\dfrac{f_0}{2K}\right)/4\left(\dfrac{kK}{U}\right)\exp(-2KL)}. \tag{14}$$

At the resonance $\omega=\Omega^*=U-f_0^2/4$, $T=|t_\omega|^2=1$, (see **Figure 2**).

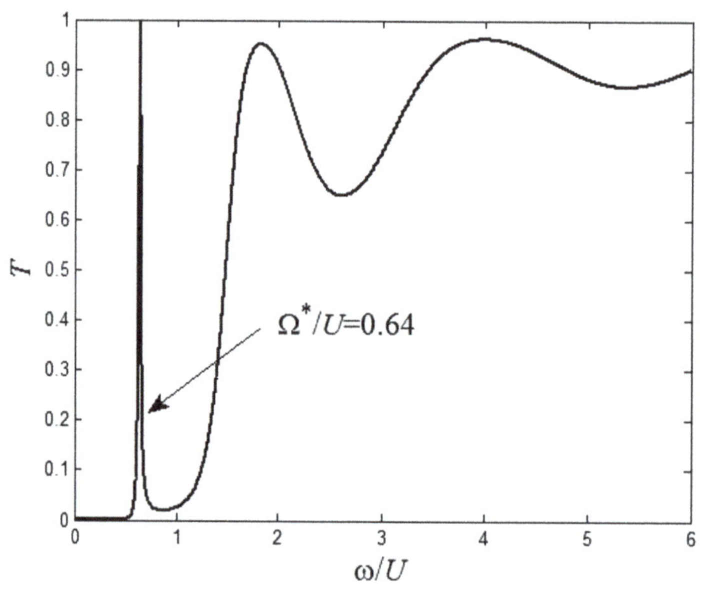

Figure 2. Transmission $T=|t_\omega|^2$ as a function of the particles energy ω for the parameters U=1, L=4 and f_0=1.2.

4. Adiabatic transition

Now, let us take a varying potential well,

$$f(t/\tau)=\frac{\lambda_0}{\tau}\exp\left[-(t/\tau)^2\right]. \tag{15}$$

In the adiabatic approximation, τ is longer than any other time-scale of the problem. The relevant time-scale (i.e., the longest one) is the resonance time or the dwelling time of the resonant state. Therefore, in the adiabatic approximation,

$$\tau U >> \exp\left(2\sqrt{U-\omega}\left(L-|x_0|\right)\right) \tag{16}$$

the potential can be regarded as stationary, and the wavefunction simply reads

$$\varphi_\omega(x)=\varphi_\omega^+(x)-\frac{\varphi_\omega^+(x_0)}{1-\tau\exp\left[(t/\tau)^2\right]/\left[\lambda_0 G_\omega^+(x_0,x_0)\right]}\frac{G_\omega^+(x,x_0)}{G_\omega^+(x_0,x_0)}. \tag{17}$$

The resonance energy of this system varies in time $\Omega^*(t)=U-f^2(t/\tau)/4$; therefore, large transmission occurs for $\omega=\Omega^*(t)$ and, in principle, can be as high as 1 when the well is located at the center of the barrier, i.e.,

When the lowest eigenenergy of the well

$$\Omega_{\min}^* \equiv \min\{\Omega^*(t)\}=U-f^2(0)/4=U-\lambda_0^2/4\tau^2 \tag{18}$$

is higher than the incoming energy ω, then the effect of the varying well is negligible; however, if the lowest eigenenergy of the well is lower than the incoming energy ω, then the eigenstate crosses the incoming energy twice. At the vicinity of the crossing time $t \cong t_0 = \pm\tau\sqrt{\ln\left(\frac{\lambda_0}{2K\tau}\right)}$, the solution has the Lorentzian shape:

$$\psi(x,t)=i\frac{\exp\left(ik(x-2L)-i\arctan\left[(K^2-k^2)/2kK\right]-i\omega t\right)}{1\mp it_0(t-t_0)/2\tau^2(kK/U)\exp(-2KL)}. \tag{19}$$

5. The general scenario

In principle, in the adiabatic approximation, the outgoing energy is equal to the incoming energy (i.e., $\omega_{out}=\omega_{in}\pm\tau^{-1}$), where τ^{-1} should be exponentially small $\tau^{-1}<<U\exp\left(-2\sqrt{U-\omega}(L-|x_0|)\right)$. However, in practice, the situation can be quite different due to the exponential decay of low energies in the tunneling process.

In general, the generic Schrödinger equation (see refs. [17, 33])

$$-\frac{\partial^2}{\partial x^2}\psi(x,t)+\left[U(x)-f(t/\tau)\delta(x-x_0)\right]\psi(x,t)=i\frac{\partial\psi(x,t)}{\partial t} \tag{20}$$

can be solved by a superposition of solutions of the type (11), namely,

$$\psi(x,t)=\varphi_{\Omega}(x)\exp(-i\Omega t)+\int_{-\infty}^{\infty}d\omega a(\omega)G_{\omega}^{+}(x,x_0)\exp(-i\omega t) \tag{21}$$

where it is taken that the incoming energy is Ω.

By substituting solution (21) into Eq. (20) (and after spatial integration), the solution can be reduced to the integral equation [17]:

$$-\frac{1}{2\pi}\varphi_{\Omega}(x_0)f(\Omega-\omega)+a(\omega)-\frac{1}{2\pi}\int_{-\infty}^{\infty}d\omega'' f(\omega''-\omega)a(\omega'')G_{\omega''}^{+}(x_0,x_0)=0 \tag{22}$$

where $f(\omega)=\int_{-\infty}^{\infty}dtf(t/\tau)\exp(-i\omega t)$ is the Fourier transform of the perturbation's amplitude.

6. Adiabatic and slow variations

In general, due to the complex structure of the Green function, this is a complex integral equation; nevertheless, as long as the spectrum of the function is mainly concentrated near the incoming energy Ω, i.e., $|a(|\omega-\Omega|\tau>>1)|<<|a(\omega=\Omega)|$, the contribution to the integral of the components $|\omega-\Omega|\tau>>1$ is negligible; therefore, we can replace $G_{\omega}^{+}(0)$ with $G_{\Omega}^{+}(0)$. In this case, the integral equation reduces to

$$-\frac{1}{2\pi}\varphi_{\Omega}(x_0)f(\Omega-\omega)+a(\omega)-\frac{1}{2\pi}\int_{-\infty}^{\infty}d\omega'' f(\omega''-\omega)a(\omega'')G_{\Omega}^{+}(x_0,x_0)=0 \tag{23}$$

which is merely a convolution equation; therefore, the inverse Fourier transform of the solution

$$a(t)=\frac{1}{2\pi}\int_{-\infty}^{\infty}d\omega a(\omega)\exp(i\omega t) \tag{24}$$

obeys [24, 33]

$$a(t)=-\frac{\varphi_{\Omega}(x_0)f(t/\tau)\exp(i\Omega t)}{1-f(t/\tau)G_{\Omega}^{+}(x_0,x_0)}. \tag{25}$$

Clearly, resonance occurs when $f(t/\tau)\Re G_{\Omega}^{+}(x_0, x_0)=1$. We will see in the next sections that nontrivial effects occur when $f(0)\Re G_{\Omega}^{+}(x_0, x_0)>1$. In this regime, it is invalid to substitute G_{Ω}^{+}

(x_0,x_0) for $G_\omega^+(x_0,x_0)$. However, in the $f(0)\Re G_\Omega^+(x_0, x_0)<1$ (where it is taken that $f(0) = \max\{f(t/\tau)\}$) regime, this approximation is still valid.

It is clear, for example, that, when $f(0)\Re G_\Omega^+(x_0, x_0)<<1$, which is equivalent to the adiabatic regime,that

$$a(t)\cong-\varphi_\Omega(x_0)f(t/\tau)\exp(i\Omega t), \tag{26}$$

and thus

$$a(\omega)\cong-\frac{1}{2\pi}\varphi_\Omega(x_0)f(\omega-\Omega),$$

which means that the spectrum broadening is exactly similar to the spectrum of the perturbation; in which case the solution is simply

$$\psi(x,t)\cong\varphi_\Omega(x)\exp(-i\Omega t)-\frac{\varphi_\Omega(x_0)}{2\pi}\int_{-\infty}^{\infty}d\omega f(\omega-\Omega)G_\omega^+(x,x_0)\exp(-i\omega t). \tag{27}$$

7. Activation

From Eq. (27), it is evident that elevation to higher energies is still a possibility even in the adiabatic and slowly varying cases, and because the Green function $G_\omega^+(x,x_0)$ increases with the energy, there is still a possibility that the mean exit energy will be considerably higher than the incoming one. The outcome depends only on the specific functional shape of the perturbation spectrum.

More importantly, when the perturbation becomes more energetic and τ decreases so that $f(0)\Re G_\Omega^+(x_0, x_0)$ approaches 1 from below (i.e., there is still no intersection), the adiabatic Eq. (25) is still approximately valid. In this case, however, one can take

$$f(t/\tau)\cong(\lambda_0/\tau)\left[1-(t/\tau)^2\right]. \tag{28}$$

Then, after substituting Eq. (28) in Eq. (25) and both of them in Eq. (24),

$$\begin{aligned}a(\omega)&=-\varphi_\Omega(x_0)\frac{\lambda_0}{\tau}\int\frac{\exp(-i(\omega-\Omega)t)}{1-(\lambda_0/\tau)G_\Omega^+(x_0,x_0)+(t/\tau)^2}dt=\\&-\varphi_\Omega(x_0)\lambda_0\sqrt{\frac{\pi}{2}}\exp\left(-\sqrt{1-\beta}\tau|\omega-\Omega|\right)\end{aligned} \tag{29}$$

where $\beta \equiv (\lambda_0/\tau)G_{\Omega}^{+}(x_0, x_0)$. Therefore,

$$\psi(x,t) \cong \varphi_{\Omega}(x)\exp(-i\Omega t) - \varphi_{\Omega}(x_0)\lambda_0\sqrt{\frac{\pi}{2}}\int_{-\infty}^{\infty} d\omega \exp\left(-\sqrt{1-\beta}\tau|\omega-\Omega|\right)G_{\omega}^{+}(x,x_0)\exp(-i\omega t). \tag{30}$$

Now, because the Green function can be written (beyond the barrier) approximately as

$$G_{\omega}^{+}(x > L, x_0) \sim \exp\left(-\sqrt{U-\omega}(L-x_0)\right) \tag{31}$$

then the exponent in the integrand, which can be regarded as an approximate evaluation of the spectrum of the outgoing wavefunction, consists of two main terms (two peaks):

$$\exp\left(-\sqrt{1-\beta}\tau|\omega-\Omega| - \sqrt{U-\omega}(L-x_0)\right) \tag{32}$$

The first peak occurs around the incoming energy Ω (suppressed activation) and the second one occurs around the barrier's height U (activation).

Each one of these peaks has a different height. The higher one will determine whether activation will occur. Therefore, the probability to tunnel through the barrier and to exit with energy ω_{out} is approximately [33]

$$P(\omega_{out}) \propto \left|a(\omega_{out})G_{\omega_{out}}^{+}(x > L)\right|^2 \propto \exp\left(-2\sqrt{1-\beta}\tau|\omega_{out}-\Omega| - 2\sqrt{U-\omega_{out}}(L-x_0)\right) \tag{33}$$

which means that the activation probability peak is proportional to

$$P(\omega_{out} \cong U) \propto \exp\left(-2\sqrt{1-\beta}\tau|U-\Omega|\right) \tag{34}$$

whereas the inactivation probability peak (i.e., the probability for suppressed activation) goes like

$$P(\omega_{out} \cong \Omega) \propto \exp\left(-2\sqrt{U-\Omega}(L-x_0)\right). \tag{35}$$

In **Figure 3**, we demonstrate the fact that the spectrum is governed by two maxima (peaks) and that Eq. (33) is a good approximation to the numerical solution at the vicinity of these maxima. Therefore, activation occurs when

$$P(\omega_{out} \cong U) \propto \exp\left(-2\sqrt{1-\beta\tau}|U-\Omega|\right) > P(\omega_{out} \cong \Omega) \propto \exp\left(-2\sqrt{U-\Omega}(L-x_0)\right). \tag{36}$$

This occurs approximately at the time scale

$$\tau_T \cong \frac{\lambda_0 + \sqrt{\lambda_0^2 + 16L^2}}{4\sqrt{U-\Omega}}, \tag{37}$$

or, for a given τ, the transition occurs for the following incoming energy:

$$\Omega_T \cong U - \frac{2\lambda_0\left(\lambda_0 + \sqrt{\lambda_0^2 + 16L^2}\right) + 16L^2}{16\tau^2}. \tag{38}$$

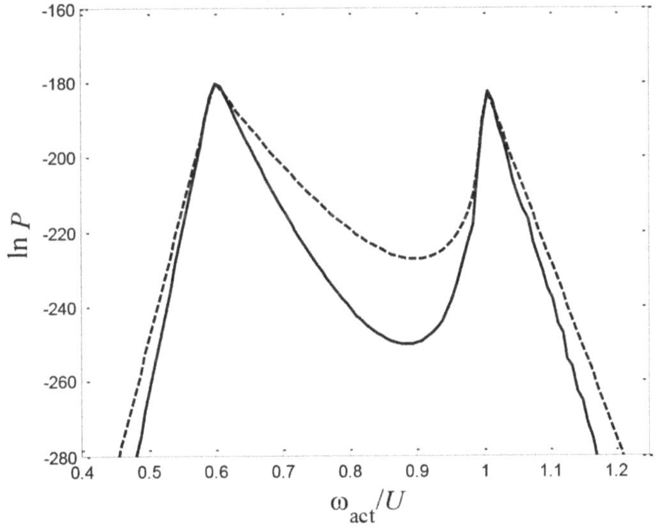

Figure 3. Comparison between the logarithm of the exact numerical solution (solid line) and the analytical approximation (dashed line), i.e., Eq. (33), for $\lambda_0=100, L\sqrt{U}=150,\ \Omega/U=0.6=\Omega_T, \tau=\tau_T=280U$.

Clearly, this energy is lower than the minimum resonance energy

$$\Omega_T < \Omega^*_{\min} = U - \frac{\lambda_0^2}{4\tau^2}. \tag{39}$$

In **Figure 4**, the dependence of the spectrum on the transition time-scale τ is presented for three different values: below τ_T where activation prevails, above τ_T when simple tunneling wins, and when they are equal, and the outgoing particle's spectrum has two equally probable outgoing energies.

It should be stressed that, because these peaks are exponentially narrow, the transition is extremely abrupt (i.e., the process resembles a phase transition). The identification of the process as a phase transition was first suggested by Azbel [17].

However, it was wrongly assumed that, if the particle's incoming energy matches the quasi-eigenstate energy, then an eigenstate-assisted activation (EAA) effect occurs (i.e., if $\Omega>\Omega_{min}^*$, then activation will definitely increase). In fact, it will be shown that this process is more complicated, and at some energies (above Ω_{min}^*), activation is "totally" suppressed.

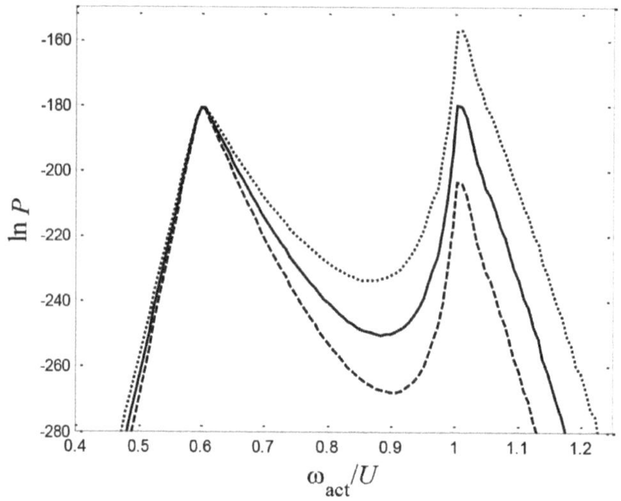

Figure 4. The (logarithm of) the exit probability as a function of the activation energy ω_{act} for the parameters $\lambda_0=100$, $L\sqrt{U}=150$, $\Omega/U=0.6$, for three different perturbation time scales: $\tau = \tau_T$ (solid line), $\tau = \tau_T + 20$ (dashed line) and $\tau_T - 20$ (dotted line).

8. Selected elevations and forbidden activations

For $\Omega>\Omega_{min}^*$, the spectrum's shape becomes more complicated. Instead of only two peaks, it has a more complex structure. There is a clear difference between the $\omega_{act}<\Omega$ (i.e., the under-activated regime) and the $\omega_{act}>\Omega$ (i.e., the activated one). The former oscillates as a function of ω_{act} but almost independent of the incoming Ω, whereas the latter oscillates as a function of the incoming Ω but has a mild dependence on the outgoing ω_{act}. In **Figure 5**, there is an illustration of this behavior, where a small change in the incoming particle's energy has an enormous effect on the activated regime. In **Figure 6**, a numerical example illustrates this phenomenon where a ~6.7% change in the incoming energy made a dramatic change from full activation to suppressed one.

As a consequence, it is clear that, for specific incoming particle's energies, the entire activated part of the spectrum is suppressed. To illustrate this point, we define the mean activated energy

$$\langle \omega_{act} \rangle \equiv \frac{\int \omega_{act} P(\omega_{act}, \Omega) d\omega_{act}}{\int P(\omega_{act}, \Omega) d\omega_{act}} \tag{40}$$

where $P(\omega_{act}\Omega)$ is the probability of an incoming particle with energy Ω to exit the barrier with the energy ω_{act}.

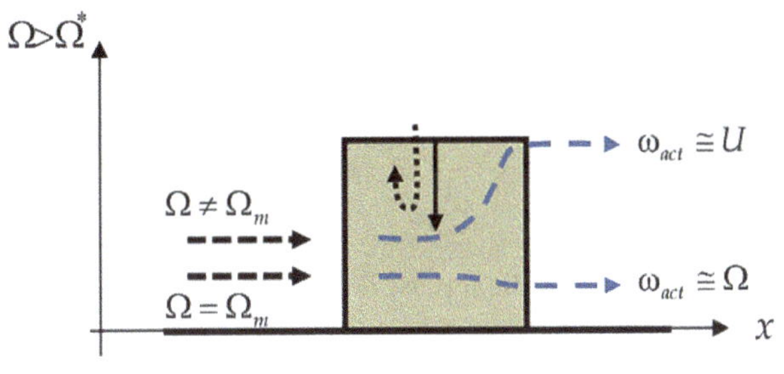

Figure 5. Schematic illustration of the suppressed activation. For most energies, activation occurs (i.e., $\omega_{act} \cong U$); however, for the specific energies (i.e., Ω=Ω_m), activation is suppressed and $\omega_{act} \cong \Omega = \Omega_m$.

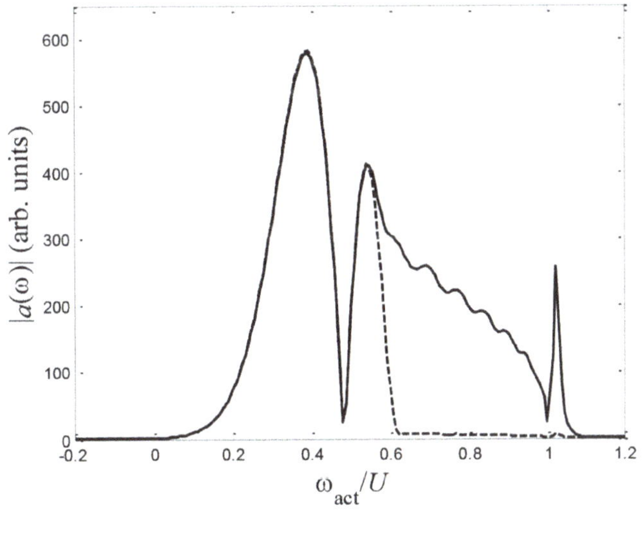

Figure 6. Absolute value of the transmission coefficient $a(\omega)$ as a function of the activation energy ω_{act}. The dashed curve corresponds to the case Ω/U = 0.6 and the solid line corresponds to Ω/U = 0.56. The other parameters are λ_0=100, $L\sqrt{U}$=10, τU=60.6.

In **Figure 7** the mean activation energy $<\omega_{act}>$ is plotted as a function of the perturbation time scale τ, and in **Figure 8**, $<\omega_{act}>$ is plotted as a function of the incoming particle's energy Ω. It is clearly seen that activation ($<\omega_{act}> \cong U$) occurs mainly below $\tau<\tau_T$. However, even below this time-scale, there are specific values of τ, for which activation is suppressed (i.e., $<\omega_{act}> \cong \Omega$). Similarly, activation occurs $<\omega_{act}> \cong U$ mainly above $\Omega>\Omega_T$; however, even in the activation regime, there are specific energies for which $<\omega_{act}> \cong \Omega$ (i.e., suppressed activation).

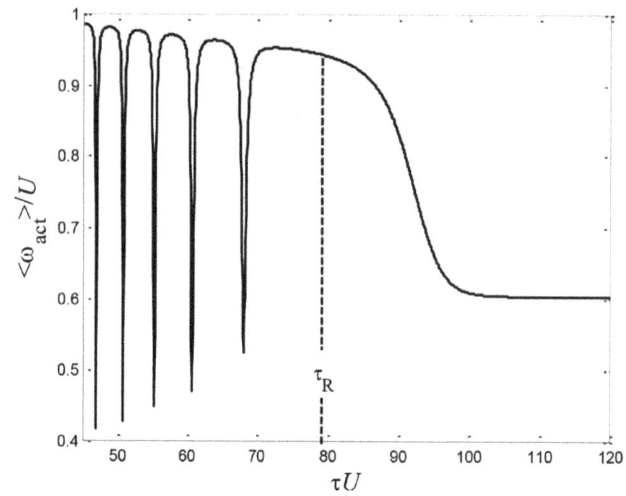

Figure 7. Mean activation energy as a function of the time-scale τ for the parameters: $\Omega/U = 0.6$, $L\sqrt{U}=6$, $\lambda_0=100$.

When $\tau<\tau_T$, two important things occur: (1) At two specific times, the particle's incoming energy is equal to the eigenenergy of the quasi-bound state of the varying well. (2) The well varies quickly enough so the particle has no time to escape from the well.

As a consequence of these two, the particle's state changes with the well's eigenstate; therefore, it is easier to excite the particle energetically. That was the logic that led Azbel to predict the EAA effect. Indeed, this effect does occur, and it is clearly seen (see **Figure 4**) that, when $\tau<\tau_T$, then, for most values of τ, the spectrum's energy is concentrated around the barrier's height ***U***. However, this process cannot last if the particle cannot dwell inside the quasi-bound state. This event occurs when there is destruction interference inside the well.

Had it been a stationary eigenstate with an eigenenergy Ω_0 the eigenstate would accumulate a linear phase [i.e., $\exp(-i\Omega_0 t)$].

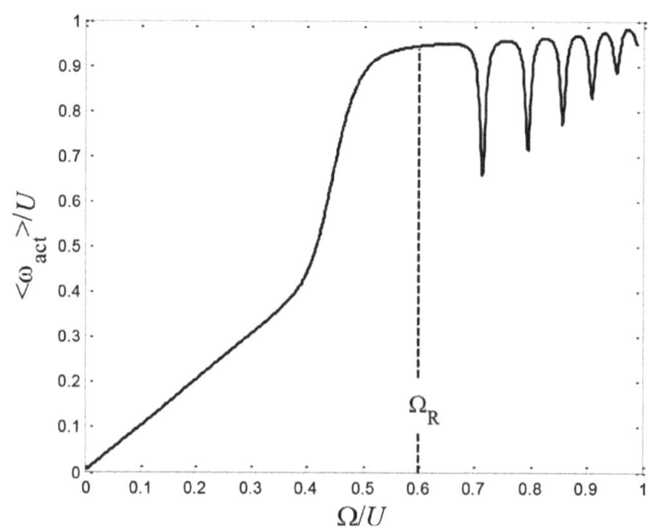

Figure 8. Mean activation energy as a function of the incoming energy Ω/U for $\tau U=79$, $L\sqrt{U}=6$, $\lambda_0=100$, Ω_R is the resonance energy

However, because the quasi-bound state evolves in time, it gains the integral

$$\exp\left(-i\int dt'\Omega^*(t')\right) \tag{41}$$

When the incoming energy is above the minimum eigenenergy (i.e., $\Omega>\Omega_{min}^*=U-\lambda_0^2/4\tau^2$), there are two times, in which $\Omega=\Omega^*(t_1)=\Omega^*(t_2)$ (see **Figure 9**), and due to the temporal symmetry of the perturbation $t_1=-t_2$. Therefore, the particle has two options to be temporally bounded to the quasi-eigenstate: it can either begin at t_1 and gain the phase $\exp\left(-i\int_{t_1}^{t} dt'\Omega^*(t')\right)$ or at t_2 and gain the phase $\exp\left(-i\Omega(t_2-t_1)-i\int_{t_2}^{t} dt'\Omega^*(t')\right)$. If the two components are out of phase and a destructive interference occurs [33], i.e.,

$$\int_{t_1}^{t_2} dt'\left[\Omega^*(t')-\Omega\right]=\pi(2m+1) \text{ for } m=0,1,2,... \tag{42}$$

the particle cannot survive within the well, and activation is frustrated.

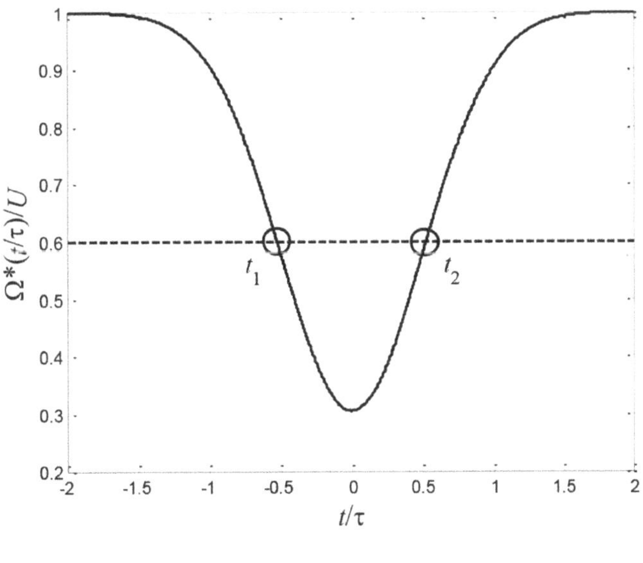

Figure 9. When the minimum of the resonance energy of the perturbation is lower than the incoming energy Ω, then the instantaneous resonance energy Ω^*t crosses the incoming energy twice (at t_1 and t_2). For successful activation, the cumulative phase between these two events must be constructive.

In our case, at the vicinity of the parabola peak,

$$\Omega^*(t)=U-\left(\frac{\lambda_0}{2\tau}\right)^2\exp\left(-\frac{2t^2}{\tau^2}\right)\cong U-\left(\frac{\lambda_0}{2\tau}\right)^2\left(1-\frac{2t^2}{\tau^2}\right) \tag{43}$$

After substituting (43) into (42), the values of the forbidden energies Ω_m, for which destructive interference occurs and the activation is suppressed, are directly given

$$\Omega_m \cong \Omega^*_{\min}+\frac{1}{2}\left[\frac{3\lambda_0}{\tau^2}\left(m+\frac{1}{2}\right)\pi\right]^{2/3} \quad \text{for } m=0,1,2,... \tag{44}$$

In each one of these energies, the activation is suppressed.

To determine these energies more accurately, we take advantage of the fact that, at the vicinity of the minimum $\Omega_{\min}{}^*$, the instantaneous resonance energy has a parabola shape; therefore, any varying potential with the same parabola should have approximately the same suppressed energies. Therefore, we replace the Gaussian with a parabolic function, that is, we choose Eq. (28) for the perturbation, namely, $f(t/\tau) \cong (\lambda_0/\tau)(1-t^2/\tau^2)$, then

$$f(\omega)\cong\int\frac{\lambda_0}{\tau}\left(1-\frac{t^2}{\tau^2}\right)\exp(-i\omega t)dt=2\pi\frac{\lambda_0}{\tau}\left[\delta(\omega)+\frac{1}{\tau^2}\delta''(\omega)\right]. \tag{45}$$

Therefore, the integral equation

$$-\frac{1}{2\pi}\varphi_\Omega(x_0)f(\Omega-\omega)+a(\omega)-\frac{1}{2\pi}\int_{-\infty}^{\infty}d\omega''\, f(\omega''-\omega)a(\omega'')G^+_{\omega''}(x_0,x_0)=0 \tag{46}$$

reduces to the differential equation

$$\frac{d^2s(n)}{dn^2}+s(n)\left[1-\frac{1}{G^+_{\omega''}(x_0,x_0)}\frac{\tau}{\lambda_0}\right]=\varphi_\Omega(x_0)\lambda_0\delta(n) \tag{47}$$

where we used the dimensionless parameters $n=(\omega-\Omega)\tau$ and .

After linearization of the Green function, Eq. (47) can be approximated to

$$\frac{d^2s(n)}{dn^2}+s(n)\left[1-2K\frac{\tau}{\lambda_0}+\frac{n}{\lambda_0 K}\right]=\varphi_\Omega(x_0)\lambda_0\delta(n) \tag{48}$$

where again $K\equiv\sqrt{U-\Omega}$.

The solution that maintain the boundary conditions that $s(n\to\infty)\to 0$ is

$$s(\xi) = -i\pi\varphi_{\Omega}(x_0)\lambda_0(2\lambda_0 K)^{1/3}\begin{cases} \mathrm{Ai}(-\xi)[\mathrm{Ai}(-\xi_0) + i\mathrm{Bi}(-\xi_0)] & \text{for} \quad \xi < \xi_0 \\ \mathrm{Ai}(-\xi_0)[\mathrm{Ai}(-\xi) + i\mathrm{Bi}(-\xi)] & \text{for} \quad \xi > \xi_0 \end{cases} \tag{49}$$

where $\xi \equiv (n + K(\lambda_0/\tau - 2K)\tau)/(2\lambda_0 K)^{1/3}$, $\xi_0 \equiv (K(\lambda_0/\tau - 2K)\tau)/(2\lambda_0 K)^{1/3}$ and Ai and Bi are the Airy functions [40].

Therefore, it is clear that activation is suppressed when

$$\mathrm{Ai}(-\xi_0) = 0\,, \tag{50}$$

which, in the slowly varying approximation (i.e., large τ), correspond to (see [40])

$$\cos^2\left[\frac{2}{3}\left(1 - \frac{2K}{\lambda_0}\tau\right)^{3/2} K\lambda_0 - \frac{\pi}{4}\right] = 0\,. \tag{51}$$

Therefore, the incoming energies for which $\Omega_{\mathrm{act}} \cong \Omega$, and thus no activation occurs, are approximately (see **Figures 10** and **11**)

$$\Omega_m = \Omega^*_{\min} + \frac{1}{2}\left[\frac{3\lambda_0}{\tau^2}\left(m + \frac{3}{4}\right)\pi\right]^{2/3}\,. \tag{52}$$

Therefore, Eq. (42) should be rewritten more accurately as

$$\int_{t_1}^{t_2} dt'\left[\Omega^*(t') - \Omega\right] = \pi\left(2m + \frac{3}{2}\right), \tag{53}$$

like destructive interference condition in the WKB approximation (see, for example, ref. [19]). Eq. (53) can be applied to any varying potential whose temporal shape has minima.

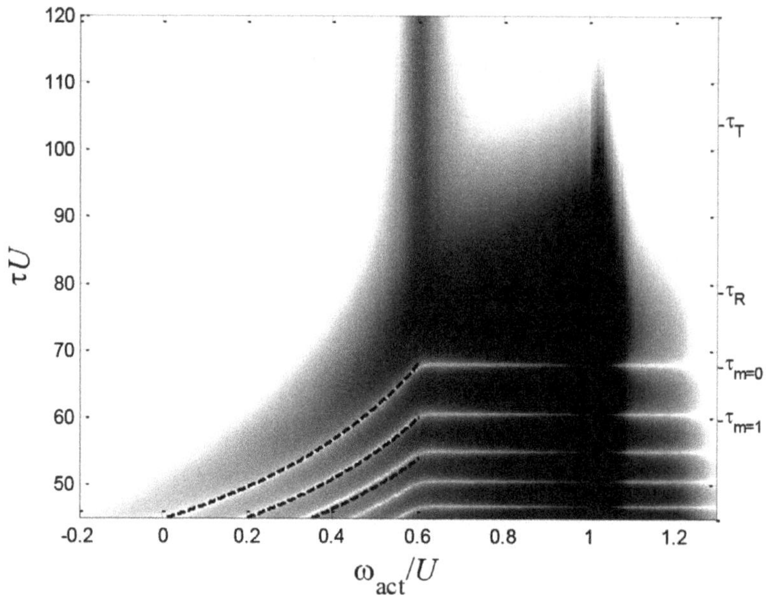

Figure 10. Approximate analytical expression of the forbidden time-scales (dashed curves) with the exact numerical solution of the probability density (the darker the spot the higher probability it represents). τ_m are the time scales for suppressed activations (52). τ_R is the minimum time scale $\tau_R \equiv -\lambda_0 G_\Omega^+(0)$, and τ_T is the transition time when activation wins. With the parameters: $\Omega / U = 0.6$, $\lambda_0 = 100$, and $L\sqrt{U} = 10$.

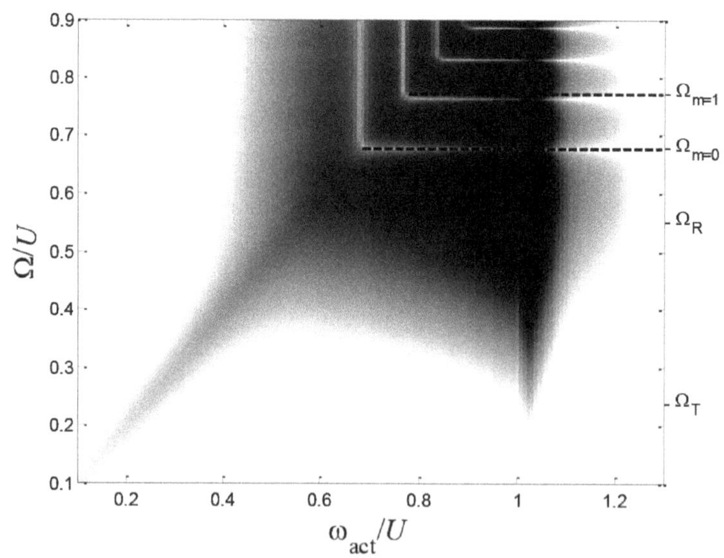

Figure 11. Presentation of the analytical suppressed activation energies Ω_m (dashed lines) on top of the numerical solution of the probability to be activated to energy ω_{act} (the darker the spot, the higher the probability it represents). Ω_T is the transition energy when activation wins, and Ω_R is the minimum resonant energy. The other parameters are $\tau U = 75$, $\lambda_0 = 100$ and $L\sqrt{U} = 10$.

Because (52) was derived for any potential, which can be approximated by a parabola [Eq. (28)], then the same conclusions and the same suppression of activation are valid to periodic potentials of the type

$$f(t/\tau)=\frac{\lambda_0}{\tau}\cos(2t/\tau)\delta(x-x_0). \tag{54}$$

See ref. [23] for a more extensive study on these potentials; nevertheless, the suppressed activation energies (52) and (53) are still valid for these potentials.

9. Instantaneous changes

Lastly, we are going to investigate the scenario in which the perturbation appears instantaneously. In this case, the Schrödinger equation reads

$$-\frac{\partial^2\psi}{\partial x^2}-f_0u(t)\delta(x-x_0)\psi+U(x)\psi=i\frac{\partial\psi}{\partial t} \tag{55}$$

where $u(t)=\begin{cases}1 & t\geq 0\\ 0 & t<0\end{cases}$ is the Heaviside step function.

It was shown in ref. [41] that the effect of such a perturbation has a universal pattern in the short time domain and in fact depends only on the product of the strength of the delta-function (f_0 in this case) and the local value of the initial wavefunction. The idea is [41] that an instantaneous delta-function perturbation is equivalent to a discontinuity in the wavefunction, and it was proven elsewhere [42, 43] that such a discontinuity has, in the short time, a universal pattern even in the presence of potentials [44]. Therefore, if the initial state was

$$\psi(x,t<0)=\varphi_\Omega^+(x)\exp(-i\Omega t) \tag{56}$$

then, after the instantaneous perturbation, the short time dynamics is simply [41]

$$\psi(x,t)\cong\varphi_\Omega^+(x)\exp(-i\Omega t)-2\sqrt{\frac{-i}{\pi}}\frac{t^{3/2}}{(x-x_0)^2}\exp\left(i\frac{(x-x_0)^2}{4t}\right)f_0\varphi_\Omega^+(0). \tag{57}$$

Because the first term is exponentially smaller than the second one, then very quickly the second term becomes dominant. In fact, it becomes dominant as early as

$$\frac{t^3}{(x-x_0)^4}\left(\frac{Uf_0^2}{U-\Omega}\right)\exp(K(L-x_0))>1. \tag{58}$$

If the delta-function well is turned on instantaneously but only for a period of 2τ, then the Schrödinger equation can be written as

$$-\frac{\partial^2\psi}{\partial x^2}-\frac{\lambda_0}{2\tau}[u(t+\tau)-u(t-\tau)]\delta(x-x_0)\psi+U(x)\psi=i\frac{\partial\psi}{\partial t}, \tag{59}$$

and the solution is

$$\psi(x,t)\cong\varphi_\Omega^+(x)\exp(-i\Omega t)-\\ \sqrt{\frac{-i}{\pi}}\frac{\lambda_0\varphi_\Omega^+(0)}{\tau(x-x_0)^2}\left\{(t+\tau)^{3/2}\exp\left(i\frac{(x-x_0)^2}{4(t+\tau)}\right)-(t-\tau)^{3/2}\exp\left(i\frac{(x-x_0)^2}{4(t-\tau)}\right)\right\} \tag{60}$$

which can be further simplified (provided $\tau<<t$) to

$$\psi(x,t)\cong\varphi_\Omega^+(x)\exp(-i\Omega t)+i\sqrt{\frac{-i}{\pi}}\frac{t^{-1/2}\lambda_0\varphi_\Omega^+(0)}{2}\exp\left(i\frac{(x-x_0)^2}{4t}\right)\mathrm{sinc}\left(\frac{\tau(x-x_0)^2}{4t^2}\right) \tag{61}$$

where $\mathrm{sinc}(x)\equiv\sin(x)/x$.

This result again suggests that, if the variation occurs quickly enough, there is no dependence on the incoming particle's energy Ω. In fact, the variation term vanishes for

$$\frac{\tau(x-x_0)^2}{4t^2}=m\pi \text{ for } m=1,2,3,... \tag{62}$$

Moreover, when $\frac{\tau(x-x_0)^2}{4t^2}<<1$, the second term has a totally universal pattern, which is even independent of τ:

$$\psi(x,t)\cong\varphi_\Omega^+(x)\exp(-i\Omega t)+i\sqrt{\frac{-i}{\pi}}\frac{t^{-1/2}\lambda_0\varphi_\Omega^+(0)}{2}\exp\left(i\frac{(x-x_0)^2}{4t}\right). \tag{63}$$

10. Applications

(A) The effect of controlled activation can be used in dynamic heterostructures, which can be used as frequency effect transistors. A schematic presentation of the system is presented in **Figure 12**. The barrier can be constructed by a semiconductor, and its potential shape can be controlled by the gate's voltage. In such a device, the current from source to drain will be governed by the frequency of the gate and therefore can be much more sensitive than any other transistor (because frequency, unlike voltage or currents, can be determined with great accuracy).

In **Figure 13**, the current $j \propto |i\psi'(x,t)\psi^*(x,t) - c.c.|$ [where $\psi(x,t)$ is the solution of Eq. (21)] of such a device is plotted as a function of the incoming energy Ω and τ. The forbidden activations values are clearly seen by the white stripes. They are exponentially narrow; therefore, the device's current can be controlled by small variations in τ.

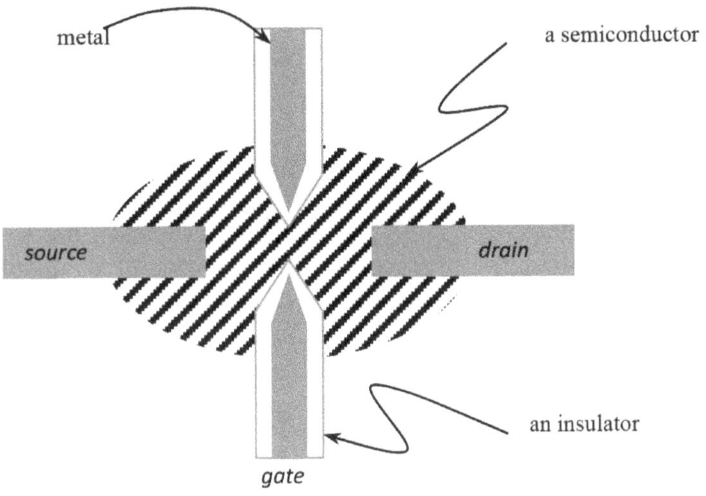

Figure 12. Schematic illustration of a frequency effect transistor. The semiconductor functions as an effective barrier, and the gate voltage oscillates with frequency ω to create the oscillating region. The insulating layer prevents current leakage from the gate to the drain.

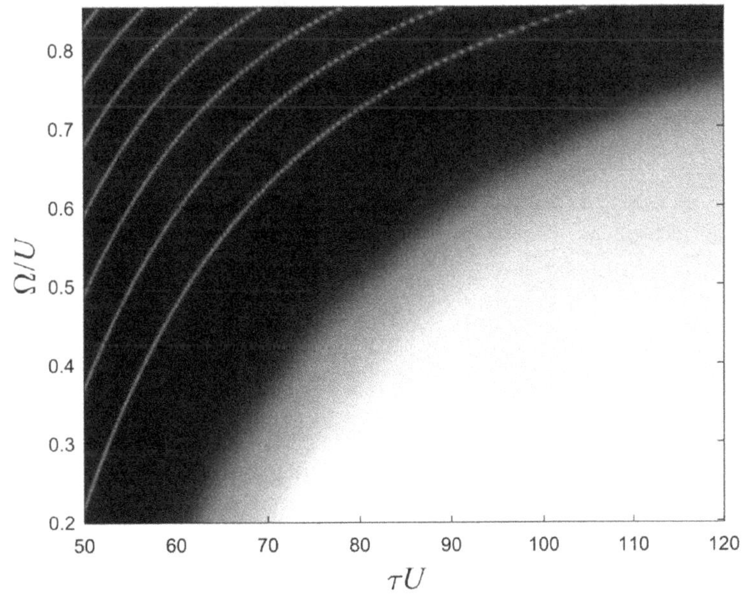

Figure 13. Plot of the current $j \propto |i\psi'(x,t)\psi^*(x,t) - c.c.|$ as a function of the Ω and τ for $\lambda_0 = 100$ and $L\sqrt{U} = 10$. The darker the color, the higher is the current.

(B) As mentioned in the Section 1, there are evidences that odor detection is governed by dynamic resonant tunneling. It is known that odor receptors are, in some sense, like resonant tunneling devices—only specific molecules, with the right chemical properties, "activate" the receptor, which sends a signal to the brain. However, it was recently recognized that the olfactory system could distinguish between two molecules, which have the same chemistry but have different mechanical properties. That is, in cases where one of the atoms in a molecule is replaced with one of its isotopes (e.g., deuterium instead of hydrogen), a different odor is detected. It was therefore suggested that the receptor actually operates as a "dynamic" resonant tunneling device [38], which is sensitive on the molecule vibrations. If this is indeed the case, an enigma still remains: how humans can distinguish between approximately 10,000 different smells while they have only few hundred receptors. If every receptor is calibrated to a specific molecule with a specific vibration, then only several hundred odors should have been detected. The dynamics, which is presented by us in this chapter by the forbidden activation energy, suggests that each molecule has a different fingerprint, and every molecule can activate several receptors in a specific combination, which characterizes only the specific molecule. From this specific combination (instead of a single specific receptor), the brain can identify the specific detected molecule (see **Figure 14**).

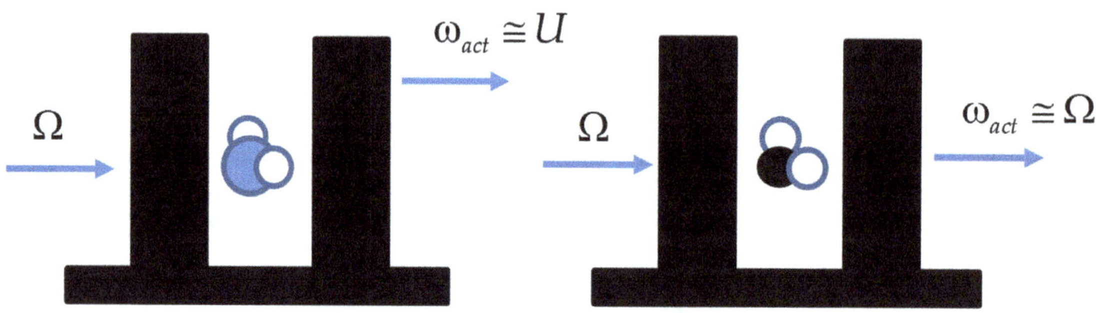

Figure 14. A receptor as a resonant tunneling system distinguishes between molecules, which are identical chemically but different mechanically. The heavier the molecule, the lower the vibrational frequencies and therefore have a different impact on each one of the receptors. Thus, every molecule would have a unique combination of activated receptors.

11. Summary

Resonant tunneling is a fascinating quantum phenomenon. In this chapter, we have focused on the transmission of a quantum particle through an opaque (mostly rectangular) barrier via a delta-function potential. We have discussed all dynamics regimes—stationary resonant tunneling, adiabatic changes, activation, forbidden activation, and instantaneous changes. The generic dynamic problem is highly complex. Each one of these regimes has its specific characteristics.

The main results are related to the domain in which the incoming particle's energy is higher than the minimum eigenenergy of the changing well and the changes are fast enough to prevent particle escape. Unlike previous predictions, the fact that the particle can be trapped in the well does not mean by itself that it will necessarily be activated. In fact, in some cases, the particle's wavefunction experiences "destructive interference" inside the well, and as a consequence, activation is suppressed. We show that all the cases in which the potential's temporal minima has a parabola shape have the same forbidden energies, and for the first time, we present a generic solution for this case.

Furthermore, the instantaneous case is also discussed for the first time. In this case, the dependence on the specific incoming energy vanishes, and a generic universal pattern appears.

Finally, we suggest testing this effect as a frequency effect transistor, which have the potential to be used as a highly accurate transistor.

Moreover, it is suggested that the effect of forbidden activation energies may explain the reason that humans can distinguish between 10,000 different odors while they have only several hundred odor receptors. According to this suggestion, every molecule can trigger different receptors, and only the combination of the activated ones creates the perception of the right smell. In any case, this research shows that dynamic resonant tunneling, in general (and forbidden activation, in particular), has a major, and totally nontrivial, role in the olfactory mechanism.

Author details

Er'el Granot* and Gilad Zangwill*

*Address all correspondence to: erelgranot@gmail.com and gzangwill@gmail.com

Department of Electrical Electronics Engineering, Ariel University, Ariel, Israel

References

[1] Bohm D., Quantum Theory (Dover Publication, New York, 1989).

[2] Ricco B. and Azbel M.Ya., Phys. Rev. B 29, 1970 (1984).

[3] Granot E., Eur. J. Phys. 27, 985–993 (2006).

[4] Kane E.O., Tunneling Phenomenon in Solids (Plenum, New York, 1969).

[5] Price P.J., IEEE Trans. Electron. Dev. 36, 2340 (1989).

[6] Stollen T.C., Goodhue W.D., Parker C.D., Tannenwald T.E., and Peck D.D., Appl. Phys. Lett. 43, 588 (1983).

[7] Sugiyama Y., Inata T., Muto S., Nakata Y., and Hiyamizu S., Appl. Phys. Lett. 52, 314 (1988).

[8] Tsu R. and Esaki L., Appl. Phys. Lett. 22, 562 (1973).

[9] Azbel M.Ya., Phys. Rev. B. 28, 4106 (1983).

[10] Singh J., Physics of Semiconductors and Their Heterostructures (McGraw-Hill, Singapore 1993).

[11] Granot E. and Azbel M.Ya., Phys. Rev. B 50, 8868 (1994).

[12] Granot E. and Azbel M.Ya., J. Phys. Condensed Matter, 11, 4031 (1999).

[13] Kwok K.Ng., Complete Guide to Semiconductor Devices, 2nd ed., pp. 75–83 (Wiley, 2010).

[14] Settnes M., Power S.R., Petersen D.H., and Jauho A.P., Phys. Rev. Lett. 112, 096801 (2014).

[15] Qi Z., Bahamon D.A., Pereira V.M., Park H.S., Campbell D.K., and Neto A.H.C., Nano. Lett. 13, 2692–2697 (2013).

[16] Kashcheyevs V. and Timoshenko J., Phys. Rev. Lett. 109, 216801 (2012).

[17] Azbel M.Ya., Phys. Rev. Lett 68, 98 (1992).

[18] Giles R., J. Math. Phys. 11, 2139–2160 (1970)

[19] Merzbacher E., Quantum Mechanics, (Wiley, 1970).

[20] Azbel M.Ya., Europhys. Lett. 18, 537 (1992).

[21] Azbel M.Ya., Phys. Rev. B. 46, 7596 (1992).

[22] Azbel M.Ya., Phys. Rev. B. 43, 6847 (1991).

[23] Granot E., Europhys. Lett. 61, 817 (2003).

[24] Granot E., Physica E 14, 397–401(2002).

[25] Kälbermann G., Phys. Rev. C. 77, 041601 (2008).

[26] Zangwill A. and Soven P., Phys. Rev. Lett. 45, 204 (1980).

[27] Ivlev B.I. and Mel'nikov V.I., Phys. Rev. Lett. 55, 1614 (1985).

[28] Fisher M.P.A., Phys. Rev. B 37, 75 (1988).

[29] Bagwell P.F. and Lake R.K., Phys. Rev B 46, 15329 (1992).

[30] Aleiner I.L. and Andreev A.V., Phys. Rev. Lett. 81, 1286 (1998).

[31] Buttiker M. and Landauer R., Phys. Rev. Lett. 49, 1739 (1982).

[32] Stefanucci G. and Almbladh C.O., Phys. Rev. B 69, 195318 (2004).

[33] Zangwil G. and Granot E., Physica B 461, 140–146 (2015).

[34] Gray H.B. and Winkler J.R., Q. Rev. Biophys. 36, 341–372 (2003).

[35] Lloyd S., J. Phys. 302, 012037 (2011).

[36] Turin L., Chem. Senses. 21, 773 (1996).

[37] Dyson G.M., Chem. Ind. 57, 647 (1938).

[38] Brookes J.C., Hartoutsiou F., Horsfield A.P., and Stoneham A.M., Phys. Rev. Lett. 98, 038101 (2007).

[39] Pavlicek N., Swart I., Niedenfuhr J., Meyer G., and Repp J., Phys. Rev. Lett. 110, 136101 (2013).

[40] Granot E. and Marchewka A., Physica B 459, 62–68 (2015).

[41] Granot E. and Marchewka A., Europhys. Lett. 72, 341–347 (2005).

[42] Marchewka A., Granot E., and Schuss Z., Opt. Spectrosc. 103, 330–335 (2007).

[43] Granot E. and Marchewka A., Phys. Rev. A 76, 012708 (2007).

Permissions

All chapters in this book were first published in RAQD, by InTech Open; hereby published with permission under the Creative Commons Attribution License or equivalent. Every chapter published in this book has been scrutinized by our experts. Their significance has been extensively debated. The topics covered herein carry significant findings which will fuel the growth of the discipline. They may even be implemented as practical applications or may be referred to as a beginning point for another development.

The contributors of this book come from diverse backgrounds, making this book a truly international effort. This book will bring forth new frontiers with its revolutionizing research information and detailed analysis of the nascent developments around the world.

We would like to thank all the contributing authors for lending their expertise to make the book truly unique. They have played a crucial role in the development of this book. Without their invaluable contributions this book wouldn't have been possible. They have made vital efforts to compile up to date information on the varied aspects of this subject to make this book a valuable addition to the collection of many professionals and students.

This book was conceptualized with the vision of imparting up-to-date information and advanced data in this field. To ensure the same, a matchless editorial board was set up. Every individual on the board went through rigorous rounds of assessment to prove their worth. After which they invested a large part of their time researching and compiling the most relevant data for our readers.

The editorial board has been involved in producing this book since its inception. They have spent rigorous hours researching and exploring the diverse topics which have resulted in the successful publishing of this book. They have passed on their knowledge of decades through this book. To expedite this challenging task, the publisher supported the team at every step. A small team of assistant editors was also appointed to further simplify the editing procedure and attain best results for the readers.

Apart from the editorial board, the designing team has also invested a significant amount of their time in understanding the subject and creating the most relevant covers. They scrutinized every image to scout for the most suitable representation of the subject and create an appropriate cover for the book.

The publishing team has been an ardent support to the editorial, designing and production team. Their endless efforts to recruit the best for this project, has resulted in the accomplishment of this book. They are a veteran in the field of academics and their pool of knowledge is as vast as their experience in printing. Their expertise and guidance has proved useful at every step. Their uncompromising quality standards have made this book an exceptional effort. Their encouragement from time to time has been an inspiration for everyone.

The publisher and the editorial board hope that this book will prove to be a valuable piece of knowledge for researchers, students, practitioners and scholars across the globe.

List of Contributors

Alexandre Coutinho Lisboa and José Roberto Castilho Piqueira
Engineering School, São Paulo University, São Paulo, Brazil

Josef Oswald
Institute of Physics, Montanuniversität Leoben, Leoben, Austria

Matteo Bonfanti
Department of Chemistry, University of Milan, Milano, Italy

Rocco Martinazzo
Department of Chemistry, University of Milan, Milano, Italy
Institute of Molecular Science and Technologies, Consiglio Nazionale delle Ricerche, Milano, Italy

Miquel Montero
Department of Fundamental Physics, University of Barcelona (UB), Barcelona, Spain

Paul Bracken
Department of Mathematics, University of Texas, Edinburg, TX, USA

Zhigang Sun
State Key Laboratory of Molecular Reaction Dynamics and Center for Theoretical Computational Chemistry, Dalian Institute of Chemical Physics, Chinese Academy of Sciences, Dalian, PR China

Takuya Machida
College of Industrial Technology, Nihon University, Narashino, Chiba, Japan

Parakhonsky A, Lebedev M, Dremin A and Kukushkin I
Institute of Solid State Physics, Russian Academy of Sciences, Chernogolovka, Moscow Region, Russia

Er'el Granot and Gilad Zangwill
Department of Electrical Electronics Engineering, Ariel University, Ariel, Israel

Index

R

S

T

U

W

www.ingramcontent.com/pod-product-compliance
Lightning Source LLC
LaVergne TN
LVHW080915210926
848485LV00004B/657
* 9 7 8 1 6 3 9 8 9 4 5 6 7 *